Modelling Methods for Energy in Buildings

Modelling Methods for Energy in Buildings

C.P. Underwood
Reader, School of the Built Environment
University of Northumbria at Newcastle

and

F.W.H. Yik
Professor, Department of Building Services Engineering
Hong Kong Polytechnic University

Blackwell
Science

Editorial offices:
Blackwell Science Ltd, 9600 Garsington Road, Oxford OX4 2DQ, UK
    Tel: +44(0)1865 776868
Blackwell Publishing Inc., 350 Main Street, Malden, MA 02148-5020, USA
    Tel: +1781 388 8250
Blackwell Science Asia Pty Ltd, 550 Swanston Street, Carlton, Victoria 3053, Australia
    Tel: +61 (0)3 8359 1011

First published 2004

Library of Congress Cataloging-in-Publication Data
Underwood, C.P.
    Modelling methods for energy in buildings / Chris Underwood & Francis Yik.
       p.   cm.
Includes bibliographical references and index.
    ISBN 0-632-05936-2 (printed case hardback : alk. paper)
1. Heating—Mathematics.   2. Air conditioning—Mathematics.   3. Ventilation—Mathematics.   4. Buildings—Energy consumption—Mathematical models.   5. Architecture and energy conservation.   I. Yik, Francis.   II. Title.

TH7226.U53 2003
696'.01'5118–dc22

                                                                                    2003022238

ISBN 0-632-05936-2

A catalogue record for this title is available from the British Library

Set in 10 on 13 pt Times
by SNP Best-set Typesetter Ltd., Hong Kong
Printed and bound in India
by Gopsons Papers Ltd, Noida

The publisher's policy is to use permanent paper from mills that operate a sustainable forestry policy, and which has been manufactured from pulp processed using acid-free and elementary chlorine-free practices. Furthermore, the publisher ensures that the text paper and cover board used have met acceptable environmental accreditation standards.

For further information on Blackwell Publishing, visit our website:
www.thatconstructionsite.com

Contents

Preface

Buildings consume between 40% and 60% of all energy use in most developed economies throughout the world, and an increasing proportion in many developing and emerging economies. Correspondingly, they are responsible for a similar proportion of humankind's carbon dioxide emissions with a consequential impact on global warming. Indeed, the International Council for Research and Innovation in Building and Construction (CIB) in 2003 established a Working Commission to stimulate and coordinate international research cooperation on the impact that buildings have on climate change as well as the impact that climate change will have on buildings in the future. Whilst the precise causes and scale of these impacts will remain uncertain and will be the subject of debate for some time to come, what is not in doubt is that our current dependency on carbon-based fossil fuels for providing comfort and amenity in buildings cannot be sustained and that alternative strategies require to be explored, designed, implemented, tested and demonstrated in the immediate years ahead. At the heart of this is the need to both extend and improve our knowledge of the processes of energy exchange in buildings whether they be the 'high tech' commercial buildings in the centres of our cities or the simpler and often naturally-ventilated buildings and residential habitats that surround them. Whilst many of the methods involved in our knowledge of these processes are grounded in well-established physical laws, in the past 30 years there have been dramatic improvements in the computational resources available for the treatment of these often very complex and detailed methods. This book concerns itself with a review and description of these methods with particular reference to heat and fluid flow problems in buildings and their related systems, plant and control.

Throughout most of the twentieth century, calculations of the thermal response of buildings have made a convenient and yet thoroughly erroneous assumption – that the boundary conditions within which the problem is cast are static. The external air temperature and moisture content, the wind speed and direction, the internal air temperature and moisture content were all treated as constants whilst the building was assumed to be unoccupied. Thus a fictitious steady-state problem could be defined which enabled a 'worst case' heating load on the building to be trivially calculated. In the later part of the twentieth century, the impact of the user and a wider range of external climate variables were introduced by assuming uniformly cyclic internal heat gains and solar irradiances transmitting through glazing and acting on internal surfaces. The impact of the thermal capacity of the building envelope material began to be considered as having an important influence on energy exchange during different times of the usage cycle, and simple analytical methods were developed and introduced to deal with the lagging effect of solar heat transfer through opaque elements in a 'quasi-static' manner, together with a more thorough treatment of the geometry of solar heat transmission through glazing and related shading systems. In the 1970s, developments in mainframe digital computers enabled a more thorough treatment of the underlying theory to be made for the first time, free from the assumption of static boundary conditions, and, with it, a more intuitive understanding of the time response of building energy flows without the need to resort to empiricism. During this time, building energy modelling methods came to be classified according to the manner in which conduction heat transfer through fabric was treated; the main approaches being the numerical solution of the governing Fourier equation using finite difference methods and a less computationally expensive approach using response factors derived analytically from the time series response to a unit pulse excitation at the surfaces of an element. Practice in North America tended to adopt the latter approach whilst European practice mainly focused on the former. In the 1980s personal computers became available and the first commercial codes for design calculations using steady-state methods of energy in buildings became available. Initially these codes addressed heat loss calculations and quasi-steady-state heat gain and peak internal summer temperature calculations mostly based on either the Response Factor Method or another analytical method based on the frequency response of building elements – the Admittance Method. Later in the 1980s, the first commercial codes for fluid flow in building services became available, enabling computer-aided pipework and ductwork designs to be done. Research progress meanwhile continued to refine thermal simulations mainly based on Response Factor and Finite-difference methods, and program codes capable of representing multiple interacting zones in a building began to emerge largely based on mainframe computers or the then UNIX workstations.

Another development started to take place in the 1980s – the introduction of models of heating and air conditioning plant and control systems. Initially, the simplest of these amounted to nothing more than control functions enabling plant energy flows to be both switched on or off and to track idealised set-points, but other developments began to take place in which detailed steady-state descriptions of plant were defined enabling the comparative design of system component options to be conducted for the first time. Later, dynamic descriptions of plant became available, though in a more bespoke context generally for the specific analysis and design of control systems. The steady-state approach to plant modelling has more or less remained the dominant method due to its computational efficiency together with the recognition that for a majority of building energy modelling problems, interest lies in the longer time-horizon situation over a complete season for which detailed dynamics of the plant are neither necessary nor of interest. Progress was also made during this period on the treatment of moisture sorption in building fabric elements though this has tended to remain as an option in the very few codes that have included it for application to the relatively few cases when it is needed.

By the mid 1990s, sophisticated and robust commercial codes were widely available on personal computers for a wide variety of building energy design and analysis problems. Research progress through the early 1990s mainly focused on the (then) neglected problem of fluid flow in and around buildings and the first computational fluid dynamics (CFD) codes for the special case dealing with the low Reynolds number flow conditions associated with building ventilation became available. Much of the later work – through to current times – on these methods (leading to commercial codes in some cases) has tended to focus on ways of treating transitory laminar-turbulent flow for the difficult problem frequently encountered in building ventilation – that of conjugate heat and fluid flow at low Reynolds number. However, the high computational demand of most CFD problems has meant that a seamless or truly conjugate and universally applicable model of energy and fluid flow in buildings, including plant and control, moisture sorption, and the stochastic influences of the user, remains elusive. On the other hand, many would argue that such an embracing and comprehensive description would in any case be unnecessary given that many building energy analyses require to address one, or a small subset, of problems. So long as the boundary conditions of the problem of interest can be sufficiently defined for an accurate solution of that immediate problem, then the method selected need be no more comprehensive than is needed to address the problem itself. And so many of the methodological developments have tended to take place in isolation which means that 'hand-shaking' between methods and codes is at best primitive at present and the issue of methodological interoperability is more or less where we are now. Much of the recent research concerning energy modelling in buildings has been attempting to develop ways of enabling methods described in a variety of program codes using a variety of data management protocols and requiring different levels of user skill and expertise to integrate with one another robustly and efficiently.

This book concerns itself with a review and descriptions of the various state-of-the-art methods for modelling energy and fluid flow problems in buildings. It is hoped that it offers something for the researcher, the scholar and the practitioner in equal measure. It is aimed at a wide readership of those engaged in the physical science of energy in buildings: building physicists, architectural scientists, building and building services engineers, HVAC engineers and building technologists.

Chapter 1 deals with methods for treating heat exchange through building elements. The transient heat conduction problem is defined and the various analytical methods for the solution of the governing partial differential equations are described and compared: conventional time domain methods; the use of Laplace transforms; approaches using unit impulse response factors (and the more computationally-efficient form of this using operator-z transfer functions); the frequency-response or 'admittance' method; first- and higher-order lumped capacitance methods. Chapter 1 concludes with a review and description of methods for dealing with absorption, transmission and reflection of solar radiation through glass.

In Chapter 2 the 'building blocks' describing conduction heat exchange through building elements are linked with procedures for dealing with the surface effects of convection and radiation leading to a closure of methods for treating the entire room space. As a lead-in to this, an alternative *numerical* approach for treating element conduction is described as a preferred choice over the analytical methods of Chapter 1 due to the flexibility and ease with which this sits alongside other numerical methods applicable to building energy. Issues concerning the explicit, implicit and semi-implicit numerical solution of the governing equations together with considerations for dealing with numerical instability are explored. Heat balances due to convection and radiation on the external surfaces of elements are then dealt with followed by methods for treating the radiation interchange between bounding internal surfaces of a room space, as well as methods for dealing with solar radiation at both external opaque surfaces and at internal surfaces after direct and diffuse transmission through glazing. The chapter concludes with a synthesis of each method for dealing with conduction through building elements with the overall building space heat balance including some simplified contemporary approaches for calculating design loads on spaces.

Chapter 3 is entirely devoted to methods for dealing with fluid flow within the built environment with a focus on air movement for ventilation. A treatment of the modelling of moisture flow through building elements is first dealt with as a linked heat and mass transfer problem. The conservation equations for mass, momentum and energy are then dealt with as a basis for the development of a computational fluid dynamics (CFD) approach to ventilation problems in buildings, including a review of the various approaches for modelling turbulence especially when linked with heat transfer at solid boundaries. Methods for discretising the governing fluid flow equations are reviewed together with common approaches to solving the resulting discretised equation sets. Chapter 3 concludes with a discussion of simplified methods of modelling ventilation flow paths using flow networks and zonal approaches.

Chapters 4 and 5 move on to look at methods for the treatment of systems, plant and control. Chapter 4 deals with two approaches to modelling plant in the steady-state – the use of curve fitting to experimental or manufacturers' data and the use of a rigorous theoretical approach as an alternative. The latter is based on the treatment of sensible and latent cooling in the air cooling process typified by an air conditioner. This represents one of the most detailed heat exchange problems encountered in building energy plant but, by relaxing the modelling assumptions, approaches to modelling other less complex sub-systems also emerge. The chapter concludes with a treatment of flow network modelling applicable to fan and pump systems. In Chapter 5, plant-modelling methods focusing on dynamic approaches with main applicability to the analysis of control systems are dealt with. Again, the sensible and latent cooling process is first treated as a generic exemplar and simplifications for less complex heat exchange processes are drawn out. There is also a treatment of the various approaches to model order reduction and model linearisation in order to improve computational efficiency without compromising accuracy. Methods for reducing problems to linearised transfer functions are dealt with for those interested in classical block diagram methods. Chapter 5 also deals with methods for the representation of control elements (i.e. control valves, dampers, drive systems) and concludes with principles for modelling conventional controllers as well as approaches to a number of 'smart' control strategies.

Chapters 6 and 7 deal with building energy modelling in practice. Chapter 6 starts by outlining the key stages in the analysis of building energy and environment problems using model-based methods and then draws out the various methods described elsewhere in the book as a methodological hierarchy. The major part of Chapter 6 then deals with a review based on the literature of some of the most recent developments of the various methods dealt with throughout the book. Chapter 7 then deals with the interrelationships between methods and their integration, again, mainly drawing from examples reported in the literature.

It is hoped that the end product provides a comprehensive text on the subject of the mathematical modelling of energy and environment in buildings both for the treatment of individual problems, as well as for the appreciation of the wider range of issues with which the individual problem might be expected to interact. To date, there are in excess of 500 program codes around the world that deal with these problems to a greater or lesser extent but probably fewer than 20 of them have sufficient rigour and detail to treat most building energy simulations with a sufficient degree of comprehensiveness needed in modern building design analysis. This book is intended both for those intending to use an existing program code with a thirst for more detailed insights into both the potential and the flaws of the methods involved as well as for those with a thirst for understanding who are inclined to develop new models from scratch.

Chapter 1
Heat Transfer in Building Elements

A key function of the envelope of a building is to act as a passive climate modifier to help maintain an indoor environment that is more suitable for habitation than the outdoors. However, besides providing shelter from stormy weather, the building envelope alone can hardly ensure that the indoor environmental conditions will always be comfortable to the occupants, or be suitable for the intended purposes of the indoor spaces, particularly during periods of unfavourable outdoor conditions, such as in the night or when the outdoor air is stagnant.

The need for building design features that would facilitate use of natural ventilation and daylight has diminished, as active means of environmental control, such as central heating, ventilating and air-conditioning (HVAC) systems and electric lights, can be used instead to maintain adequate indoor thermal and visual environmental conditions, and air quality. This has also helped to remove the restrictions imposed on the design of buildings, particularly to maximisation of the amount of floor area that can be built upon a given piece of land.

The increased reliance on HVAC and lighting systems for active control over the indoor environmental conditions, however, has made buildings the dominant energy consumers in modern cities worldwide. This has not only accelerated the depletion of the limited reserve of fossil fuels on earth; it has also exacerbated global warming and environmental pollution due to the emissions of combustion products resulting from burning of fuels for generating electricity, steam, hot water or chilled water for use in buildings. Buildings also contribute to other environmental problems, such as the use of CFC as refrigerants in HVAC plants and halons as fire extinguishing agents, which are causes of the depletion of the stratospheric ozone layer. Therefore, besides fulfilment of the functional needs and aesthetics, the environmental performance of buildings, which includes energy efficiency, has become an essential attribute of environmentally friendly buildings.

Measures that can be used to enhance the energy efficiency of a building include the adoption of building design features that can help reduce the frequency and intensity of use of the HVAC plants and the lighting installations, and the use of more efficient HVAC and lighting system designs and equipment. For instance, the cooling or heating load due to heat transmission through the building envelope can be reduced through the use of thermal insulation at external walls and roofs, and high performance glazing and shading devices at windows. The use of energy-efficient boilers and chillers, variable speed motor drives in heating and air-conditioning systems, and energy-efficient lamps and electronic ballasts can lead to very significant energy saving.

In the design process, the effectiveness of individual energy efficiency enhancement measures, particularly the possible energy and running cost saving, would need to be quantified. The financial benefit, derived from the difference in the annual energy consumption of the building with and without a particular measure, would be essential input to a financial appraisal for determining whether or not to adopt individual measures, and for selecting the most viable choices.

Quantification of the annual energy use in a building requires prediction of the space cooling loads of individual rooms in the building that would arise at different times in the operating periods throughout the year. This involves determination of the heat and mass transfer through the building envelope that are significant parts of the heat and moisture gains or losses of an indoor space. The other sources of heat and moisture gains include occupants, equipment and appliances present within the air-conditioned spaces, and infiltration.

This chapter reviews the methods for modelling the heat and mass transfer through the building envelope, which is a key starting point in the prediction of the annual energy consumption in a building. In most such analyses, the mass transfer modelled would be limited to the bulk air transport into or out of buildings through infiltration and exfiltration, while the moisture transport through the building fabric elements would be ignored.

1.1 Heat and mass transfer processes in buildings

The range of heat and mass transfer processes that would take place in buildings is as illustrated in Fig.1.1, which shows a perimeter room on an intermediate floor in a multi-storey office building. The room is separated from the outdoors by an external wall and a window, and from adjoining rooms at the sides by internal partitions, and at above and below by a ceiling and a floor slab. The room is equipped with a HVAC system that would supply heating or cooling to the room by circulating air between the room and the air-handling unit via the supply and return air ducts.

As shown in Fig.1.1, the heat and mass transfer processes that would take place in a building include:

(a) conduction heat transfer through the building fabric elements, including the external walls, roof, ceiling and floor slabs and internal partitions;
(b) solar radiation transmission and conduction through window glazing;
(c) infiltration of outdoor air and air from adjoining rooms;
(d) heat and moisture dissipation from the lighting, equipment, occupants and other materials inside the room; and
(e) heating or cooling and humidification or dehumidification provided by the HVAC system.

The conduction heat transfer through an opaque building fabric element, such as an external wall as shown in Fig. 1.2, is the effect of the convective heat that the surface at each side of the element is exchanging with the surrounding air, and the radiant heat exchanges with other surfaces that the surface is exposed to. For an external wall or a roof, the radiant heat exchange at

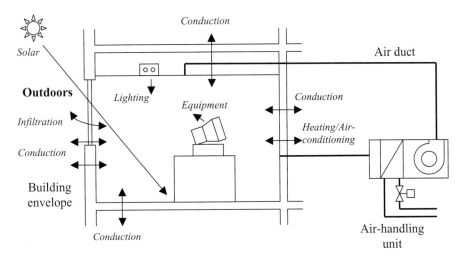

Fig. 1.1   Heat and mass transfer processes involved in building energy simulation.

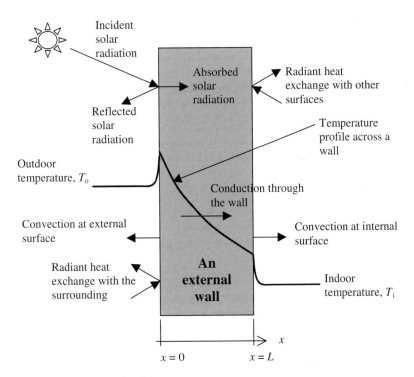

Fig. 1.2   Heat transfer at an external wall.

the external side includes the absorbed solar radiation, including both direct and diffuse radiation.

The heat transfer through a window is shown in Fig.1.3. The window glass will transmit part of the incident solar radiation into the indoor space. While the solar radiation penetrates

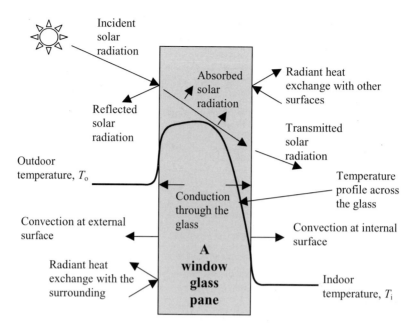

Fig. 1.3   Heat transfer at a window glass pane.

the glass pane, some of the energy will be absorbed by the glass, leading to an increase in the glass temperature, which, in turn, will cause heat to flow in both the indoor and the outdoor directions, first by conduction within the glass and then by convection and radiation at the surfaces at both sides.

The heat flows through a building fabric element resulting from the absorbed solar radiation and the outdoor to indoor temperature difference are often treated together through the use of an equivalent outdoor air temperature, called 'sol-air temperature', that will, in the absence of the radiant heat exchange, cause the same amount of conduction and convection heat flow through the element. A similar parameter, called 'environmental temperature', is used to account for the combined effects of the convective heat transfer from the internal surface to the room air and the radiant energy gain at the surface.

The transmitted solar radiation from the windows will be imparted to the indoor air and become cooling load only after it has been absorbed by the internal surfaces. Consequently, the temperature at such surfaces will rise, leading to convective heat flow from the surfaces into the room air. It is this convective heat flow that will affect the indoor air temperature and constitutes a component of the space cooling load. This cooling load component will differ in magnitude and in the time of occurrence of its peak value from those of the radiant heat gain, as shown in Fig. 1.4, which is the result of the thermal capacitance of the fabric elements or furniture materials that are subject to thermal irradiation. Besides the short wave radiation from the sun, radiant heat gains from the lighting and equipment and the long wave radiation exchange among the internal surfaces within the space will need to undergo a similar process to become a cooling load.

When there are air movements into and out of an indoor space, heat and moisture will be brought into or out of the room if the airs that enter the space are at thermodynamic states

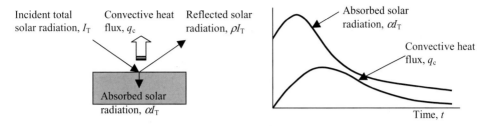

Fig. 1.4   Radiant heat gain and the resultant cooling load.

different from that of the indoor air. The air movements can be set up by pressure differences between the room and the adjoining rooms and the outdoor, due to wind or stack effect, or imbalance in the supply and extract flow rates maintained by the ventilation system.

The thermodynamic state of the air within the room would vary with the net heat and moisture gains experienced by the room air, resulting from heat and moisture exchanges with the enclosure surfaces, air transport into or out of the room bringing with it heat and moisture, heat and moisture gains from sources present within the room, and heating, cooling, humidification or dehumidification provided by the HVAC system serving the room. These heat and moisture transfer processes would need to be modelled for the prediction of the indoor air condition or the rate of heating or cooling, and humidification or dehumidification required for maintaining the indoor air condition at the set point state.

1.2 Heat transfer through external walls and roofs

The equation that governs the heat transfer through a homogeneous material can be derived through taking account of:

(a)   the rate of heat transfer across each boundary surface of an elemental control volume within the material that would arise corresponding to the temperature gradient that exists at the surface;
(b)   the rate of heat generation or removal by internal heat sources or sinks present within the control volume; and
(c)   the change in internal energy of the material in the control volume, which is reflected by the change in temperature of the material.

The resultant equation is:

$$\rho c \frac{\partial T}{\partial t} - \nabla \cdot k \nabla T - \dot{q} = 0 \tag{1.1}$$

where ρ, c and k are respectively the density (kg m^{-3}), specific heat (kJ kg^{-1} K^{-1}) and thermal conductivity (kW m^{-1} K^{-1}) of the material; T is the temperature (K) and $\dot{q}$ the net rate of heat gain from the internal source or sink within the material (kW m^{-3}); and t is time (s).

For applications to analysis of the heat transfer through walls and slabs in buildings, Equation 1.1 can be simplified by making the following assumptions:

(a)   only the heat transfer in the direction across the thickness of each wall or slab (assumed to be in the x-direction, Fig.1.2) needs to be modelled whereas heat transfer in the other two directions (the y- and z-directions) can be ignored;

(b)   heat transfer in the material is isotropic;

(c)   properties of the material (ρ, c and k) are independent of temperature; and

(d)   no internal heat source or sink exists within the material ($\dot{q} = 0$).

The governing equation can then be simplified to:

$$\frac{\partial T}{\partial t} = \alpha \frac{\partial^2 T}{\partial x^2} \tag{1.2}$$

where $\alpha = k/\rho c$ is the thermal diffusivity of the material. Equation 1.2 is a one-dimensional (in space) partial differential equation (PDE) governing the transient heat conduction in a homogeneous slab with no internal heat source or sink.

Boundary conditions

Equation 1.2 needs to be solved in conjunction with the equations that define the boundary conditions that apply to the two surfaces of the fabric element. Typical boundary conditions that apply to building heat transfer problems include the following.

(a)   The temperature at either surface is known, i.e.:

$$T(x,t) = f(t) \qquad (\text{for } x = 0, L) \tag{1.3}$$

where $f(t)$ is a known function that relates the surface temperature to time, and L is the thickness of the fabric element (m). This includes the condition that the surface temperature is a steady temperature, i.e. $f(t) = \text{constant}$.

(b)   The surface is exposed to an ambient air at a known temperature and is exchanging radiant heat with the surroundings. In this case, the heat balance at the surface can be described by:

$$q(0,t) = -k\frac{\partial T(x,t)}{\partial x} = h_0\left(T_{\text{Amb},0} - T(x,t)\right) + q_{\text{r,net},0} \qquad (\text{for } x = 0) \tag{1.4a}$$

$$q(L,t) = -k\frac{\partial T(x,t)}{\partial x} = h_L\left(T(x,t) - T_{\text{Amb},L}\right) - q_{\text{r,net},L} \qquad (\text{for } x = L) \tag{1.4b}$$

where h_0 and h_L are the convective heat transfer coefficients ($\text{kW m}^{-2}\text{K}^{-1}$), $T_{\text{Amb},0}$ and $T_{\text{Amb},L}$ are the ambient air temperatures (K) and $q_{\text{r,net},0}$ and $q_{\text{r,net},L}$ are the net radiant heat gains per unit area of the surfaces (kW m^{-2}) at the surfaces at $x=0$ and $x=L$ respectively.

Note that both the ambient air temperatures and the net radiant heat gains could be complex functions of time, such as the outdoor air temperature and the solar radiation absorbed by the external surface of an external wall or roof of a building.

1.3 Analytical methods for solving the one-dimensional transient heat conduction equation

Conventional time-domain analysis

The one-dimensional transient heat conduction equation shown in Equation 1.2 can be solved in various ways; for instance, by using the conventional method of separation of variables to decompose the PDE into two ordinary differential equations (ODEs), followed by integration of the ODEs to yield a solution. The solution to Equation 1.2 for a wall, corresponding to the initial (at $t=0$) and boundary (at $x=0$ and $x=L$, for $t>0$) conditions as given in Equations 1.5 to 1.7, is shown in Equation 1.8:

$$T(x,0) = 0 \quad (\text{for } 0 \le x \le L) \tag{1.5}$$

$$T(0,t) = u(t) = \begin{bmatrix} 0 & (\text{for } t < 0) \\ 1 & (\text{for } t \ge 0) \end{bmatrix} \tag{1.6}$$

$$T(L,t) = 0 \quad (\text{for } t > 0) \tag{1.7}$$

where $u(t)$, given in Equation 1.6, is a unit step function, as shown in Fig.1.5:

$$T(x,t) = 1 - \frac{x}{L} - \frac{2}{\pi}\sum_{n=1}^{\infty}\frac{1}{n}\exp\left\{-\alpha\left(\frac{n\pi}{L}\right)^{2}t\right\}\sin\left(\frac{n\pi}{L}x\right) \tag{1.8}$$

Since the conduction heat flux through a cross sectional plane within the wall, $q(x,t)$, is related to the temperature by Equation 1.9, the equation for the heat flux, in response to the unit temperature step excitation at $x = 0$, can be found from Equation 1.8, as shown in Equation 1.10:

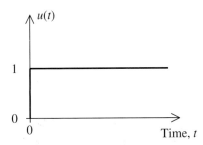

Fig. 1.5   A unit step function.

$$q(x,t) = -k \frac{\partial T(x,t)}{\partial x} \tag{1.9}$$

$$q(x,t) = \frac{k}{L} + \frac{2k}{L} \sum_{n=1}^{\infty} \exp\left\{-\alpha\left(\frac{n\pi}{L}\right)^2 t\right\} \cos\left(\frac{n\pi}{L}x\right) \tag{1.10}$$

And, at $x=L$, the heat flux is:

$$q(L,t) = \frac{k}{L} + \frac{2k}{L} \sum_{n=1}^{\infty} (-1)^n \exp\left\{-\alpha\left(\frac{n\pi}{L}\right)^2 t\right\} \tag{1.11}$$

Although the above unit step response of the wall would hardly be directly applicable to solving actual problems, it is useful because its time derivative is the response of the wall to a unit impulse excitation, which is the basis for determining the response of the wall to an arbitrary excitation.

The unit impulse function, $\delta(t)$, as shown in Fig. 1.6, is defined as a function of time that has the following characteristics:

$$\delta(t) = \lim_{\Delta t \to 0} \frac{1}{\Delta t} \qquad \text{for } 0 \le t \le \Delta t \tag{1.12}$$

$$\delta(t) = 0 \qquad \text{for } t < 0 \text{ and } t > \Delta t \tag{1.13}$$

The area under the unit impulse function is:

$$\int_{-\infty}^{+\infty} \delta(t)dt = \int_{-\infty}^{+\infty} \lim_{\Delta t \to 0} \frac{1}{\Delta t} dt = \lim_{\Delta t \to 0} \frac{1}{\Delta t} \int_{0}^{\Delta t} dt = \lim_{\Delta t \to 0} \frac{\Delta t}{\Delta t} = 1 \tag{1.14}$$

For an impulse function $p(t)$ that equals $A/\Delta t$ from $0 < t \le \Delta t$, with Δt tending to zero, the area under the curve of the impulse function is:

$$\int_{-\infty}^{+\infty} p(t)dt = \lim_{\Delta t \to 0} \frac{A}{\Delta t} \int_{0}^{\Delta t} dt = A \lim_{\Delta t \to 0} \frac{1}{\Delta t} \int_{0}^{\Delta t} dt = A \int_{-\infty}^{+\infty} \delta(t)dt = A \tag{1.15}$$

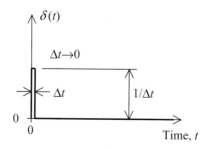

Fig. 1.6   A unit impulse function.

By comparing the above equation with Equation 1.14, it can be seen that the impulse function $p(t)$ has an area under the curve A times as large as the unit impulse function. This area (A in this case) is called the strength of the impulse function, and it follows that the strength of a unit impulse function is one, which is the reason for calling it a unit impulse function.

The conduction heat flux through the wall in response to a unit temperature impulse applied to the side at $x=0$ can be derived by differentiating Equation 1.10 with respect to time, which yields:

$$q(x,t) = -\frac{2\alpha k\pi^2}{L^3} \sum_{n=1}^{\infty} n^2 \exp\left\{-\alpha\left(\frac{n\pi}{L}\right)^2 t\right\} \cos\left(\frac{n\pi}{L}x\right) \tag{1.16}$$

And, at $x=L$, the conduction heat flux is:

$$q(L,t) = \frac{2\alpha k\pi^2}{L^3} \sum_{n=1}^{\infty} (-1)^n n^2 \exp\left\{-\alpha\left(\frac{n\pi}{L}\right)^2 t\right\} \tag{1.17}$$

Since the PDE that governs the conduction heat transfer through the wall is linear, the response of the wall to an impulse of strength different from unity will be the unit impulse response of the wall scaled by the impulse strength. For instance, if the heat flux through the wall at $x=L$ is regarded as the 'output' of a system (the wall in this case), denoted as $y(t)$, in response to an 'input' (the temperature excitation at $x = 0$) to the system, and let $y_0(t)$ be the system's output in response to the unit impulse input $\delta(t)$ (given in Equation 1.17), the output of the system to an impulse input of strength A will be:

$$y(t) = Ay_0(t) \tag{1.18}$$

If the impulse is applied at a time greater than zero instead, say at $t = \tau$, the output will become:

$$y(t) = Ay_0(t - \tau) \tag{1.19}$$

Assume that the wall is subject to an arbitrary time-varying temperature excitation at $x = 0$, denoted by an excitation function $f(t)$; the value of $f(t)$ at $t = \tau$, is $f(\tau)$. The wall's response to an impulse of strength that equals $f(\tau)$, applied to the wall at $t = \tau$, is:

$$y(t) = f(\tau)y_0(t - \tau) \tag{1.20}$$

Note that:

$$\int_{-\infty}^{+\infty} f(t)\delta(t - \tau)dt = f(\tau) \tag{1.21}$$

Equation 1.21 shows that the excitation applied to the wall at time τ, given by the value $f(\tau)$, can be represented by an impulse of strength $f(t)$ that takes place at $t = \tau$. The excitation, as described by the time function $f(t)$, can, therefore, be represented by a consecutive sequence of impulses, each with a strength that equals the value of the function at the time of its occurrence, as shown in Fig.1.7. It follows that the response of the wall to this excitation function, at an arbitrary time t, can be determined by summing the responses of the wall to each of the sequence of impulses that took place from $\tau = 0$ to $\tau = t$.

This means that the response of the wall to a particular impulse applied to the wall at time τ, as shown in Equation 1.20, will only be an incremental part of the overall response of the wall, denoted as $\Delta y(t)$, when the effect of an additional impulse, taking place between $t = \tau$ and $t = \tau + \Delta\tau$, has been accounted for. Taking the limit that $\Delta\tau$ tends to zero:

$$\underset{\Delta\tau \to 0}{\text{Lim}} \frac{\Delta y(t)}{\Delta\tau} = \frac{dy(t)}{d\tau} = f(\tau)y_0(t-\tau) \tag{1.22}$$

Therefore, the overall response of the wall observable at time t to all the impulses taking place from $\tau = 0$ to $\tau = t$ can be determined by integrating Equation (1.22) over this lapsed time as follows:

$$y(t) = \int_0^t f(\tau)y_0(t-\tau)d\tau \tag{1.23}$$

Note that since $y_0(t)$ is the response to a unit impulse $\delta(t)$, $y_0(t)$ will have zero value for $t < 0$. Likewise, $y_0(t-\tau)$ will have zero value for $t < \tau$. The upper limit of integration can, therefore, be extended to infinity without affecting the result. Hence, Equation 1.23 can be re-written as:

$$y(t) = \int_0^\infty f(\tau)y_0(t-\tau)d\tau \tag{1.24}$$

Equation 1.24 is called a convolution integral, which allows the output of a system to an arbitrary excitation to be found from the unit impulse response of the system.

Laplace transformation

A more widely adopted method for solving the PDE governing the conduction heat transfer through a slab of material is to apply Laplace transformation to the PDE to yield an ODE in the Laplace domain. The solution to the ODE can then be inverse-transformed to yield the

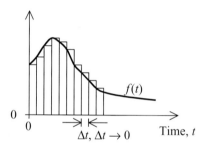

Fig. 1.7   Representing an arbitrary function by a sequence of unit impulses.

solution in the time domain. One of the major advantages of using Laplace transformation to solve the PDE is that a far greater variety of situations can be dealt with and many important observations can be made in the analysis in the Laplace domain.

The Laplace transform of a function $f(t)$, denoted by $L\{f(t)\}$, which would yield a function in the Laplace domain, denoted as $\phi(s)$, is defined as:

$$L\{f(t)\} = \phi(s) = \int_0^\infty f(t)\exp(-st)dt \qquad (1.25)$$

On the basis of this definition, it can be shown that:

$$L\{\partial f(t)/\partial t\} = s\phi(s) - f(0) \qquad (1.26)$$

For avoiding the need to assign two different symbols to denote each function and its Laplace transform, the following nomenclature is defined for representing the Laplace transform of the temperature $T(x,t)$ and heat flux $q(x,t)$:

$$L\{T(x,t)\} = T(x,s) \qquad (1.27)$$

$$L\{q(x,t)\} = q(x,s) \qquad (1.28)$$

Here, the substitution of the argument t by s in the functions means the Laplace transform of the functions rather than a simple substitution of the variable t by s in the function. The same convention will apply also to other time functions and their respective Laplace transforms. Note, however, that where zero appears as the time argument, the function remains the time domain function, for instance:

$$T(x,0) = T(x,t)|_{t=0} \qquad (1.29)$$

and:

$$q(x,0) = q(x,t)|_{t=0} \qquad (1.30)$$

Following the convention defined above, Equation 1.2 when transformed, becomes:

$$\frac{\partial^2 T(x,s)}{\partial x^2} = \frac{s}{\alpha} T(x,s) - \frac{1}{\alpha} T(x,0) \qquad (1.31)$$

Note that Equation 1.31 is an ODE in x, which can be solved when the boundary conditions are defined.

Equation 1.9 can also be transformed into:

$$q(x,s) = -k \frac{\partial T(x,s)}{\partial x} \qquad (1.32)$$

The initial and boundary conditions assumed in the preceding time domain analysis were:

$$T(x,0) = 0 \tag{1.33}$$

$$T(0,t) = f(0,t) \tag{1.34}$$

$$T(L,t) = 0 \tag{1.35}$$

Equations 1.34 and 1.35 are transformed to:

$$T(0,s) = f(0,s) \tag{1.36}$$

$$T(L,s) = 0 \tag{1.37}$$

Based on the initial condition as defined in Equation 1.33, Equation 1.31 then becomes:

$$\frac{\partial^2 T(x,s)}{\partial x^2} = \frac{s}{\alpha} T(x,s) \tag{1.38}$$

Equation 1.38 is an ODE in x that has the following solution:

$$T(x,s) = c_1 \exp\left(\sqrt{s/\alpha} \cdot x\right) + c_2 \exp\left(-\sqrt{s/\alpha} \cdot x\right) \tag{1.39}$$

where c_1 and c_2 are coefficients to be evaluated from the defined boundary conditions. Applying the boundary conditions as defined in Equations 1.36 and 1.37:

$$f(0,s) = c_1 + c_2 \tag{1.40}$$

$$0 = c_1 \exp\left(\sqrt{s/\alpha} \cdot L\right) + c_2 \exp\left(-\sqrt{s/\alpha} \cdot L\right) \tag{1.41}$$

The coefficients c_1 and c_2, solved from the above equations, are:

$$c_1 = -\frac{\exp\left(-\sqrt{s/\alpha} \cdot L\right)}{\exp\left(\sqrt{s/\alpha} \cdot L\right) - \exp\left(-\sqrt{s/\alpha} \cdot L\right)} f(0,s) \tag{1.42}$$

$$c_2 = \frac{\exp\left(\sqrt{s/\alpha} \cdot L\right)}{\exp\left(\sqrt{s/\alpha} \cdot L\right) - \exp\left(-\sqrt{s/\alpha} \cdot L\right)} f(0,s) \tag{1.43}$$

Substituting c_1 and c_2 given in Equations 1.42 and 1.43 into Equation 1.39 yields:

$$T(x,s) = \frac{\exp\left(\sqrt{s/\alpha} \cdot (L-x)\right) - \exp\left(-\sqrt{s/\alpha} \cdot (L-x)\right)}{\exp\left(\sqrt{s/\alpha} \cdot L\right) - \exp\left(-\sqrt{s/\alpha} \cdot L\right)} f(0,s)$$

The above can be expressed more succinctly as:

$$T(x,s) = \frac{\sinh\left(\sqrt{s/\alpha} \cdot (L-x)\right)}{\sinh\left(\sqrt{s/\alpha} \cdot L\right)} f(0,s) \tag{1.44}$$

By substituting Equation 1.44 into Equation 1.32 the heat flux equation in the Laplace domain can be derived as:

$$q(x,s) = k\sqrt{\frac{s}{\alpha}} \left\{ \frac{\cosh\left(\sqrt{s/\alpha} \cdot (L-x)\right)}{\sinh\left(\sqrt{s/\alpha} \cdot L\right)} \right\} f(0,s) \tag{1.45}$$

Note that if the temperature excitation at $x = 0$, $f(0,t)$, is an unit impulse function, its Laplace transform, $f(0,s)$, equals 1. The right hand side of Equation 1.45 will then become the Laplace transform of the right hand side of Equation 1.16.

Through similar procedures, the temperature and heat flux equations corresponding to the initial and boundary conditions given in Equations 1.46 to 1.48 can be derived, which are given in Equations 1.49 and 1.50:

$$T(x,0) = 0 \tag{1.46}$$

$$T(0,t) = 0 \tag{1.47}$$

$$T(L,t) = f(L,t) \tag{1.48}$$

$$T(x,s) = \frac{\sinh\left(\sqrt{s/\alpha} \cdot x\right)}{\sinh\left(\sqrt{s/\alpha} \cdot L\right)} f(L,s) \tag{1.49}$$

$$q(x,s) = -k\sqrt{\frac{s}{\alpha}} \left\{ \frac{\cosh\left(\sqrt{s/\alpha} \cdot x\right)}{\sinh\left(\sqrt{s/\alpha} \cdot L\right)} \right\} f(L,s) \tag{1.50}$$

Let:

$$G_{q,0}(x,s) = k\sqrt{\frac{s}{\alpha}} \left\{ \frac{\cosh\left(\sqrt{s/\alpha} \cdot (L-x)\right)}{\sinh\left(\sqrt{s/\alpha} \cdot L\right)} \right\} \tag{1.51}$$

and:

$$G_{q,L}(x,s) = -k\sqrt{\frac{s}{\alpha}} \left\{ \frac{\cosh\left(\sqrt{s/\alpha} \cdot x\right)}{\sinh\left(\sqrt{s/\alpha} \cdot L\right)} \right\} \tag{1.52}$$

It follows that:

$$G_{q,0}(0,s) = k\sqrt{\frac{s}{\alpha}} \left\{ \frac{\cosh\left(\sqrt{s/\alpha} \cdot L\right)}{\sinh\left(\sqrt{s/\alpha} \cdot L\right)} \right\} \tag{1.53}$$

$$G_{q,0}(L,s) = k\sqrt{\frac{s}{\alpha}} \left\{ \frac{1}{\sinh(\sqrt{s/\alpha} \cdot L)} \right\} \tag{1.54}$$

$$G_{q,L}(0,s) = -k\sqrt{\frac{s}{\alpha}} \left\{ \frac{1}{\sinh(\sqrt{s/\alpha} \cdot L)} \right\} \tag{1.55}$$

$$G_{q,L}(L,s) = -k\sqrt{\frac{s}{\alpha}} \left\{ \frac{\cosh(\sqrt{s/\alpha} \cdot L)}{\sinh(\sqrt{s/\alpha} \cdot L)} \right\} \tag{1.56}$$

With reference to Equations 1.45, 1.50 and 1.53 to 1.56, the equations for the heat fluxes at $x = 0$ and at $x = L$, when the wall is subject to excitations $f(0,s)$ at $x = 0$ and $f(L,s)$ at $x = L$, can be written as:

$$q(0,s) = G_{q,0}(0,s)f(0,s) + G_{q,L}(0,s)f(L,s) \tag{1.57}$$

$$q(L,s) = G_{q,0}(L,s)f(0,s) + G_{q,L}(L,s)f(L,s) \tag{1.58}$$

which can be expressed in the following matrix form:

$$\begin{Bmatrix} q(0,s) \\ q(L,s) \end{Bmatrix} = \begin{bmatrix} G_{q,0}(0,s) & G_{q,L}(0,s) \\ G_{q,0}(L,s) & G_{q,L}(L,s) \end{bmatrix} \begin{Bmatrix} f(0,s) \\ f(L,s) \end{Bmatrix} \tag{1.59}$$

Equation 1.59 allows the heat fluxes at the two sides of the wall (the 'output') in response to the temperature excitations at the two wall surfaces (the 'input') to be determined. The coefficient matrix is, therefore, the transfer function of the wall (the 'system') relating the input to the output of the system.

Through inverting the coefficient matrix in the equation, the following equation can be obtained to relate the temperature response of the wall that can be observed at the surfaces at the two sides to heat inputs at the two surfaces:

$$\begin{Bmatrix} T(0,s) \\ T(L,s) \end{Bmatrix} = \begin{bmatrix} G_{q,0}(0,s) & G_{q,L}(0,s) \\ G_{q,0}(L,s) & G_{q,L}(L,s) \end{bmatrix}^{-1} \begin{Bmatrix} q(0,s) \\ q(L,s) \end{Bmatrix} \tag{1.60}$$

Note that $f(0,s)$ and $f(L,s)$ have been replaced by $T(0,s)$ and $T(L,s)$ respectively in the above equation, to reflect more appropriately that the temperatures are the 'responses' rather than the 'forcing functions'. In the following, $T(0,s)$ and $T(L,s)$ are used to denote the Laplace transforms of the temperature functions for the two surfaces of the wall, irrespective of whether they represent forcing functions or responses to heat input.

Other useful forms of the equation can be derived by algebraic manipulations with Equations 1.57 and 1.58, for instance:

$$\begin{Bmatrix} T(0,s) \\ q(0,s) \end{Bmatrix} = \begin{bmatrix} A(s) & B(s) \\ C(s) & D(s) \end{bmatrix} \begin{Bmatrix} T(L,s) \\ q(L,s) \end{Bmatrix} \tag{1.61}$$

where:

$$A(s) = -\frac{G_{q,L}(L,s)}{G_{q,0}(L,s)}$$

$$B(s) = \frac{1}{G_{q,0}(L,s)}$$

$$C(s) = G_{q,L}(0,s) - \frac{G_{q,L}(L,s)G_{q,0}(0,s)}{G_{q,0}(L,s)}$$

$$D(s) = \frac{G_{q,0}(0,s)}{G_{q,0}(L,s)}$$

Substituting Equations 1.53 to 1.56 into the above:

$$A(s) = \cosh\left(\sqrt{s/\alpha} \cdot L\right) \tag{1.62}$$

$$B(s) = \frac{\sinh\left(\sqrt{s/\alpha} \cdot L\right)}{k\sqrt{s/\alpha}} \tag{1.63}$$

$$C(s) = k\sqrt{s/\alpha} \sinh\left(\sqrt{s/\alpha} \cdot L\right) \tag{1.64}$$

$$D(s) = \cosh\left(\sqrt{s/\alpha} \cdot L\right) \tag{1.65}$$

Note that the determinant of the coefficient matrix in Equation 1.61 $(= A(s)D(s) - B(s)C(s))$ equals one. The equation can also be inverted into:

$$\begin{Bmatrix} T(L,s) \\ q(L,s) \end{Bmatrix} = \begin{bmatrix} D(s) & -B(s) \\ -C(s) & A(s) \end{bmatrix} \begin{Bmatrix} T(0,s) \\ q(0,s) \end{Bmatrix} \tag{1.66}$$

Equations 1.61 and 1.66 are particularly useful, as each of them relays the excitations at one side of the wall to the response of the wall at the other side. Therefore, when a composite wall involves N layers of different materials, the response of the wall can be modelled by recognising that the output of one layer becomes the input to the next layer. Hence, the transfer function matrices of the layers can be multiplied in sequence to provide an overall transfer function:

$$\begin{Bmatrix} T(0,s) \\ q(0,s) \end{Bmatrix} = \begin{bmatrix} A_1(s) & B_1(s) \\ C_1(s) & D_1(s) \end{bmatrix} \begin{bmatrix} A_2(s) & B_2(s) \\ C_2(s) & D_2(s) \end{bmatrix} \cdots$$

$$\cdots \begin{bmatrix} A_{N-1}(s) & B_{N-1}(s) \\ C_{N-1}(s) & D_{N-1}(s) \end{bmatrix} \begin{bmatrix} A_N(s) & B_N(s) \\ C_N(s) & D_N(s) \end{bmatrix} \begin{Bmatrix} T(L,s) \\ q(L,s) \end{Bmatrix}$$

$$= \begin{bmatrix} A_M(s) & B_M(s) \\ C_M(s) & D_M(s) \end{bmatrix} \begin{Bmatrix} T(L,s) \\ q(L,s) \end{Bmatrix} \tag{1.67}$$

where:

$$L = \sum_{i=1}^{N} l_i \tag{1.68}$$

and l_i is the thickness of the ith layer.

Note also that in the overall matrix in Equation 1.67, $A_M(s)$ will not be equal to $D_M(s)$ as in the single layer equations, but its determinant will still be equal to one.

So far, the heat transfer equations for the wall have been derived based solely on the temperature and heat flux at the surfaces. The convective and radiant heat exchange at the surfaces would need to be properly treated for applications to buildings.

Recall Equations 1.4a and 1.4b, with the convective heat transfer coefficients at the outdoor (assumed to be at $x=0$) and the indoor (at $x=L$) sides denoted respectively as h_o and h_i, the time varying ambient air temperatures at the two sides as $T_o(t)$ and $T_i(t)$, and the time varying net radiant heat gains at the two side as $q_{r,o}(t)$ and $q_{r,i}(t)$:

$$q(0,t) = h_o(T_o(t) - T(0,t)) + q_{r,o}(t) \tag{1.69a}$$

$$q(L,t) = h_i(T(L,t) - T_i(t)) - q_{r,i}(t) \tag{1.69b}$$

When the two surfaces are each treated as one layer of a composite wall, the heat flow into the wall (at the surface at $x = 0$) or out of the wall (at the surface at $x = L$) (in the positive x-direction), which are given respectively by the right hand side of Equations 1.69a and 1.69b, should now be regarded as the heat flow into and out of the respective layers. Denoting these heat flows as $q_o(t)$ and $q_i(t)$:

$$q(0,t) = q_o(t) \tag{1.70a}$$

$$q(L,t) = q_i(t) \tag{1.70b}$$

The above equations reflect the property of the 'surface layers' that each layer has zero heat capacity ($\rho c = 0$), and thus any heat that flows in must flow out at the same rate.

It follows that:

$$q_o(t) = h_o(T_o(t) - T(0,t)) + q_{r,o}(t) \tag{1.71a}$$

$$q_i(t) = h_i(T(L,t) - T_i(t)) - q_{r,i}(t) \tag{1.71b}$$

Solving for the wall surface temperatures:

$$T(0,t) = T_o(t) + \frac{1}{h_o} q_{r,o}(t) - \frac{1}{h_o} q_o(t) \tag{1.72a}$$

$$T(L,t) = T_i(t) + \frac{1}{h_i} q_{r,i}(t) + \frac{1}{h_i} q_i(t) \tag{1.72b}$$

The indoor/outdoor temperature and the net absorbed radiant heat terms in the above equations can be combined and denoted by the environmental temperature $T_{e,i}$ and the sol-air temperature $T_{e,o}$, allowing Equation 1.72 to be simplified to:

$$T(0,t) = T_{e,o}(t) - \frac{1}{h_o} q_o(t) \tag{1.73a}$$

$$T(L,t) = T_{e,i}(t) + \frac{1}{h_i} q_i(t) \tag{1.73b}$$

The Laplace transforms of Equations 1.70 and 1.73 are:

$$T(0,s) = T_{e,o}(s) - \frac{1}{h_o} q_o(s) \tag{1.74}$$

$$q(0,s) = q_o(s) \tag{1.75}$$

$$T(L,s) = T_{e,i}(s) + \frac{1}{h_i} q_i(s) \tag{1.76}$$

$$q(L,s) = q_i(s) \tag{1.77}$$

For the surface layer at $x = L$, the heat flow out of the wall and the surface temperature of the wall can be related to the heat flow out of the surface layer and the indoor environmental temperature as follows:

$$\begin{Bmatrix} T(L,s) \\ q(L,s) \end{Bmatrix} = \begin{bmatrix} 1 & 1/h_i \\ 0 & 1 \end{bmatrix} \begin{Bmatrix} T_{e,i}(s) \\ q_i(s) \end{Bmatrix} \tag{1.78}$$

For the surface at $x = 0$, Equations 1.74 and 1.75 are re-arranged as:

$$T_{e,o}(s) = T(0,s) + \frac{1}{h_o} q(0,s) \tag{1.79}$$

$$q_o(s) = q(0,s) \tag{1.80}$$

The corresponding matrix form of the two equations is:

$$\begin{Bmatrix} T_{e,o}(s) \\ q_o(s) \end{Bmatrix} = \begin{bmatrix} 1 & 1/h_o \\ 0 & 1 \end{bmatrix} \begin{Bmatrix} T(0,s) \\ q(0,s) \end{Bmatrix} \tag{1.81}$$

Equations (1.78) and (1.81) can now be substituted into Equation 1.67 to yield:

$$\begin{Bmatrix} T_{e,o}(s) \\ q_o(s) \end{Bmatrix} = \begin{bmatrix} 1 & 1/h_o \\ 0 & 1 \end{bmatrix} \begin{bmatrix} A_M(s) & B_M(s) \\ C_M(s) & D_M(s) \end{bmatrix} \begin{bmatrix} 1 & 1/h_i \\ 0 & 1 \end{bmatrix} \begin{Bmatrix} T_{e,i}(s) \\ q_i(s) \end{Bmatrix} \tag{1.82}$$

which can be simplified to:

$$\begin{Bmatrix} T_{e,o}(s) \\ q_o(s) \end{Bmatrix} = \begin{bmatrix} A_O(s) & B_O(s) \\ C_O(s) & D_O(s) \end{bmatrix} \begin{Bmatrix} T_{e,i}(s) \\ q_i(s) \end{Bmatrix} \tag{1.83}$$

Equation 1.83 can be re-arranged (taking note that $A_O(s)D_O(s) - B_O(s)C_O(s) = 1$), to yield the following, which would be more directly applicable for the prediction of the heat transfer across a wall when the wall is subject to variations in the indoor and outdoor temperatures and radiant heat gains:

$$\begin{Bmatrix} q_o(s) \\ q_i(s) \end{Bmatrix} = \begin{bmatrix} \dfrac{D_O(s)}{B_O(s)} & -\dfrac{1}{B_O(s)} \\ \dfrac{1}{B_O(s)} & -\dfrac{A_O(s)}{B_O(s)} \end{bmatrix} \begin{Bmatrix} T_{e,o}(s) \\ T_{e,i}(s) \end{Bmatrix} \tag{1.84}$$

The Laplace transformation analysis reviewed above is a common basis for several modelling methods that are in common use nowadays. These include the Response Factor Method, Transfer Function Method and the Admittance Method. A review of the theoretical basis of each of these methods is given in the following sections.

Response Factor Method

The Response Factor Method was first proposed by Stephenson and Mitalas (1967). The elements in the matrix in Equation 1.59, which relates the heat fluxes at the two surfaces of a homogeneous slab of material to the temperature excitations at the two surfaces (denoted as surfaces 0 and 1, and the temperature and heat flux at these surfaces denoted by subscripts 0 and 1 respectively), can be expressed in terms of the transfer functions $A(s)$, $B(s)$ and $D(s)$ shown in Equations 1.62 to 1.65 as follows:

$$\begin{Bmatrix} q_0(s) \\ q_1(s) \end{Bmatrix} = \begin{bmatrix} \dfrac{D(s)}{B(s)} & -\dfrac{1}{B(s)} \\ \dfrac{1}{B(s)} & -\dfrac{A(s)}{B(s)} \end{bmatrix} \begin{Bmatrix} T_0(s) \\ T_1(s) \end{Bmatrix} \tag{1.85}$$

It can be seen from Equation 1.85 that the Laplace transforms of the heat flux at the two surfaces of the layer of material due to the temperature excitation $T_0(s)$ alone are given by:

$$q_0(s) = \frac{D(s)}{B(s)} T_0(s) \tag{1.86}$$

$$q_1(s) = \frac{1}{B(s)} T_0(s) \tag{1.87}$$

For clarity, $q_0(s)$ and $q_1(s)$ in Equations 1.86 and 1.87 are denoted as $q_{0,0}(s)$ and $q_{1,0}(s)$ respectively, to denote that they are the heat fluxes due to excitation $T_0(s)$.

If the excitation is a unit impulse, $T_0(s) = 1$. Therefore, the unit impulse responses of the slab of material in terms of the surface heat fluxes in the time-domain that will arise due to this unit impulse excitation, denoted as $x(t)$ and $y(t)$ respectively, can be found as follows:

$$x(t) = L^{-1}\left\{\frac{D(s)}{B(s)}\right\}$$
(1.88)

$$y(t) = L^{-1}\left\{\frac{1}{B(s)}\right\}$$
(1.89)

If, instead of using a sequence of impulses, each with time duration that tends to zero, a sequence of pulses, each of duration of one time unit ($\Delta t = 1$) and a strength of $T_0(t) \cdot \Delta t$, is used to approximate the temperature excitation function, as shown in Fig. 1.8, the response of the wall to this excitation function at $t = n \cdot \Delta t = n$, can be approximately determined by:

$$q_{0,0}(n) = x(0)T_0(n) + x(1)T_0(n-1) + x(2)T_0(n-2) + \dots$$

$$= \sum_{j=0}^{\infty} x(j)T_0(n-j)$$
(1.90)

$$q_{1,0}(n) = y(0)T_0(n) + y(1)T_0(n-1) + y(2)T_0(n-2) + \dots$$

$$= \sum_{j=0}^{\infty} y(j)T_0(n-j)$$
(1.91)

Similarly, the heat flux at the two wall surfaces at time $n \cdot \Delta t$, in response to the temperature excitation at surface 1 ($T_1(t)$), denoted as $q_{0,1}(n)$ and $q_{1,1}(n)$ respectively, can be approximated by:

$$q_{0,1}(n) = w(0)T_1(n) + w(1)T_1(n-1) + w(2)T_1(n-2) + \dots$$

$$= \sum_{j=0}^{\infty} w(j)T_1(n-j)$$
(1.92)

$$q_{1,1}(n) = z(0)T_1(n) + z(1)T_1(n-1) + z(2)T_1(n-2) + \dots$$

$$= \sum_{j=0}^{\infty} z(j)T_1(n-j)$$
(1.93)

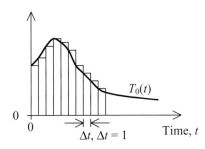

Fig. 1.8   Representing an arbitrary function by a sequence of rectangular pulses.

where:

$$w(t) = L^{-1}\left\{-\frac{1}{B(s)}\right\} \tag{1.94}$$

$$z(t) = L^{-1}\left\{-\frac{A(s)}{B(s)}\right\} \tag{1.95}$$

Comparing Equations 1.94 and 1.89 reveals that $w(t) = -y(t)$. Hence, only the impulse responses $x(t)$, $y(t)$ and $z(t)$ would need to be known.

Since the impulse responses are used in the above method, it will yield accurate results only if the time step, Δt, used in the calculation is very small. The reasons for this include that the response of a system to a pulse input of a particular strength will approach that to an impulse of the same strength only if the duration of the pulse tends to zero. Furthermore, the discrepancy between the excitation function and the pulses used to represent it will become larger the longer the duration of the pulse used. Using a very small time step, however, means that the calculation will need to cover a large number of previous time steps for determining the heat fluxes at a given time.

A better way of representing an excitation function is to use triangular pulses, as shown in Fig. 1.9. A unit triangular pulse function $f_T(t)$ with pulse strength of one time unit (Δt) can be constructed from a ramp function $f_R(t)$, as shown in Fig. 1.10.

The unit triangular pulse, therefore, can be expressed in terms of the ramp function as:

$$f_T(t) = \frac{f_R(t + \Delta t) - 2f_R(t) + f_R(t - \Delta t)}{\Delta t} \tag{1.96}$$

Hence, if $\xi(t)$ is the wall's response to the ramp function, the wall's response to the unit triangular pulse function, say $X(t)$, can be found as:

$$X(t) = \frac{\xi(t + \Delta t) - 2\xi(t) + \xi(t - \Delta t)}{\Delta t} \tag{1.97}$$

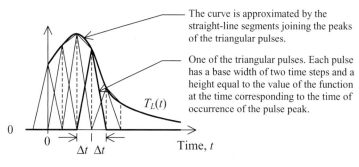

The curve is approximated by the straight-line segments joining the peaks of the triangular pulses.

One of the triangular pulses. Each pulse has a base width of two time steps and a height equal to the value of the function at the time corresponding to the time of occurrence of the pulse peak.

$T_L(t)$

Time, t

Fig. 1.9   Representing an arbitrary function by a sequence of triangular pulses.

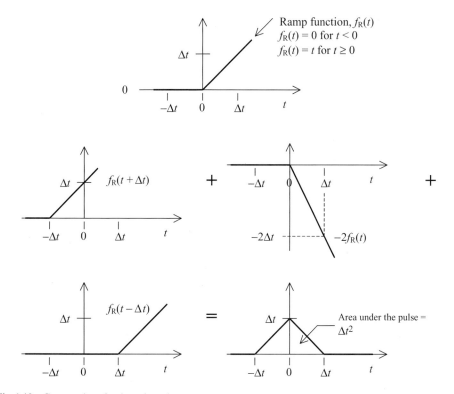

Fig. 1.10 Construction of a triangular pulse.

Note that the above result holds because the ramp function has been defined such that its value is 0 for $t<0$.

The Laplace transform of $f_R(t)$ is:

$$L\{f_R(t)\} = f_R(s) = \frac{1}{s^2} \tag{1.98}$$

Let $\xi_{i,j}(t)$ be the heat flux at surface i due to the application of a temperature input at surface j, which is a ramp function, according to Equations 1.85 and 1.98:

$$\xi_{0,0}(t) = L^{-1}\left\{\frac{D(s)}{s^2 B(s)}\right\} \tag{1.99}$$

$$\xi_{1,0}(t) = L^{-1}\left\{\frac{1}{s^2 B(s)}\right\} \tag{1.100}$$

$$\xi_{0,1}(t) = L^{-1}\left\{-\frac{1}{s^2 B(s)}\right\} = -\xi_{1,0}(t) \tag{1.101}$$

$$\xi_{1,1}(t) = L^{-1}\left\{-\frac{A(s)}{s^2 B(s)}\right\} = -\xi_{0,0}(t) \tag{1.102}$$

Since $A(s) = D(s)$ for a homogeneous slab (see Equations 1.62 and 1.65), only the inverse Laplace transforms in Equations 1.99 and 1.100 need to be evaluated. By using the residue theorem, it can be shown that the inverse Laplace transforms are:

$$\xi_{0,0}(t) = L^{-1}\left\{\frac{D(s)}{s^2 B(s)}\right\} = \frac{D(0)}{B(0)}t + \left[\frac{d}{ds}\left(\frac{D(s)}{B(s)}\right)\right]_{s=0}$$

$$+ \sum_{m=1}^{\infty} \frac{D(\beta_m)}{\beta_m^2 (dB(s)/ds)_{s=\beta_m}} \exp(\beta_m t) \tag{1.103}$$

$$\xi_{1,0}(t) = L^{-1}\left\{\frac{1}{s^2 B(s)}\right\} = \frac{1}{B(0)}t + \left[\frac{d}{ds}\left(\frac{1}{B(s)}\right)\right]_{s=0}$$

$$+ \sum_{m=1}^{\infty} \frac{1}{\beta_m^2 (dB(s)/ds)_{s=\beta_m}} \exp(\beta_m t) \tag{1.104}$$

where β_m for $m = 1, 2, \ldots$ are the roots of the transfer function $B(s)$ such that $B(\beta_m) = 0$.

For the transfer functions $B(s)$ and $D(s)$ for a homogeneous slab of material, as shown in Equations 1.63 and 1.65, it can be shown that (see Appendix A for details of the steps involved in the derivation):

$$\xi_{0,0} = \frac{k}{L}t + \frac{kL}{3\alpha} - \frac{kL}{\alpha}\frac{2}{\pi^2}\sum_{m=1}^{\infty}\frac{1}{m^2}\exp\left(-\frac{m^2\pi^2\alpha}{L^2}t\right) \tag{1.105}$$

$$\xi_{1,0} = \frac{k}{L}t - \frac{kL}{6\alpha} - \frac{kL}{\alpha}\frac{2}{\pi^2}\sum_{m=1}^{\infty}\frac{(-1)^m}{m^2}\exp\left(-\frac{m^2\pi^2\alpha}{L^2}t\right) \tag{1.106}$$

Having determined the ramp function responses, the unit triangular pulse responses can be found as follows:

$$X(t) = \frac{\xi_{0,0}(t + \Delta t) - 2\xi_{0,0}(t) + \xi_{0,0}(t - \Delta t)}{\Delta t} \tag{1.107}$$

$$Y(t) = \frac{\xi_{1,0}(t + \Delta t) - 2\xi_{1,0}(t) + \xi_{1,0}(t - \Delta t)}{\Delta t} \tag{1.108}$$

$$Z(t) = X(t) \tag{1.109}$$

where $X(t)$ is the heat flux at surface 0 due to a unit triangular temperature pulse imposed onto surface 0, and $Y(t)$ is the corresponding heat flux at surface 1. It follows that the heat flux at surface 0 due to a unit triangular temperature pulse applied to surface 1 will be $-Y(t)$. The heat flux at surface 1 in response to a unit triangular temperature pulse applied to surface 1, however, is traditionally denoted as $-Z(t)$. $Z(t)$ will then have the same numerical value as $X(t)$, as shown in Equation 1.109, which applies only to a homogeneous slab of material. $X(t)$ and $Z(t)$ will have different values for a composite wall.

Note that the triangular pulse excitation comprises three ramp excitations, applied at times separated from each other by one time unit. At $t = 0$, only the ramp excitation with a slope of

+1 has been applied for one time unit. The response of the slab at $t = 0$ (the discussion below makes reference to $X(t)$ only), therefore, is:

$$X(0) = \frac{\xi_{0,0}(\Delta t)}{\Delta t} \tag{1.110}$$

At $t = \Delta t$, the response will include that due to the first ramp excitation for two time units and that due to the ramp excitation with a slope of -2 for one time unit. Hence:

$$X(\Delta t) = \frac{\xi_{0,0}(2\Delta t) - 2\xi_{0,0}(\Delta t)}{\Delta t} \tag{1.111}$$

Starting from $t = 2\Delta t$, the responses due to all the three ramp excitations that make up the triangular pulse will be obtained. Hence, for $j \geq 2$:

$$X(j\Delta t) = \frac{\xi_{0,0}((j+1)\Delta t) - 2\xi_{0,0}(j\Delta t) + \xi_{0,0}((j-1)\Delta t)}{\Delta t} \tag{1.112}$$

The starting values of $Y(j)$ and $Z(j)$ for $j = 0$ and 1 are to be evaluated from the respective ramp function responses in a similar manner.

The values of the unit triangular pulse responses $X(j)$, $Y(j)$ and $Z(j)$ for $j = 0, 1, 2, \ldots$, are called the response factors of a wall or slab. Based on Equations 1.105 to 1.112 and by choosing a time step of one unit ($\Delta t = 1$) such that $t = j \cdot \Delta t = j$, the following expressions for determining the response factors can be obtained:

$$X(0) = \frac{kL}{\alpha} \left\{ \frac{\alpha}{L^2} + \frac{1}{3} - \frac{2}{\pi^2} \sum_{m=1}^{\infty} \frac{1}{m^2} \phi(m) \right\} \tag{1.113a}$$

$$X(1) = \frac{kL}{\alpha} \left\{ -\frac{1}{3} - \frac{2}{\pi^2} \sum_{m=1}^{\infty} \frac{1}{m^2} \left[(\phi(m))^2 - 2\phi(m) \right] \right\} \tag{1.113b}$$

$$X(j) = -\frac{2kL}{\alpha\pi^2} \sum_{m=1}^{\infty} \frac{1}{m^2} \left[\phi(m)^{j+1} - 2(\phi(m))^j + (\phi(m))^{j-1} \right] \quad \text{for } j = 2, 3, 4, \ldots \tag{1.113c}$$

$$Y(0) = \frac{kL}{\alpha} \left\{ \frac{\alpha}{L^2} - \frac{1}{6} - \frac{2}{\pi^2} \sum_{m=1}^{\infty} \frac{(-1)^m}{m^2} \phi(m) \right\} \tag{1.114a}$$

$$Y(1) = \frac{kL}{\alpha} \left\{ \frac{1}{6} - \frac{2}{\pi^2} \sum_{m=1}^{\infty} \frac{(-1)^m}{m^2} \left[(\phi(m))^2 - 2\phi(m) \right] \right\} \tag{1.114b}$$

$$Y(j) = -\frac{2kL}{\alpha\pi^2} \sum_{m=1}^{\infty} \frac{(-1)^m}{m^2} \left[(\phi(m))^{j+1} - 2(\phi(m))^j + (\phi(m))^{j-1} \right] \quad \text{for } j = 2, 3, 4, \ldots \tag{1.114c}$$

where

$$\phi(m) = \exp\left(-\frac{m^2\pi^2\alpha}{L^2}\right) \tag{1.115}$$

Figure 1.11 shows the response factors calculated using the above equations for a slab of concrete.

The heat flux at the surfaces of the slab of material at time t can then be determined as follows:

$$q_{0,0}(n) = X(0)T_0(n) + X(1)T_0(n-1) + X(2)T_0(n-2) + \ldots$$

$$= \sum_{j=0}^{\infty} X(j)T_0(n-j) \tag{1.116}$$

$$q_{1,0}(n) = Y(0)T_0(n) + Y(1)T_0(n-1) + Y(2)T_0(n-2) + \cdots$$

$$= \sum_{j=0}^{\infty} Y(j)T_0(n-j) \tag{1.117}$$

$$q_{0,1}(n) = -Y(0)T_1(n) - Y(1)T_1(n-1) - Y(2)T_1(n-2) + \ldots$$

$$= -\sum_{j=0}^{\infty} Y(j)T_1(n-j) \tag{1.118}$$

$$q_{1,1}(n) = -Z(0)T_1(n) - Z(1)T_1(n-1) - Z(2)T_1(n-2) + \ldots$$

$$= -\sum_{j=0}^{\infty} Z(j)T_1(n-j) \tag{1.119}$$

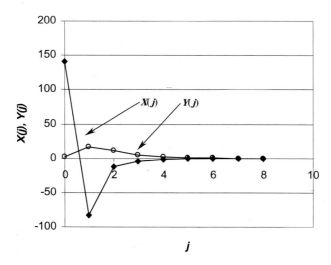

Fig. 1.11   Response factors for a 200 mm thick concrete slab.

Note that, as a consequence of the definition of $Z(j)$, all $Z(j)$ terms in Equation 1.119 are assigned with a negative sign to yield the correct heat flow direction.

Equations 1.113 to 1.119 can be summarised as:

$$\begin{Bmatrix} q_0(n) \\ q_1(n) \end{Bmatrix} = \sum_{j=0}^{\infty} \begin{bmatrix} X(j) & -Y(j) \\ Y(j) & -Z(j) \end{bmatrix} \begin{Bmatrix} T_0(n-j) \\ T_1(n-j) \end{Bmatrix} \tag{1.120}$$

For a composite wall, instead of combining the transfer functions of individual layers and then inverse-transforming the resultant combined transfer functions, the overall response factors may be found from the response factors of individual layers. Appendix B shows the derivation of the equations for calculating the overall response factors for a wall comprising two layers of homogeneous materials, as shown in Fig. 1.12. Chen and Li (1987) derived the following abbreviated versions of Equations B.16a to B.16c in Appendix B:

$$X(j) = X_A(j) - \sum_{i=0}^{j} Y_A(i)TR_A(j-i) \tag{1.121}$$

$$Y(j) = \sum_{i=0}^{j} Y_B(i)TR_A(j-i) \tag{1.122}$$

$$Z(j) = Z_B(j) - \sum_{i=0}^{j} Y_B(i)TR_B(j-i) \tag{1.123}$$

which are based on the following recursive formulae:

$$TR_A(j) = \frac{1}{X_B(0) + Z_A(0)} \left\{ Y_A(j) - \sum_{i=1}^{j} (X_B(i) + Z_A(i))TR_A(j-i) \right\} \tag{1.124}$$

$$TR_B(j) = \frac{1}{X_B(0) + Z_A(0)} \left\{ Y_B(j) - \sum_{i=1}^{j} (X_B(i) + Z_A(i))TR_B(j-i) \right\} \tag{1.125}$$

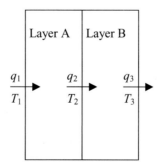

Fig. 1.12 A wall comprising two layers of homogeneous materials.

where $X(j)$, $Y(j)$ and $Z(j)$ are the response factors for the two-layered composite wall, $X_A(j)$, $Y_A(j)$ and $Z_A(j)$ and $X_B(j)$, $Y_B(j)$ and $Z_B(j)$ are respectively the response factors of the two homogeneous layers of materials. Note that $TR_A(j)$ and $TR_B(j)$ both equal zero when $j < 0$.

The above procedure can be applied repeatedly to yield the overall response factors for a multi-layer wall, including the 'surface' layers at the two sides of the wall. The response factors for several wall materials and a composite wall, calculated using the method described above, are summarised in Table 1.1.

The method, however, may not provide accurate enough results, especially when the number of layers involved is large, due to accumulation of errors introduced by linear interpolation of the boundary temperatures between layers. More accurate results require inverse-transformation of the combined transfer functions. Continued efforts are being made by researchers to develop enhanced methods for finding response factors for multi-layer walls (e.g. Wang and Chen 2003; Zhang and Ding 2003).

In using the response factor method to calculate the heat transfer through an envelope element at a particular time step, the summation process needs to go back in time for a sufficiently long period in order to get accurate results. It follows that response factors for a long series of time steps have to be available.

Note that when j is large, from Equation 1.113c:

$$X(j) = -\frac{2kL}{\alpha\pi^2} \sum_{m=1}^{\infty} \frac{1}{m^2} \left[\phi(m)^{j+1} - 2(\phi(m))^j + (\phi(m))^{j-1} \right] \approx -\frac{2kL}{\alpha\pi^2} (\phi(1))^{j-1} \qquad (1.126)$$

Therefore,

$$C = \frac{X(j+1)}{X(j)} \approx \frac{(\phi(1))^j}{(\phi(1))^{j-1}} = \phi(1) = \exp\left(-\frac{\pi^2\alpha}{L^2}\right) \qquad (1.127)$$

where C is called the common ratio of the response factors for large j values. Note that the other response factors $Y(j)$ and $Z(j)$ have the same common ratio.

Therefore, values of response factors for building fabric elements are often given only up to a large enough value of j, and the response factors corresponding to higher j values can be determined from the common ratio.

Based on the above, the summation processes (Equations 1.116 to 1.119) involved in calculating the heat fluxes can be simplified as follows.

Let:

$$q(n) = \sum_{j=0}^{\infty} Y(j)T(n-j) \qquad (1.128)$$

The summation process in Equation 1.128 may be split into two processes, with the first from $j = 0$ to k and the second from $j = k + 1$ to infinity. If k is a large enough value, Equation 1.128 can be written in terms of the common ratio C as follows:

$$q(n) = \sum_{j=0}^{k} Y(j)T(n-j) + \sum_{j=1}^{\infty} C^j \cdot Y(k)T(n-k-j) \qquad (1.129)$$

Table 1.1 Response factors for several building materials and a composite wall.

(a) Properties of materials

Material	k $\text{W m}^{-1}\text{K}^{-1}$	k^* $\text{kJ h}^{-1}\text{m}^{-1}\text{K}^{-1}$	ρ kg m^{-3}	c $\text{kJ kg}^{-1}\text{K}^{-1}$
Plaster	0.72	2.592	1860	0.84
Concrete	2.16	7.776	2400	0.84
Gypsum	0.38	1.368	1120	0.84

(b) Response factors for envelope layers

j	100 mm thick Concrete $X(j)$	$Y(j)$	150 mm thick concrete $X(j)$	$Y(j)$	200 mm thick concrete $X(j)$	$Y(j)$	13 mm thick plaster $X(j)$	$Y(j)$	10 mm thick gypsum $X(j)$	$Y(j)$
0	144.052328	45.0676675	141.337021	12.7077343	141.279649	2.7726787	206.155015	195.999415	139.936	135.232
1	-65.404822	31.8048318	-80.272411	29.9428598	-82.588178	17.1833557	-6.7704	3.3852	-3.136	1.568
2	-0.8677871	0.86778209	-7.5289968	7.49383305	-12.3248	11.4570125	-3.451E-42	3.4508E-42	-8.989E-63	8.989E-63
3	-0.0192805	0.01928054	-1.3833472	1.38330675	-4.5998085	4.58052792	-2.893E-84	2.8931E-84	-4.24E-125	4.238E-125
4	-0.0004284	0.00042838	-0.2547581	0.2547581	-1.7723841	1.77195573	-2.43E-126	2.426E-126	-2E-187	1.998E-187
5	-9.518E-06	9.5177E-06	-0.0469171	0.04691711	-0.6842033	0.6841938	-2.03E-168	2.034E-168	-9.42E-250	9.423E-250
6	-2.115E-07	2.1147E-07	-0.0086404	0.00864041	-0.2641552	0.26415496	-1.7E-210	1.705E-210	0	0
7	-4.698E-09	4.6984E-09	-0.0015912	0.00159125	-0.1019849	0.10198487	-1.43E-252	1.429E-252	0	0
8	-1.044E-10	1.0439E-10	-0.000293	0.00029305	-0.0393743	0.03937427	-1.2E-294	1.198E-294	0	0
C	0.02221811		0.18416334		0.38607955		8.384E-43		4.715E-63	

(c) Response factors for a wall comprising a 13 mm thick plaster layer, 100 mm thick concrete and 10 mm gypsum board

j	$X(j)$	$Y(j)$	$Z(j)$
0	94.17374	12.2612485	74.1975353
1	-148.28022	15.9130757	-88.453671
2	-41.479725	18.7774082	-26.951989
3	-18.668312	9.16710639	-10.389376
4	-10.098314	3.95942326	-4.2743595
5	-4.9396557	1.66995028	-1.7843373
6	-2.2441537	0.70084519	-0.7472162
7	-0.9803058	0.2938221	-0.313117
8	-0.4197096	0.12315436	-0.1312288
C	0.41914603		

*Value used in calculation of response factors due to the use of a time step of 1 h.
Symbols: C = common ratio; k = thermal conductivity; ρ = density; c = heat capacity.

For the $(n-1)^{th}$ time step:

$$q(n-1) = \sum_{j=0}^{k} Y(j)T(n-1-j) + \sum_{j=1}^{\infty} C^j \cdot Y(k)T(n-1-k-j) \qquad (1.130)$$

It follows that:

$$q(n) - C \cdot q(n-1) = Y(0)T(n) + \sum_{j=1}^{k} \{Y(j) - C \cdot Y(j-1)\}T(n-j) \qquad (1.131)$$

Hence, the summation process can be limited to k terms as follows:

$$q(n) = C \cdot q(n-1) + Y(0)T(n) + \sum_{j=1}^{k} \{Y(j) - C \cdot Y(j-1)\}T(n-j) \qquad (1.132)$$

Transfer Function Method

The Transfer Function Method (TFM) (Stephenson & Mitalas 1971) was a further development of the Response Factor Method, which will allow the heat transfer through building envelopes to be determined more efficiently than using the Response Factor Method described in the preceding section. TFM is based on the z-transform, which is itself based on the Laplace transform.

Recall that the Laplace transform of a time function $f(t)$ is defined as:

$$f(s) = L\{f(t)\} = \int_{0}^{\infty} f(t)\exp(-st)dt \qquad (1.133)$$

The following summation process may be used to approximate the integral in the above Laplace transform:

$$f(s) \approx \Delta t \sum_{j=0}^{n} f(j\Delta t)\exp(-js\Delta t)$$
$$= \Delta t\{f(0) + f(\Delta t)\exp(-s\Delta t) + f(2\Delta t)\exp(-2s\Delta t) + \ldots + f(n\Delta t)\exp(-ns\Delta t)\} \quad (1.134)$$

With the complex variable z as defined in Equation 1.135 and setting Δt equals 1 (time unit), Equation 1.134 can be re-written as shown in Equation 1.136:

$$z = \exp(s\Delta t) \qquad (1.135)$$

$$f(s) \approx f(0) + f(1)z^{-1} + f(2)z^{-2} + \cdots + f(n)z^{-n} \qquad (1.136)$$

The right hand side of Equation 1.136 may be regarded as the definition of the z-transform of the time function $f(t)$. Let $f(z)$ and $g(z)$ be the z-transform of the time functions $f(t)$ and $g(t)$ respectively, then:

$$f(z) = f(0) + f(1)z^{-1} + f(2)z^{-2} + \ldots + f(n)z^{-n} \qquad (1.137)$$

$$g(z) = g(0) + g(1)z^{-1} + g(2)z^{-2} + \ldots + g(n)z^{-n} \tag{1.138}$$

If $f(t)$ is the input function to a system while $g(t)$ is the corresponding output of the system, the relation between the two can be written as:

$$\frac{g(z)}{f(z)} = K(z) \tag{1.139}$$

The function $K(z)$ is then the z-transfer function of the system, which is the z-transform of the unit impulse response of the system. Assuming that $K(z)$ is given by:

$$K(z) = \frac{a_0 + a_1 z^{-1} + a_2 z^{-2} + a_3 z^{-3} + \ldots}{1 + b_1 z^{-1} + b_2 z^{-2} + b_3 z^{-3} + \ldots} \tag{1.140}$$

substituting Equations 1.137, 1.138 and 1.140 into Equation 1.139 yields.

$$\begin{aligned}
(f(0) + f(1)z^{-1} &+ f(2)z^{-2} + f(3)z^{-3} + \ldots)(a_0 + a_1 z^{-1} + a_2 z^{-2} + a_3 z^{-3} + \ldots) \\
&= (g(0) + g(1)z^{-1} + g(2)z^{-2} + g(3)z^{-3} + \ldots)(1 + b_1 z^{-1} + b_2 z^{-2} + b_3 z^{-3} + \ldots)
\end{aligned} \tag{1.141}$$

Note that multiplying the two series at each side of the above equation will yield a polynomial of z. Furthermore, the coefficient of z of a specific power, say z^{-n}, at one side of the equation must equal the coefficient of z of the same power at the other side, which is as shown below:

$$\begin{aligned}
a_0 f(n) + a_1 f(n-1) &+ a_2 f(n-2) + a_3 f(n-3) + \ldots \\
&= g(n) + b_1 g(n-1) + b_2 g(n-2) + b_3 g(n-3) + \ldots
\end{aligned}$$

It follows that the response of the system at time step n can be found from a weighted sum of the past responses and the current and past inputs as follows:

$$\begin{aligned}
g(n) = a_0 f(n) &+ a_1 f(n-1) + a_2 f(n-2) + a_3 f(n-3) + \ldots \\
&- [b_1 g(n-1) + b_2 g(n-2) + b_3 g(n-3) + \ldots]
\end{aligned} \tag{1.142}$$

The coefficients $a_{0\ldots j}$ and $b_{0\ldots j}$ in Equation 1.142 are called the weighting factors. Recall that the heat flux at surface 0 of a homogeneous slab of material in response to a unit triangular impulse temperature excitation applied to surface 0 is as given in Equation 1.116, which is reproduced below with the subscripts for the heat flux and temperature dropped for simplicity:

$$q(n) = X(0)T(n) + X(1)T(n-1) + X(2)T(n-2) + \ldots \tag{1.143}$$

If this heat flux is regarded as the 'output', and the temperature series as the 'input', Equation 1.142 may be re-written, for various time steps j for $j = 0, 1, 2, \ldots$, as follows:

For $j = 0$:

$$q(0) = a_0 T(0)$$

Comparing the above with Equation 1.143, it can be seen that:

$$a_0 = X(0) \tag{1.144a}$$

For $j = 1$:

$$q(1) = a_0 T(1) + a_1 T(0) - b_1 q(0)$$
$$= a_0 T(1) + a_1 T(0) - b_1 X(0) T(0)$$

$$q(1) = a_0 T(1) + (a_1 - b_1 X(0)) T(0)$$

It follows that:

$$a_1 - b_1 X(0) = X(1)$$
$$a_1 = X(1) + b_1 X(0) \tag{1.144b}$$

For $n = 2$:

$$q(2) = a_0 T(2) + a_1 T(1) + a_2 T(0) - b_1 q(1) - b_2 q(0)$$
$$= a_0 T(2) + a_1 T(1) + a_2 T(0) - b_1 (X(0) T(1) + X(1) T(0)) - b_2 X(0) T(0)$$

$$q(2) = a_0 T(2) + (a_1 - b_1 X(0)) T(1) + (a_2 - b_1 X(1) - b_2 X(0)) T(0)$$

It follows that:

$$a_2 - b_1 X(1) - b_2 X(0) = X(2)$$

$$a_2 = X(2) + b_1 X(1) + b_2 X(0) \tag{1.144c}$$

By continuing with the process for greater values of j, it can be shown that:

$$a_j = X(j) + b_1 X(j-1) + b_2 X(j-2) + \ldots + b_j X(0) \tag{1.145}$$

This shows that the coefficients $a_{0\ldots j}$ can be determined as a weighted sum of the response factors for the current and past time steps. The weights $(b_{0\ldots j})$ are the coefficients of the polynomial in z at the denominator of the z-transfer function $K(z)$ shown in Equation 1.140.

Note that the heat flux to temperature relation, in Laplace transform, is:

$$q(s) = \frac{D(s)}{B(s)} T(s)$$

The equivalent z-transform of the above expression is:

$$q(z) = \frac{D(z)}{B(z)} T(z) = K(z) T(z)$$

The transfer function $B(z)$ can be expressed as:

$$B(z) = 1 + b_1 z^{-1} + b_2 z^{-2} + b_3 z^{-3} + b_4 z^{-4} \dots$$
$$= (1 - z_1 z^{-1})(1 - z_2 z^{-2})(1 - z_3 z^{-3})(1 - z_4 z^{-4}) \dots \qquad (1.146)$$

It can be seen that z_j for $j = 1, 2, \dots$, are the roots of the polynomial expression for $B(z)$ such that $B(z_j) = 0$. From Equations 1.135 and 1.136, it can be seen that, when z_j is a root of $B(z)$, the corresponding value of s (Equation 1.135), denoted as β_j, will also be a solution of $B(s)$.

The coefficient b_j will therefore be the coefficient of z^{-j} in the expression obtained by multiplying out Equation 1.146.

From the above analysis, it can be seen that the heat fluxes at the two surfaces of a wall in response to the temperature excitations applied to the surfaces can be expressed in matrix form as follows:

$$\begin{Bmatrix} q_0(n) \\ q_1(n) \end{Bmatrix} = \sum_{j=0}^{n} \begin{bmatrix} E(j) & -F(j) \\ F(j) & -G(j) \end{bmatrix} \begin{Bmatrix} T_0(n-j) \\ T_1(n-j) \end{Bmatrix} - \sum_{j=1}^{n} \begin{Bmatrix} H(j)q_0(n-j) \\ H(j)q_1(n-j) \end{Bmatrix} \qquad (1.147)$$

$$E(j) = \sum_{j=0}^{n} X(j)H(n-j) \qquad (1.148)$$

$$F(j) = \sum_{j=0}^{n} Y(j)H(n-j) \qquad (1.149)$$

$$G(j) = \sum_{j=0}^{n} Z(j)H(n-j) \qquad (1.150)$$

where $X(j)$, $Y(j)$ and $Z(j)$ are the response factors as defined before, and $E(j)$, $F(j)$ and $G(j)$ and $H(j)$ are the weighting factors for the wall in the TFM. Furthermore, $H(j)$ will be the coefficient of z^{-j} in the following polynomial of z, obtained by expanding the right hand side of the equation:

$$H(0) + H(1)z^{-1} + H(2)z^{-2} + H(3)z^{-3} + H(4)z^{-4} \dots$$
$$= (1 - z_1 z^{-1})(1 - z_2 z^{-2})(1 - z_3 z^{-3})(1 - z_4 z^{-4}) \dots \qquad (1.151)$$

where:

$$z_j = \exp(\beta_j \Delta t) \qquad (1.152)$$

As shown in Equation 1.84, the four transfer functions in the matrix share the same denominator $B(s)$. Hence the same set of $H(j)$ values apply to the heat fluxes at the two surfaces of a wall.

The widest application of the TFM has been for design cooling load calculation for buildings, particularly for generation of the factors needed in the simplified one-step cooling load calculation method of ASHRAE (2001). For this purpose, the assumption is made that the indoor temperature would be fixed and the key concern is the heat flow into the indoor space due to the external climate conditions. Hence, the heat gain from an external wall of area A_w can be expressed as:

$$Q(n) = A_W\left[\sum_{j=0}^{n}F(j)T_{e,o}(n-j)\right] - \sum_{j=1}^{n}H(j)Q(n-j) - A_W T_{e,i}\sum_{j=0}^{n}G(j) \qquad (1.153)$$

Frequency response functions

It has been shown that the temperature and heat flux at one side of a homogeneous slab of material (surface 0) can be related to the temperature and heat flux at the other side (surface 1) by Equation 1.61, as shown once again below:

$$\begin{Bmatrix}T(0,s)\\q(0,s)\end{Bmatrix} = \begin{bmatrix}A(s) & B(s)\\C(s) & D(s)\end{bmatrix}\begin{Bmatrix}T(L,s)\\q(L,s)\end{Bmatrix}$$

For a system the input ($x(s)$) and output ($y(s)$) of which are related by:

$$y(s) = G(s)x(s)$$

If the excitation function $x(t)$ is a sinusoidal function of fixed amplitude and frequency, it can be shown that the steady response will also be sinusoidal with the same frequency, but with a different fixed amplitude and a phase shift from the excitation function. Furthermore, the amplitude ratio and the phase shift between the input and the output time functions can be obtained by replacing the variable s in the transfer function $G(s)$ by $i\omega$ as follows:

$$\frac{|y(t)|}{|x(t)|} = |G(i\omega)| \qquad (1.154)$$

$$y(t+\phi) = |G(i\omega)|x(t) \qquad (1.155)$$

$$\phi = \angle G(i\omega) \qquad (1.156)$$

It can, therefore, be seen that if the excitations to a wall or roof were periodic functions, prediction of the heat transfer through the wall or roof would be a much simpler task, since the response can be determined directly from the transfer function. Note also that the matrix operations for finding the overall transfer function for composite walls equally applies to the frequency analysis.

The frequency analysis as described above is the basis of the Admittance Method (Milbank & Harrington-Lynn 1974) that CIBSE has adopted as the standard method of cooling load prediction (CIBSE 1988). In this method, the sol-air temperature is expressed as a sinusoidal wave of a period of 24 hours. Furthermore, the heat flow rate is decomposed into two components: a steady component and a fluctuating component as follows:

$$Q(t+\phi) = AU\left(\overline{T}_{e,o} + \overline{T}_{e,i}\right) + AU\left(T_{e,o} - \overline{T}_{e,o}\right)f \qquad (1.157)$$

where A is area of the wall or roof, U is the overall heat transfer coefficient of the wall or roof, $\bar{T}_{e,o}$ is the daily mean outdoor sol-air temperature, $\bar{T}_{e,i}$ is the indoor environmental temperature (assumed to be constant), f is called the decrement factor, and ϕ is called the time-lag.

If the sol-air temperature is not a pure sinusoidal function, it can be represented by a set of sinusoidal functions of different frequencies as long as it is periodic. By the superposition theorem for linear systems, the output of a system can be found by summing the response to each of the harmonic components of the input.

Due to the assumption of periodic excitations, the Admittance Method is useful for determining design cooling loads based on a fixed daily pattern of outdoor conditions, but would be cumbersome to use for predicting cooling loads of buildings based on actual weather conditions.

1.4 Lumped capacitance methods

In certain special cases, modelling of the thermal response of building spaces must be done at high time resolution. Examples include: studies on the transient response of secondary HVAC plant; analysis of control system response; control system optimisation; and the design analysis of smart controllers. The previous methods, while suitable for the longer time horizon studies associated for instance with energy prediction, are not generally well suited to these essentially short time horizon problems. A method that gives acceptable accuracy over short time horizon simulations and, through its simplicity, exhibits low computational demands, can be arrived at by treating each building element as one or a small number of 'lumps' in which uniform thermal response is assumed.

First-order lumped capacitance method

It is instructive to view a single lumped element as an equivalent resistive-capacitive electrical network. Figure 1.13 shows this, in which T_i and T_o are the internal and external surface temperatures, r_i and r_o are the inner and outer thermal resistances, T_m is the uniform material temperature, C_m is the thermal capacity of the material per unit area and q_r is a radiant source term.

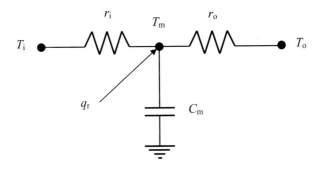

Fig. 1.13   Lumped capacitance construction element.

Recognising that most construction elements are built up from several series layers of material (Fig. 1.12), arriving at a lumped equivalent element for a multiple layer construction element requires the inner and outer material resistances to be weighted in some way with respect to the overall element thermal resistance. Laret (1980) proposed the following for the calculation of the inner and outer material thermal resistances:

$$r_i = a r_m \tag{1.158}$$

and:

$$r_o = (1-a)r_m \tag{1.159}$$

in which a is an 'accessibility factor' given by:

$$a = 1 - \frac{1}{r_m C_m} \sum_{k=1}^{k=n} r_k^* C_k \tag{1.160}$$

where:

$$r_k^* = \sum_{j=1}^{j=n-1} r_j + r_k/2$$

in which r_m and C_m are the overall thermal resistance and overall thermal capacitance per unit area ($\mathrm{m^2\,K\,W^{-1}}$ and $\mathrm{J\,m^{-2}\,K^{-1}}$ respectively) of the multiple layer element, and n is the number of layers forming the element.

Thus the equation of heat conduction based on this first-order description of a construction element can be written as follows, in which q_r is a radiant source term:

$$\frac{dT_m}{dt} = \frac{1}{C_m}\left[q_r + \frac{(T_i - T_m)}{r_i} - \frac{(T_m - T_o)}{r_o} \right] \tag{1.161}$$

Second-order lumped capacitance method

Modelling based on first-order lumped capacitance elements is useful for certain types of simplified problem solving but breaks down in two areas. First, to complete the heat balance of a building space, surface temperatures of all construction elements are required and the uniform lumped element temperature, T_m, does not give an adequate approximation to the temperature at the internal surface of an element. Simplified modelling procedures get round this by approximating the mean radiant temperature of a space to its dry bulb air temperature, T_i, but there are clear accuracy implications associated with this. Second, though the response in heat transfer to a step change in T_i or T_o by the first-order element can be shown to agree well with a fully distributed element, disturbances in surface heat transfer due to radiant inputs do not yield responses equivalent to rigorously modelled elements. This is due to inaccurate modelling of the energy split between conduction and surface convection, again due to the uniformity of material temperature inherent in the method. The error associated with this is especially evident in high thermal capacity construction elements.

One approach to remedy this has been proposed by Tindale (1993) through the introduction of a fictitious 'rad-air' temperature node but, though this method gives improved performance over the first-order room approach, problematical results are reported for cases involving high thermal capacity elements. Another approach, based on second-order element descriptions with optimally tuned parameters, has been proposed by Gouda *et al.* (2002) and is discussed briefly here. Each construction element is treated as a second-order lumped capacitance giving two material body temperatures, one of which can be used as a surface temperature prediction (Fig. 1.14).

The five parameters $(r_i, r_i', r_o, c_i, c_o)$ of the second-order element may be adjusted such that the response of the element is nearly identical to that of a fully distributed element with equivalent overall thermal resistance, r_m, and thermal capacitance, C_m. A fully distributed element may be envisaged as one with a large number of resistive-capacitive lumps. Gouda *et al.* (2002) suggest a minimum of 20 for a typical multiple layer construction element featuring materials with high thermal capacity). The parameter adjustment problem then becomes a straightforward optimisation problem that can be stated as follows.

If the overall element resistance and capacitance can be expressed as:

$$r_i = x_1 r_m$$
$$r_i' = x_2 r_m$$
$$r_o = x_3 r_m \qquad\qquad (1.162a,b,c)$$

$$C_i = y_1 C_m$$
$$C_o = y_2 C_m \qquad\qquad (1.163a,b)$$

then:

$$\text{minimise}_\varepsilon : f(\varepsilon)$$
$$\text{subject to}: x_1 + x_2 + x_3 = 1$$
$$x_1; x_2; x_3 \geq 0$$
$$y_1 + y_2 = 1$$
$$y_1; y_2 \geq 0$$
$$L_{\text{bound}} \leq \varepsilon \leq U_{\text{bound}} \qquad\qquad (1.164)$$

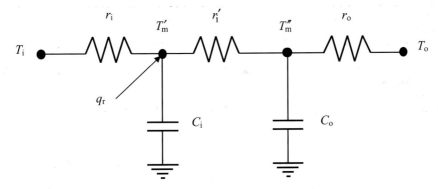

Fig. 1.14   Second-order lumped capacitance element.

in which the single objective function, $f(\varepsilon)$, is the square root of the sum-square-error between the distributed and second-order elements and L_{bound}, U_{bound} are the lower and upper tolerance bounds respectively on ε.

Various schemes are available for conducting optimisations like this. In the work reported by Gouda *et al.* (2002) the Kuhn–Tucker system of Lagrange multipliers was used as a basis for cancelling gradients between the objective function and active constraints at solution points (Mijangos and Nabona 2001). Though each construction element type needs to be uniquely parameter-fitted to use this method, results for a wide range of construction types show only moderate variation in results. Results from two sets of contrasting construction element types ranging from typical 'high' thermal capacity (masonry-based) to typical 'low' thermal capacity (timber-based) are given in Table 1.2 and these results would appear to cover a wide range of practical modelling problems.

The equations for this second-order approach to element modelling are as follows, in which T'_m represents the effective surface temperature of the element:

$$\frac{dT'_m}{dt} = \frac{1}{y_1 C_m}\left[q_r + \frac{(T_i - T'_m)}{x_1 r_m} - \frac{(T'_m - T''_m)}{x_2 r_m}\right] \tag{1.165}$$

$$\frac{dT''_m}{dt} = \frac{1}{y_2 C_m}\left[\frac{(T'_m - T''_m)}{x_2 r_m} - \frac{(T''_m - T_o)}{x_3 r_m}\right] \tag{1.166}$$

Or, in the more convenient state-space form:

$$\begin{bmatrix} \dot{T}'_m \\ \dot{T}''_m \end{bmatrix} = \begin{bmatrix} -1/[r_m C_m (x_1 y_1 + x_2 y_1)] & 1/(x_2 y_1 r_m C_m) \\ -1/(x_2 y_2 r_m C_m) & -1/[r_m C_m (x_2 y_2 + x_3 y_2)] \end{bmatrix} + \dots$$

$$\dots \begin{bmatrix} 1/(r_m C_m x_1 y_1) & 1/(y_1 C_m) & 0 \\ 0 & 0 & 1/(r_m C_m x_3 y_2) \end{bmatrix} \times \begin{bmatrix} T_i \\ q_r \\ T_o \end{bmatrix} \tag{1.167}$$

Table 1.2   Typical resistance and capacitance fractions.

Construction element type	x_1	x_2	x_3	y_1	y_2
External wall:					
Brick/insulation/concrete block/plaster – 'high'	0.111	0.506	0.383	0.186	0.814
External wall:					
Metalclad/air/insulation/aerated concrete block/air/plasterboard – 'low'	0.022	0.380	0.598	0.188	0.812
Internal partition:					
Plaster/concrete block/plaster – 'high'	0.107	0.424	0.469	0.090	0.910
Internal partition:					
Plasterboard/air/plasterboard – 'low'	0.111	0.477	0.412	0.198	0.802
Internal floor:					
Carpet/concrete screed/cast concrete – 'high'	0.087	0.267	0.646	0.126	0.874
Internal floor:					
Timber/air/plasterboard – 'low'	0.100	0.482	0.418	0.211	0.789

1.5 Heat transfer through glazing

Apart from external walls and roofs, windows and skylights are the key components of the building envelope through which heat transfer between indoors and outdoors takes place. Since window glasses are typically much thinner than the external walls and, being transparent elements that allow penetration of solar radiation, windows, in fact, often dominate the heat gain or loss of a heated or air-conditioned space in a building. As shown in Fig. 1.15, the heat transfer processes that may take place at a piece of window glass include:

- Reflection, absorption and transmission of direct and diffuse solar radiation;
- Conduction and convection of the absorbed solar radiation to the ambient air;
- Conduction and convection due to a temperature difference between indoors and outdoors;
- Radiant heat exchanges with the internal surfaces of other fabric elements and with lighting, appliances and other heat sources/sinks in the room.

Because the window glasses used in buildings are typically thin sheets of glasses, and the thermal conductivity of the glass is high compared with that of other fabric materials, the thermal resistance of the glass pane is often ignored in calculating the heat transfer through windows, which will largely simplify the analysis.

Reflectance, absorption and transmission of radiation through glass

When a beam of direct solar radiation strikes the external surface of a piece of glass, part of the incident radiation will be reflected and the remainder transmitted, as shown in Fig. 1.15. The angle of incidence and the angle of reflection are equal and, according to Snell's law, are related to the angle of refraction as given in Equation 1.168:

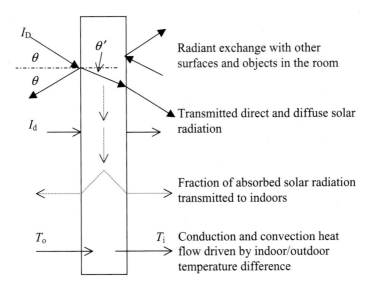

Fig. 1.15   Heat transfer processes at a piece of window glass.

$$n = \sin\theta/\sin\theta' \tag{1.168}$$

where θ = angle of reflection, θ' = angle of refraction, and n is a constant for a particular type of glass.

Let the fraction of the incident intensity that is reflected be r; it can be shown that:

$$r = \frac{1}{2}\left(\frac{\tan^2(\theta-\theta')}{\tan^2(\theta+\theta')} + \frac{\sin^2(\theta-\theta')}{\sin^2(\theta+\theta')}\right) \tag{1.169}$$

Let I_D be the intensity of the incident direct radiation and I_0 be the intensity of radiation that enters the glass:

$$I_0 = (1-r)I_D \tag{1.170}$$

While the beam of solar radiation penetrates through the glass, the intensity of the radiation will gradually drop, as the glass will absorb part of its energy. Consider the piece of glass as shown in Fig. 1.16. If the thickness of the glass is L, the horizontal distance from the external surface of the glass is y and the linear distance along the path of the sun's ray is x, which is measured from the point it enters the glass:

$$x = \frac{y}{\cos\theta'}$$

The fraction of the radiation intensity that will be absorbed when the sunray passes through the glass can be determined from the following equation:

$$\frac{dI}{dx} = -kI \tag{1.171}$$

where k is the absorption coefficient, which is a characteristic of the glass (≈ 0.02/mm for typical glass).

Integrating Equation 1.171 using the boundary conditions that at $x=0$, $I=I_0$, the intensity of the radiation at an intermediate position x within the glass is given by:

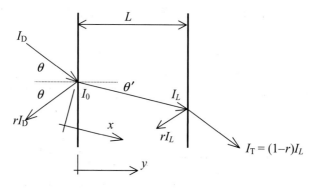

Fig. 1.16   Solar radiation through a piece of glass.

$$I = I_0 \exp(-kx) \tag{1.172}$$

At $x = L/\cos\theta'$:

$$I_L = I_0 \exp(-kL/\cos\theta') \tag{1.173}$$

Hence, the amount of radiation absorbed by the glass is:

$$I_0 - I_L = I_0(1 - \exp(-kL/\cos\theta')) \tag{1.174}$$

and the absorbed fraction, a, is:

$$a = \frac{I_0 - I_L}{I_0} = 1 - \exp(-kL/\cos\theta') \tag{1.175}$$

It follows that:

$$I_L = (1-a)I_0 \tag{1.176}$$

and the intensity of solar radiation that is transmitted through the glass pane is:

$$I_T = (1-r)I_L \tag{1.177}$$

It can be seen from Equations 1.169 and 1.175 that both r and a are dependent on the incident angle of the direct solar radiation.

The above describes the processes that a sunray will undergo when it passes through the thickness of the glass once. As there is a portion of the radiation energy that will be reflected back into the glass every time a ray hits a surface, the ray will undergo repeated reflection, absorption and transmission until the intensity has been completely dissipated due to absorption of the glass or transmission in or out of the glass, as shown in Fig. 1.17.

Table 1.3 summarises the fractional intensities of a sunray as it bounces back and forth between the two surfaces of the glass plane.

The total amount of solar radiation that will ultimately be reflected back to the outdoors, absorbed by the glass plane and transmitted into the indoor space can, therefore, be calculated

Table 1.3 Fractional intensity of a sunray as it bounces back and forth between the two surfaces of a glass pane.

(1)	(2)	(3)	(4)	(5)	(6)
Outdoor surface			Indoor surface		
Incident	To outside	To inside	Incident	Reflected	Transmitted
1 (from outside)	r	$1-r$	$(1-r)(1-a)$	$r(1-r)(1-a)$	$(1-r)^2(1-a)$
$r(1-r)(1-a)^2$	$r(1-r)^2(1-a)^2$	$r^2(1-r)(1-a)^2$	$r^2(1-r)(1-a)^3$	$r^3(1-r)(1-a)^3$	$r^2(1-r)^2(1-a)^3$
$=(1-a)\times(5)*$	$=(1-r)\times(1)*$	$=r\times(1)*$	$=(1-a)\times(3)*$	$=r\times(4)*$	$=(1-r)\times(4)*$

* (N) means the value in column N in the above row.

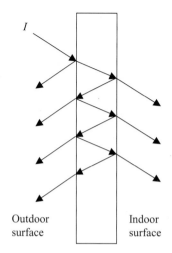

Fig. 1.17 Repeated reflection, absorption and transmission processes in glass.

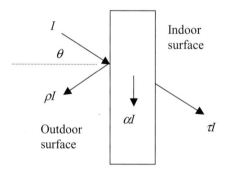

Fig. 1.18 Simplified reflection, absorption and transmission processes in glass.

by summing up the corresponding fractions of energy that leave the outer surface to the outdoors, are absorbed by the glass, and leave the inner surface to the indoor space for each pass
of the ray through the thickness of the glass.

Let ρ, α and τ be the reflectivity, absorptivity and transmissivity of the glass respectively,
which are the overall fractions of the incident intensity that are reflected, absorbed and transmitted; the heat transfer processes can be regarded simply as a one-step process as shown in
Fig. 1.18.

From the above analysis on multiple reflection, absorption and transmission of the sun's
ray in a glass pane, ρ, α and τ can be determined as follows:

$$\alpha = a(1-r) + ar(1-r)(1-a) + ar^2(1-r)(1-a)^2 + \dots$$
$$= a(1-r)\left[1 + r(1-a) + r^2(1-a)^2 + r^3(1-a)^3 + \dots\right]$$
$$= \frac{a(1-r)}{1-r(1-a)} \tag{1.178}$$

$$\rho = r + r(1-a)^2(1-r)^2 + r^3(1-a)^4(1-r)^2 + \ldots$$
$$= r\left\{1 + (1-a)^2(1-r)^2\left[1 + r^2(1-a)^2 + \ldots\right]\right\}$$
$$= r\left(1 + \frac{(1+a)^2(1-r)^2}{1 - r^2(1-a)^2}\right) \tag{1.179}$$

$$\tau = (1-r)^2(1-a) + r^2(1-r)^2(1-a)^3 + r^4(1-r)^2(1-a)^5 + \ldots$$
$$= (1-r)^2(1-a)\left[1 + r^2(1-a)^2 + \ldots\right]$$
$$= \frac{(1-r)^2(1-a)}{1 - r^2(1-a)^2} \tag{1.180}$$

Influence of solar incidence

Since r and a are both dependent on the angle of incidence (θ) of the direct radiation, ρ, α and τ are all functions of θ, as illustrated in Fig. 1.19.

The optical properties of a glass pane can be obtained using Snell's law and the Fresnel reflection equation (i.e. Equations 1.168 and 1.169). However, the refractive indices and other essential details for special glass types such as coated glass are frequently unavailable. Where normal or near-normal transmissivities are known, Karlsson and Roos (2000) recognised that angle-dependent transmissivity decreases monotonically with increasing incidence angle. They propose the following polynomial-based model for its prediction in either single, multiple, coated or uncoated glass panes:

$$\tau(\theta) = \tau_0\left(1 - A\theta^d - B\theta^e - C\theta^f\right) \tag{1.181}$$

(in which $A + B + C = 1$ and θ' is the normalised angle of incidence, i.e. $\theta/90$). The following are recommended for the constants and exponents (Karlsson and Roos 2000):

$$A = 8; \; B = 0.25/G; \; C = 1 - a - b \tag{1.182a,b,c}$$

$$d = 5.2 + 0.7/p; \; e = 2; \; f = (5.26 + 0.06p) + G \times (0.73 + 0.04p) \tag{1.183a,b,c}$$

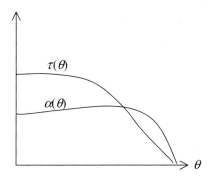

Fig. 1.19   Dependence of the transmissivity and absorptivity on solar incidence for glass.

(where p is the number of panes of glass and G is a category parameter whose value lies in the range $1 \leq G \leq 10$).

Data on basic optical properties for a range of glass types can be found in Hutchins *et al.* (2001).

Transmission of diffuse radiation

Since diffuse radiation comes from all directions, the transmissivity and absorptivity for diffuse radiation can be found by integrating the directional transmissivity and absorptivity over a hemisphere of unit radius for the range of incident angle from $0°$ to $90°$ as follows:

$$\tau_d = 2 \int_0^{\pi/2} \tau_\theta \sin \theta \cos \theta d\theta \tag{1.184}$$

$$\alpha_d = 2 \int_0^{\pi/2} \alpha_\theta \sin \theta \cos \theta d\theta \tag{1.185}$$

ASHRAE's method for window glazing heat transfer

In ASHRAE's cooling load calculation method (ASHRAE 2001), a type of glass called DSA (double strength, heat absorbing) glass is used as a standard reference for quantifying the solar properties of all type of glasses, through the use of a parameter called Shading Coefficient (SC) (see later discussions). For the standard DSA glass (ASHRAE 2001):

$$\alpha_\theta = \sum_{j=0}^{5} \alpha_j (\cos \theta)^j \tag{1.186}$$

$$\tau_\theta = \sum_{j=0}^{5} \tau_j (\cos \theta)^j \tag{1.187}$$

The value of α_j and τ_j for $j=0$ to 5 are as summarised in Table 1.4.

Using Equations 1.186 and 1.187 for determining the absorptivity and transmissivity for diffuse radiation for DSA glass,

Table 1.4   Absorptivity and transmissivity data for DSA glass.

j	α_j	τ_j
0	0.01154	−0.00885
1	0.77674	2.71235
2	−3.94657	−0.62062
3	8.57881	−7.07329
4	−8.38135	9.75995
5	3.01188	−3.89922

Source: (ASHRAE Handbook of Fundamentals (2001). © American Society of Heating, Refrigerating and Air-Conditioning Engineers, Inc., www.ashrae.org.)

$$\alpha_d = 2 \int_0^{\pi/2} \left(\sum_{j=0}^{5} \alpha_j (\cos\theta)^j \right) \sin\theta \cos\theta \, d\theta$$

$$= 2 \sum_{j=0}^{5} \frac{\alpha_j}{j+2} \tag{1.188}$$

$$\tau_d = 2 \sum_{j=0}^{5} \frac{\tau_j}{j+2} \tag{1.189}$$

For a particular incident angle of solar radiation at a particular time, the total intensity of solar radiation that will be absorbed is:

$$\alpha I_T = \alpha_\theta I_D + \alpha_d I_d \tag{1.190}$$

where I_D, I_d and I_T are respectively the direct, diffuse and total intensity of solar radiation incident upon a glass surface. This absorbed energy will flow out of the glass in both directions.

For a glass pane exposed at both sides to air of the same temperature (T_a), as shown in Fig. 1.20, the convective heat transfers to the outdoors (q_{out}) and to the indoors (q_{in}) at the two surfaces are given by:

$$q_{in} = h_i (T_G - T_a) \tag{1.191}$$

$$q_{out} = h_o (T_G - T_a) \tag{1.192}$$

where T_G is the temperature of the glass plane, which is assumed to be uniform across the thickness of the glass. These convective heat transfers are sustained by the absorbed solar energy. Hence:

$$\alpha I_T = h_i (T_G - T_a) + h_o (T_G - T_a)$$
$$= (h_i + h_o)(T_G - T_a) \tag{1.193}$$

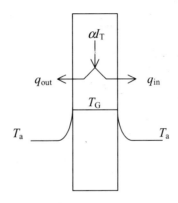

Fig. 1.20 Flow of the absorbed solar heat.

From Equations 1.191 and 1.192:

$$q_{in} = \frac{h_i}{h_i + h_o} \alpha I_T \tag{1.194}$$

If there exists a difference between the indoor and outdoor temperatures, as shown in Fig. 1.21, the resultant rate of heat transfer ($q_{\Delta T}$) will be (assuming negligible thermal resistance from the glass pane):

$$q_{\Delta T} = \frac{h_i h_o}{h_i + h_o} (T_o - T_i) \tag{1.195}$$

Therefore, the total heat flow into the indoor space (heat gain) caused by solar radiation incident upon a window is:

$$q_G = \tau_\theta I_D + \tau_d I_d + \frac{h_i}{h_i + h_o} (\alpha_\theta I_D + \alpha_d I_d) + \frac{h_i h_o}{h_i + h_o} (T_o + T_i) \tag{1.196}$$

Note that the first two components in Equation 1.196 are the solar heat gains of the space, which will become the cooling load of the space only after the energy is absorbed by the internal surfaces, causing convective heat transfer from such surfaces to the room air. The last two components, however, will be instantaneous cooling load components.

In ASHRAE's cooling load calculation method, the heat gain from a window as given in Equation 1.196 is simplified into:

$$q_G = \left(\tau + \frac{h_i}{h_i + h_o} \alpha \right) I_T + \frac{h_i h_o}{h_i + h_o} (T_o - T_i) \tag{1.197}$$

or:

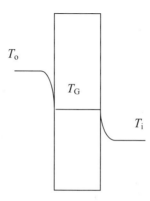

T_o

T_G

T_i

Fig. 1.21   Heat transfer through a glass pane due to the temperature difference of the ambient air at the two sides.

$$q_{G} = \left(\tau + \frac{U\alpha}{h_{o}} \right)I_{T} + U(T_{o} - T_{i})$$
(1.198)

The DSA glass is used as a standard reference and the heat gains due to transmission and absorption through a sunlit DSA glass are designated as the solar heat gain factors (*SHGF*).

The heat gain for a particular type of glass will be determined with reference to *SHGF* and *SC* of the glass as follows:

$$q_{G} = SHGF \cdot SC + U(T_{o} - T_{i})$$
(1.199)

And the definition of *SC* is:

$$SC = \frac{\text{Solar heat gain of fenestration}}{\text{Solar heat gain of reference glass}}$$
(1.200)

References

ASHRAE (2001) *Handbook of Fundamentals,* American Society of Heating, Refrigerating and Air Conditioning Engineers, Inc., Atlanta, GA.

CIBSE (1988) *CIBSE Guide Volume A: Design Data,* Chartered Institution of Building Services Engineers, London.

Chen, T.Y. & Li, S.L. (1987) Simulation study of a passive solar heating apartment building. *Tianjin University Acta Energiae Solaris Sinica* 8(3) pp.238–45.

Gouda, M.M., Danaher, S. & Underwood, C.P. (2001) Building thermal model reduction using non-linear constrained optimisation. *Building and Environment* 37 pp.1255–65.

Hutchins, M.G., Topping, A.J., Anderson, C., *et al.* (2001) Measurement and prediction of angle-dependent optical properties of coated glass products: Results of an inter-laboratory comparison of spectral transmittance and reflectance. *Thin Solid Films* 392 pp.269–75.

Karlsson, J. & Roos, A. (2000) Modelling the angular behaviour of the total solar energy transmittance of windows. *Solar Energy* 69(4) pp.321–9.

Laret, L. (1980) Use of general models with a small number of parameters: Part 1 – theoretical analysis. *Proceedings of 7th International Congress of Heating and Air Conditioning CLIMA 2000.* Budapest.

Mijangos, E. & Nabona, N. (2001) On the first-order estimation of multipliers from Kuhn-Tucker systems. *Computers and Operations Research* 28 pp.243–70.

Milbank, N.O. & Harrington-Lynn, J. (1974) Thermal response and the admittance procedure. *Building Services Engineer* 42 pp.38–47.

Stephenson, D.G. & Mitalas, G.P. (1967) Room thermal response factors. *ASHRAE Transactions* 73(2) pp. III.2.1–III.2.10.

Stephenson, D.G. & Mitalas, G.P. (1971) Calculation of heat conduction transfer functions for multi-layer slabs. *ASHRAE Transactions* 77(2) pp.117–26.

Tindale, A. (1993) Third-order lumped-parameter simulation method. *Building Services Engineering Research & Technology* 14(3) pp.87–97.

Wang, S. & Chen, Y. (2003) Transient heat flow calculation for multilayer constructions using a frequency-domain regression method. *Building and Environment* 38 pp.45–61.

Zhang, C. & Ding, G. (2003) A stable series expansion method for calculating thermal response factors of multi-layer walls. *Building and Environment* 38 pp.699–705.

Chapter 2
Modelling Heat Transfer in Building Envelopes

The method of treatment of conduction heat transfer through the elements that form building envelopes has generally been used as the main classifier of a building thermal model. One reason for this is the range of fundamentally differing approaches to treating conduction through building elements that formed the basis of Chapter 1. Nonetheless, heat transfer in building elements is merely a segment of the wider issues concerned with the modelling of energy and environment of a building space and, for that matter, an entire building of interacting spaces.

In this chapter, consideration is given to the synthesis of the various heat and mass transfer exchanges that contribute to the thermal environment of a building space. First, consideration will be given to an alternative approach to dealing with the transient heat conduction equation. In Chapter 1 there was an emphasis on analytical approaches to dealing with this problem. These approaches have sound intuitive appeal and enjoy computational efficiency but they can severely complicate the seamless solution of the ultimate building space model, which might include both heat and mass transfer, forced and natural convection with field-dependent laminar and turbulent characteristics, and an interaction with plant. An alternative approach is to treat the transient heat conduction equation as a nodal field problem and solve it numerically, thus enabling this segment of the problem to participate as part of a wider building envelope field problem. Consideration is then given to the treatment of heat exchange at room surfaces, including the treatment of both long wave and short wave radiant exchanges, and this is followed by the closure of surface, source and room air exchanges resulting in a room space heat balance. Next, procedures are described for the synthesis of the resulting room space heat balance with the various approaches to treating heat transfer through building elements, including both analytical (Chapter 1) and numerical approaches. Finally, methodologies for the treatment of latent loads are discussed.

2.1 Finite Difference Method – a numerical method for solving the heat conduction equation

The Response Factor Method, Transfer Function Method and Admittance Method described in Chapter 1 are all analytical methods for solving the partial differential equation (PDE) that governs the one-dimensional transient heat conduction through a slab of material. An alternative approach that is also widely used in modelling heat transfer in buildings is to solve the PDE using numerical methods, such as the Finite Difference Method and the Finite Element Method. Since it is comparatively simpler to formulate a building heat transfer model based on the Finite Difference Method than the Finite Element Method, and the former is adopted in most building energy simulation programmes that are based on numerical methods, only the application of the Finite Difference Method to solving the one-dimensional heat conduction equation is described here.

The Finite Difference Method for solving a differential equation is based on the assumption that when the spatial domain is divided into a finite number of slices (subdivisions), and the temporal domain into finite time steps, at a particular time step, the state of the material, here the temperature, within each slice can be represented by the state at one point, referred to as a node; for instance a point on the mid-plane within the slice. Also, the variation of the state between each pair of adjoining slices is assumed to be linear, as shown in Fig. 2.1.

Various methods can be used to derive the finite difference equations for describing the time evolution of the nodal temperatures. In the following, the energy balance approach is adopted, as it is a simple and direct method that is easily comprehensible from basic heat transfer considerations for each subdivision.

Let the temperatures at the $(i-1)^{th}$, i^{th} and $(i+1)^{th}$ nodes, which are at the mid-planes of the $(i-1)^{th}$, i^{th} and $(i+1)^{th}$ slices of a homogeneous piece of material, be T_{i-1}, T_i and T_{i+1} respectively, and these slices are of equal thickness Δx, as shown in Fig. 2.2, the temperature gradient between the $(i-1)^{th}$ and the ith nodes and between the ith and the $(i+1)^{th}$ nodes are respectively:

$$\frac{T_i - T_{i-1}}{\Delta x} \quad \text{and} \quad \frac{T_{i+1} - T_i}{\Delta x}$$

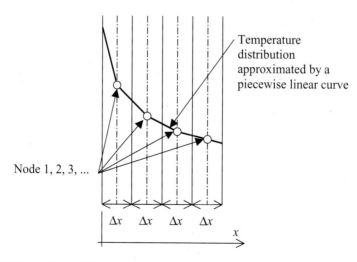

Fig. 2.1   Subdivision of the spatial domain into a finite number of slices.

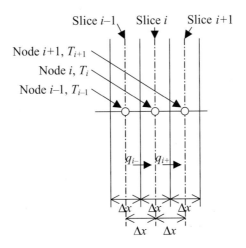

Fig. 2.2   Conduction heat transfer at an internal slice.

Since the rate of conduction heat transfer through a plane is proportional to the temperature gradient at the plane, the heat flux through the interface between the $(i-1)^{th}$ and the ith slices (q_{i-}) and between the ith and the $(i+1)^{th}$ slices (q_{i+}) are:

$$q_{i-} = -k\frac{T_i - T_{i-1}}{\Delta x}$$

$$q_{i+} = -k\frac{T_{i+1} - T_i}{\Delta x}$$

From heat balance on the ith slice, assuming that there are no internal heat sources or heat sinks:

$$\rho c \Delta x \frac{\Delta T_i}{\Delta t} = -k\frac{T_i - T_{i-1}}{\Delta x} + k\frac{T_{i+1} - T_i}{\Delta x} \tag{2.1}$$

The right hand side of Equation 2.1 describes the net heat gain of the ith slice due to conduction heat transfer between the slice and each of the two adjoining slices. The left hand side describes the rate of change in the internal energy of the ith slice of the slab resulting from this net conduction heat gain, which will be reflected by a rise in temperature ΔT_i at node i over the time interval Δt.

Equation 2.1 can be re-arranged as:

$$\frac{\Delta T_i}{\Delta t} = \alpha \frac{T_{i-1} - 2T_i + T_{i+1}}{\Delta x^2} \tag{2.2}$$

Recall that the one-dimensional transient heat conduction equation is:

$$\frac{\partial T}{\partial t} = \alpha \frac{\partial^2 T}{\partial x^2} \tag{2.3}$$

It can, therefore, be seen that the partial derivatives in the equation are approximated by the finite difference terms as follows:

$$\frac{\partial T}{\partial t} \approx \frac{\Delta T_i}{\Delta t} \tag{2.4}$$

$$\frac{\partial^2 T}{\partial x^2} \approx \frac{T_{i-1} - 2T_i + T_{i+1}}{\Delta x^2} \tag{2.5}$$

Instead of the heat balance approach as described above, Equation 2.2 can be derived directly from Equation 2.3 by replacement of the derivatives with the respective finite difference approximations. The finite difference approximation for the time derivative is as shown in Equation 2.4. The second order derivative in the x-dimension can be approximated as follows:

$$\frac{\partial^2 T}{\partial x^2} = \frac{\partial}{\partial x}\left(\frac{\partial T}{\partial x}\right) \approx \frac{\left(\dfrac{\partial T}{\partial x}\right)_+ - \left(\dfrac{\partial T}{\partial x}\right)_-}{\Delta x} \approx \frac{\dfrac{T_{i+1} - T_i}{\Delta x} - \dfrac{T_i - T_{i-1}}{\Delta x}}{\Delta x} = \frac{T_{i+1} - 2T_i + T_{i-1}}{\Delta x^2}$$

Note that in the finite difference approximation for the second order derivative as shown above, the temperature gradients are evaluated from the temperature difference between node i and each of the adjacent nodes at the two sides of node i. This scheme is, therefore, called a central-in-space difference scheme.

Since Equation 2.2 relates the change in the temperature at node i to the temperatures at the nodes at both sides of node i, it applies only to the interior nodes within a slab of material.

Boundary conditions

The finite difference equation for the outermost slice of a slab of material can also be derived from the heat balance approach. Using T_e to denote the sol-air temperature or environmental temperature of the surrounding, the convective and radiant heat flux at the surface can be described by:

$$q_o = h_o(T_e - T_s) \tag{2.6}$$

From a heat balance on the outermost slice at the left, denoted as slice 1 as shown in Fig. 2.3:

$$\rho c \Delta x \frac{\Delta T_1}{\Delta t} = h_o(T_e - T_s) + k\frac{T_2 - T_1}{\Delta x} \tag{2.7}$$

Note that in Equations 2.6 and 2.7, the surface heat flux is related to the surface temperature. The surface temperature, therefore, needs to be related to the temperature of slice 1, which can be obtained by considering:

$$h_o(T_e - T_s) = k\frac{T_s - T_1}{\Delta x/2}$$

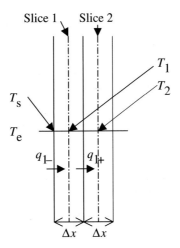

Fig. 2.3   Heat transfer between the outermost slice and the external surroundings.

The surface temperature can be solved from the above as:

$$T_s = \frac{(h_o \Delta x / 2k)T_e + T_1}{1 + (h_o \Delta x / 2k)}$$ (2.8)

Equation 2.8 can now be substituted into Equation 2.7 to yield:

$$\rho c \Delta x \frac{\Delta T_1}{\Delta t} = \frac{h_o}{1 + (h_o \Delta x / 2k)}(T_e - T_1) + k \frac{T_2 - T_1}{\Delta x}$$ (2.9)

The use of this method requires an additional calculation step for determining the surface temperature based on the environmental temperature and the temperature at node 1 determined for each time step.

An alternative method that allows the surface temperature to be directly evaluated is to define the outmost slice to be half as thick as the interior slices, as shown in Fig. 2.4. Heat balance on slice 1 shown in Fig. 2.4 can then be described by:

$$\rho c \frac{\Delta x}{2} \frac{\Delta T_1}{\Delta t} = h_o (T_e - T_1) + k \frac{T_2 - T_1}{\Delta x}$$ (2.10)

If the outermost slice at the other side of the slab is the Nth slice, the corresponding finite difference equation for that slice is:

$$\rho c \frac{\Delta x}{2} \frac{\Delta T_N}{\Delta t} = k \frac{T_{N-1} - T_N}{\Delta x} - h_o (T_N - T_e)$$ (2.11)

For a composite wall comprising multiple layers of different materials, a composite slice made partly of the material of one layer and partly of the material of the adjoining layer can be defined to describe the heat transfer through the interface between two adjoining layers, as shown in Fig. 2.5.

Heat balance on the composite slice, slice i in Fig. 2.5, yields:

$$\left(\rho_a c_a \frac{\Delta x_a}{2} + \rho_b c_b \frac{\Delta x_b}{2}\right)\frac{\Delta T_i}{\Delta t} = -k_a \frac{T_i - T_{i-1}}{\Delta x_a} + k_b \frac{T_{i+1} - T_i}{\Delta x_b} \tag{2.12}$$

Forward difference scheme

Equation 2.2 is re-arranged into the following to allow the change in the temperature at an interior node from the nth to $(n + 1)^{th}$ time step to be evaluated from the temperature of the node and its adjacent nodes:

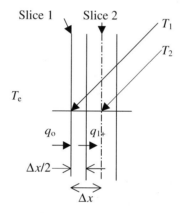

Fig. 2.4   Heat transfer at the outmost slice.

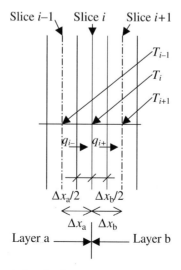

Fig. 2.5   Conduction heat transfer at the interface between two layers of different materials in a composite wall.

$$\Delta T_i = \frac{\alpha \Delta t}{\Delta x^2}(T_{i-1} - 2T_i + T_{i+1}) \tag{2.13}$$

If all the nodal temperatures used to estimate the rise in temperature at the ith node are the known values at time step n, the temperature at the ith node at time step $n+1$ can be determined from Equation 2.13 as follows:

$$\Delta T_i = T_i^{n+1} - T_i^n = \frac{\alpha \Delta t}{\Delta x^2}(T_{i-1}^n - 2T_i^n + T_{i+1}^n)$$

Then:

$$T_i^{n+1} = \frac{\alpha \Delta t}{\Delta x^2}\left(T_{i-1}^n - \left(2 - \frac{\Delta x^2}{\alpha \Delta t}\right)T_i^n + T_{i+1}^n \right) \tag{2.14}$$

(Note here that the superscript of T, (i.e. n), denotes the time-step in the calculation process.)
Using the following definition of Fourier number (Fo):

$$Fo = \frac{\alpha \Delta t}{\Delta x^2} \tag{2.15}$$

Equation 2.14 can be simplified to:

$$T_i^{n+1} = Fo\left(T_{i-1}^n - \left(2 - \frac{1}{Fo}\right)T_i^n + T_{i+1}^n \right) \tag{2.16}$$

The corresponding equations for a boundary node and an interface node are:
for the boundary nodes:

$$T_1^{n+1} = 2Fo\left(BiT_{e,1}^n - \left(1 + Bi - \frac{1}{2Fo}\right)T_1^n + T_2^n \right) \tag{2.17}$$

$$T_N^{n+1} = 2Fo\left(T_{N-1}^n - \left(1 + Bi - \frac{1}{2Fo}\right)T_N^n + BiT_{e,N}^n \right) \tag{2.18}$$

for the interface node between two layers of different materials:

$$T_i^{n+1} = 2Fo_a T_{i-1}^n - (2Fo_a + 2Fo_b - 1)T_i^n + 2Fo_b T_{i+1}^n \tag{2.19}$$

where:

$$Bi = \frac{h_o \Delta x}{k} = \text{Biot number} \tag{2.20}$$

$$Fo_a = \frac{k_a \Delta t / \Delta x_a}{\rho_a c_a \Delta x_a + \rho_b c_b \Delta x_b} = \text{Fourier number for layer a} \qquad (2.21)$$

$$Fo_b = \frac{k_b \Delta t / \Delta x_b}{\rho_a c_a \Delta x_a + \rho_b c_b \Delta x_b} = \text{Fourier number for layer b} \qquad (2.22)$$

For a wall subdivided into N slices, Equations 2.16 to 2.19 can be combined and written in matrix form as follows:

$$\begin{Bmatrix} T_1^{n+1} \\ T_2^{n+1} \\ \vdots \\ T_{N-1}^{n+1} \\ T_N^{n+1} \end{Bmatrix} = \begin{bmatrix} a_{1,1} & a_{1,2} & 0 & \cdots & 0 \\ a_{2,1} & a_{2,2} & a_{2,3} & \cdots & 0 \\ \vdots & \vdots & \vdots & \vdots & \vdots \\ 0 & \cdots & a_{N-1,N-2} & a_{N-1,N-1} & a_{N-1,N} \\ 0 & \cdots & 0 & a_{N,N-1} & a_{N,N} \end{bmatrix} \begin{Bmatrix} T_1^n \\ T_2^n \\ \vdots \\ T_{N-1}^n \\ T_N^n \end{Bmatrix} + \begin{Bmatrix} 2FoBiT_{e,1}^n \\ 0 \\ \vdots \\ 0 \\ 2FoBiT_{e,N}^n \end{Bmatrix} \qquad (2.23)$$

When the finite difference equations for all the nodes are available, the evolution of the nodal temperatures can be predicted from the known initial conditions, and the boundary conditions at various time steps, with the outcome of calculation for a time step taken as the values for the current time step in the next round of calculation. This method of estimating the state of the material at the next time step based solely on the known state values of the current time step, as depicted in Equation 2.23, is called the forward difference scheme. Since the node temperatures at the next time step can be calculated explicitly, this scheme is also called the explicit scheme.

Stability considerations in forward difference scheme

One important concern of using the forward difference scheme is the limits on the time step Δt and the size of the subdivision Δx that can be used without making the numerical scheme unstable. The criterion for Equations 2.16 to 2.19 to be stable is to have the value of the coefficient of the temperature for the node under concern to be positive. Hence, with reference to Equations 2.16 to 2.19, the stability criteria for the internal, boundary and interface nodes are:

for an internal node:

$$-\left(2 - \frac{1}{Fo}\right) \geq 0 \text{ or } Fo \leq \frac{1}{2} \qquad (2.24)$$

for the boundary nodes:

$$-\left(1 + Bi - \frac{1}{2Fo}\right) \geq 0 \text{ or } Fo \leq \frac{1}{2(1 + Bi)} \qquad (2.25)$$

for an interface node between two layers of different materials:

$$-(2Fo_a + 2Fo_b - 1) \geq 0 \text{ or } (Fo_a + Fo_b) \leq \frac{1}{2} \qquad (2.26)$$

Since the Fourier numbers, as shown in Equations 2.15, 2.21 and 2.22, are related to the time step Δt and the size of the subdivision Δx, the requirement that the Fourier number should not exceed a certain value means that the thinner the slab slice, the smaller the time step needed for stability. In most cases, the spatial subdivisions will be determined first, followed by determination of the time step using the criteria as given in Equations 2.24 to 2.26. The time step that satisfies the most stringent requirement among all the nodes can then be chosen for the solution scheme.

Proof of the stability criterion is omitted here but can be found in many textbooks of numerical methods. However, the implication of not observing this requirement can be seen by noting that Equations 2.16 to 2.19 can each be re-written in the following form:

$$T_i^{n+1} = AT_{i-}^n - (A + B - 1)T_i^n + BT_{i+}^n \tag{2.27}$$

which can be re-arranged into:

$$T_i^{n+1} = A(T_{i-}^n - T_i^n) + B(T_{i+}^n - T_i^n) + T_i^n \tag{2.28}$$

where the coefficients A and B are both positive and finite. For an interior node, T_{i-} and T_{i+} denote the temperatures of the adjacent nodes. For a boundary node, they denote the temperature of one adjacent node and the environmental temperature. Note that the sum of the coefficients of the three temperatures at the right hand side of Equation 2.27 equals one. The stability criterion requires that:

$$-(A + B - 1) \geq 0$$

which in turn requires that:

$$A + B \leq 1$$

If the above criterion is not satisfied, $A + B$ will be greater than one. It can then be seen from Equation 2.28 that for the case where the adjacent temperatures T_{i-}^n and T_{i+}^n are both higher than T_i^n by δT, the value of T_i^{n+1} will then be:

$$T_i^{n+1} = (A + B)\delta T + T_i^n$$

It can be seen from Fig. 2.6 that when $A + B$ is greater than one, T_i^{n+1} is higher than T_{i-}^n and T_{i+}^n, which is physically impossible, and this will give rise to oscillations in the prediction of T_i in subsequent time steps.

The effects of complying and violating the stability criteria are illustrated in Fig. 2.7, which shows the calculation results for a homogeneous slab of material with five equally spaced nodes. The temperature at one of the boundary nodes, node 1, was maintained steadily at 4 units while that at the other boundary node, node 5, was fixed at 0 units. The initial temperatures of nodes 2 to 4 were all 0 units. It is expected that in the steady state, the temperatures at nodes 1 to 5 will become 4, 3, 2, 1 and 0 units. The properties of the material, the spacing between the nodes and the time step size were assumed to have values such that when

Equation 2.28 is used to predict the temperature changes over time, the values of the coefficients A and B are both 0.4 for the case shown in Figs. 2.7a and 2.7b, and are both 0.6 for the other case shown in Figs. 2.7c and 2.7d. From the results shown in Figs. 2.7(a–d), it is clear that the solution process was stable for the case with $(A + B)$ less than 1 but unstable when $(A + B)$ was greater than 1.

Backward difference scheme

If, instead of using the known temperatures at the current time step, the temperature changes at the nodes from the current to the next time step are estimated based solely on the temperatures of the nodes at the next time step, the finite difference equations will become:

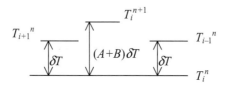

Fig. 2.6   Change in temperature in a time step for an unstable scheme.

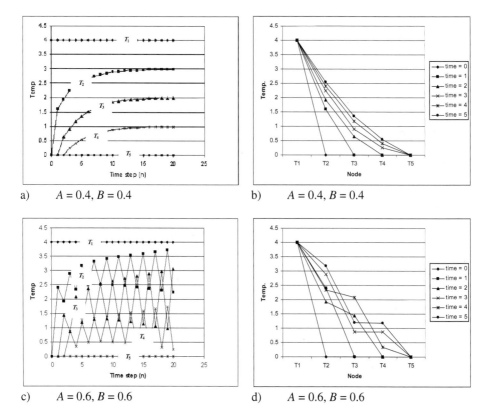

a) $A = 0.4, B = 0.4$

b) $A = 0.4, B = 0.4$

c) $A = 0.6, B = 0.6$

d) $A = 0.6, B = 0.6$

Fig. 2.7   Sample calculation results for a stable and an unstable scheme.

for the interior nodes:

$$-FoT_{i-1}^{n+1} + (2Fo+1)T_i^{n+1} - FoT_{i+1}^{n+1} = T_i^n \tag{2.29}$$

for the boundary nodes:

$$(2Fo + 2FoBi + 1)T_1^{n+1} - 2FoT_2^{n+1} = T_1^n + 2FoBiT_e^{n+1} \tag{2.30}$$

$$-2FoT_{N-1}^{n+1} + (2Fo + 2FoBi + 1)T_N^{n+1} = T_N^n + 2FoBiT_e^{n+1} \tag{2.31}$$

for an interface node between two layers of different materials:

$$-2Fo_aT_{i-1}^{n+1} + (1 + 2Fo_a + 2Fo_b)T_i^{n+1} - 2Fo_bT_{i+1}^{n+1} = T_i^n \tag{2.32}$$

It can be seen that, when the backward difference scheme is used, all the temperatures at the left hand side of the finite difference equations are unknowns that need to be solved simultaneously with the temperatures at other nodes. The problem then becomes an implicit problem and, therefore, the scheme is also called the implicit scheme.

For a N-layer wall, the equations for all the nodes can be written in matrix form as follows:

$$\begin{bmatrix} a_{1,1} & a_{1,2} & 0 & .. & 0 \\ a_{2,1} & a_{2,2} & a_{2,3} & .. & 0 \\ : & : & : & : & : \\ 0 & .. & a_{N-1,N-2} & a_{N-1,N-1} & a_{N-1,N} \\ 0 & .. & 0 & a_{N,N-1} & a_{N,N} \end{bmatrix} \begin{Bmatrix} T_1^{n+1} \\ T_2^{n+1} \\ : \\ T_{N-1}^{n+1} \\ T_N^{n+1} \end{Bmatrix} = \begin{Bmatrix} T_1^n + 2FoBiT_{e,1}^{n+1} \\ T_2^n \\ : \\ T_{N-1}^n \\ T_N^n + 2FoBiT_{e,N}^{n+1} \end{Bmatrix} \tag{2.33}$$

Compared with the explicit scheme, many more calculation steps are involved in the implicit scheme for progressing from one time step to the next. This solution scheme, however, is not subject to a limit on the time step, and thus allows a much larger time step to be used. The results, however, will be less accurate.

Crank–Nicholson scheme

For achieving higher accuracy without subject to the stability limit for the time step, a weighted average of the changes in temperature predicted by the explicit and by the implicit schemes can be used. Let λ denote the weighting factor; the finite difference equation for an interior node will be:

$$T_i^{n+1} - T_i^n = Fo[\lambda(T_{i-1}^{n+1} - 2T_i^{n+1} + T_{i+1}^{n+1}) + (1-\lambda)(T_{i-1}^n - 2T_i^n + T_{i+1}^n)]$$

which can be re-arranged into:

$$-\lambda FoT_{i-1}^{n+1} + (2\lambda Fo + 1)T_i^{n+1} - \lambda FoT_{i+1}^{n+1}$$
$$= (1-\lambda)FoT_{i-1}^n - (2Fo(1-\lambda) - 1)T_i^n + (1-\lambda)FoT_{i+1}^n \tag{2.34}$$

For the boundary nodes (node 1 and node N):

$$(2\lambda FoBi + 2\lambda Fo + 1)T_1^{n+1} - 2\lambda FoT_2^{n+1}$$
$$= 2FoBi\big((1 - \lambda)T_e^n + \lambda T_e^{n+1}\big) - (2(1 - \lambda)Fo(Bi + 1) - 1)T_1^n + 2(1 - \lambda)FoT_2^n \qquad (2.35)$$

$$-2\lambda FoT_{N-1}^{n+1} + (2\lambda FoBi + 2\lambda Fo + 1)T_N^{n+1}$$
$$= 2(1 - \lambda)FoTN_{N-1}^n - (2(1 - \lambda)Fo(Bi + 1) - 1)T_N^n + 2FoBi\big((1 - \lambda)T_e^n + \lambda T_e^{n+1}\big) \qquad (2.36)$$

For an interface node:

$$-2\lambda Fo_a T_{i-1}^{n+1} + (2\lambda Fo_a + 2\lambda Fo_b + 1)T_i^{n+1} - 2\lambda Fo_b T_{i+1}^{n+1}$$
$$= 2(1 - \lambda)Fo_a T_{i-1}^n - (2(1 - \lambda)(Fo_a + Fo_b) - 1)T_i^n + (1 - \lambda)Fo_b T_{i+1}^n \qquad (2.37)$$

For a N-layer wall, the equations for all the nodes can be written in matrix form as follows:

$$\begin{bmatrix} a_{1,1} & a_{1,2} & 0 & .. & 0 \\ a_{2,1} & a_{2,2} & a_{2,3} & .. & 0 \\ \vdots & \vdots & \vdots & \vdots & \vdots \\ 0 & .. & a_{N-1,N-2} & a_{N-1,N-1} & a_{N-1,N} \\ 0 & .. & 0 & a_{N,N-1} & a_{N,N} \end{bmatrix} \begin{Bmatrix} T_1^{n+1} \\ T_2^{n+1} \\ \vdots \\ T_{N-1}^{n+1} \\ T_N^{n+1} \end{Bmatrix} = \begin{Bmatrix} d_1 \\ d_2 \\ \vdots \\ d_{N-1} \\ d_N \end{Bmatrix} \qquad (2.38)$$

where the vector elements at the right hand side are quantities that can be determined from the known values at the nth time step, as shown below:

$$\begin{Bmatrix} d_1 \\ d_2 \\ \vdots \\ d_{N-1} \\ d_N \end{Bmatrix} = \begin{bmatrix} b_{1,1} & b_{1,2} & 0 & .. & 0 \\ b_{2,1} & b_{2,2} & b_{2,3} & .. & 0 \\ \vdots & \vdots & \vdots & \vdots & \vdots \\ 0 & .. & b_{N-1,N-2} & b_{N-1,N-1} & b_{N-1,N} \\ 0 & .. & 0 & b_{N,N-1} & b_{N,N} \end{bmatrix} \begin{Bmatrix} T_1^n \\ T_2^n \\ \vdots \\ T_{N-1}^n \\ T_N^n \end{Bmatrix} + \begin{Bmatrix} e_1 \\ 0 \\ \vdots \\ 0 \\ e_N \end{Bmatrix} \qquad (2.39)$$

It can be seen from Equations 2.34 to 2.37 that, if $\lambda = 0$, the equations will be identical to the corresponding equations for the explicit method and will become those for the implicit method if $\lambda = 1$. In the Crank–Nicholson scheme, $\lambda = 0.5$, which will be unconditionally stable.

Tri-diagonal matrix algorithm

In using the backward or the Crank–Nicholson scheme, the temperatures at the next time step at various nodes have to be determined by solving simultaneously a set of linear equations, as shown in Equations 2.33 and 2.38. Note, however, that the coefficient matrix in each of these equations is a tri-diagonal matrix, which can be solved efficiently using the tri-diagonal matrix algorithm (TDMA) described below.

In the tri-diagonal matrix shown in Equation 2.40, the elements in the diagonal are denoted by $d_{1\ldots N}$ and those above and below the diagonal by $a_{1\ldots N-1}$ and $b_{1\ldots N}$ respectively. Elements in the constant vector at the right hand side of the equation are denoted by $c_{1\ldots N}$:

$$
\begin{bmatrix}
d_1 & a_1 & 0 & .. & 0 \\
b_2 & d_2 & a_2 & .. & 0 \\
\vdots & \vdots & \vdots & \vdots & \vdots \\
0 & .. & b_{N-1} & d_{N-1} & a_{N-1} \\
0 & .. & 0 & b_N & d_N
\end{bmatrix}
\begin{Bmatrix}
T_1^{n+1} \\
T_2^{n+1} \\
\vdots \\
T_{N-1}^{n+1} \\
T_N^{n+1}
\end{Bmatrix}
+
\begin{Bmatrix}
c_1 \\
c_2 \\
\vdots \\
c_{N-1} \\
c_N
\end{Bmatrix}
\tag{2.40}
$$

The TDMA can be implemented as follows:

for $i = 1$:

$$
c_i' = c_i \\
d_i' = d_i
\tag{2.41}
$$

for $i = 2, 3, \ldots, N$:

$$
c_i' = c_i - \frac{b_i}{d_{i-1}'} c_{i-1}'
$$

$$
d_i' = d_i - \frac{b_i}{d_{i-1}'} a_{i-1}
\tag{2.42}
$$

where c_i', d_i' denote the updated values of the coefficients c_i and d_i respectively.

Having evaluated c_i', d_i' for $i = 1, 2, 3, \ldots, N$, the nodal temperatures T_i^{n+1} at time step $n + 1$ can be determined by the following back substitutions:

for $i = N$:

$$
T_i^{n+1} = \frac{c_i'}{d_i'}
\tag{2.43}
$$

for $i = N-1, N-2, \ldots, 1$:

$$
T_i^{n+1} = \frac{c_i' - a_i T_{i+1}^{n+1}}{d_i'}
\tag{2.44}
$$

2.2 Heat transfer in building spaces

The heat gains of an indoor space include those from the enclosing walls and roofs, solar heat gain through windows, heat gains due to infiltration and heat gains from the internal sources.

The heat gains from walls and roofs can be determined by using one of the methods discussed in the preceding sections, which include the Response Factor Method, the Transfer Function Method, the Admittance Method and the Finite Difference Method, or other appropriate analytical and numerical methods. The method for determining the heat gain from windows, including the transmitted solar radiation and the heat flow into the indoor space due to the part of solar radiation that is absorbed by the window glazing, has also been described. In this section, these methods will be re-visited and integrated to account for the interactions among the fabric components and the internal sources of a room.

The convective heat transfer from the internal surfaces of the envelope to the indoor air will immediately become a part of the cooling load that the air-conditioning system should offset in order to maintain the indoor temperature at the set-point level. The radiant heat gains, however, will not contribute instantaneously to the cooling load but will undergo a series of absorption and reflection processes at the surfaces of the building fabric and of other objects within the space, and will become a component of the cooling load only when the heat energy is transferred to the indoor air through convection from those surfaces that have absorbed the energy. Energy re-distribution will also take place through the radiation heat exchanges among the fabric surfaces, and with the internal heat sources, such as lighting and equipment inside the space.

The conversion of the heat gains into cooling load in a room can be modelled by considering the overall heat balance among the composing elements of the room, and thus the method is called the Heat Balance Method. Due to the complexity of the processes involved, simplifications are required in the treatment of some of the processes. This section describes the practical methods for modelling the cooling load of an air-conditioned space.

Heat transfer at the exterior surface of walls and roofs

In the preceding sections, the effect of radiation heat transfer at the exterior surfaces of walls and roofs was accounted for by defining a fictitious temperature, the 'sol-air temperature'. This is the outdoor air temperature that would, in the absence of any radiation heat transfer at the surface, cause the same rate of heat transfer across the surface as would the actual outdoor air temperature and the net radiation energy absorbed by the exterior surface. A more detailed treatment of the heat transfer processes involved is given below.

The rate of convection heat transfer from the outdoor air to the wall surface is described by:

$$q_{W,i,0,\text{Conv}}(t) = h_o\left(T_o(t) - T_{W,i,0}(t)\right) \tag{2.45}$$

where:

$q_{W,i,0,\text{Conv}}(t)$ = convective heat flux at the external surface of wall i at time t
$T_{W,i,0}(t)$ = external surface temperature of wall i at time t
$T_o(t)$ = outdoor temperature at time t
h_o = convective heat transfer coefficient at the external wall surface

The value of h_o is dependent on the outdoor air velocity and direction of wind relative to the wall, and the temperature difference between the wall surface and the outdoor air.

McClellan & Pedersen (1997) compared the methods used in different building energy simu-lation models and recommended the use of the MoWiTT (Mobile Window Thermal Test; Yazdanian & Klems 1994) model as follows:

$$h_{\mathrm{o}} = \sqrt{\left[C_{\mathrm{t}}(\Delta T)^{1/3}\right]^2 + \left[aV_{\mathrm{o}}^b\right]^2} \tag{2.46}$$

where:

C_{t} = turbulent natural convection constant
ΔT = temperature difference between the exterior surface and the outside air
a and b = constants
V_{o} = wind speed at standard conditions

The values of the coefficients and constants for the MoWiTT model are summarised in Table 2.1.

For certain applications, such as design cooling load estimations, a constant value for h_{o} is often used, and appropriate values can be found in the literature (see for example Loveday & Taki 1996).

The rate of heat that the external surface of a wall or roof (denoted as wall i) would absorb from the incident solar radiation can be estimated by:

$$q_{\mathrm{W},i,0,\mathrm{Sol}}(t) = \alpha_{\mathrm{W},i,0} \cdot I_{\mathrm{T},i}(t) \tag{2.47}$$

where:

$q_{\mathrm{W},i,0,\mathrm{Sol}}(t)$ = solar heat gain at the external surface of wall i at time t
$I_{\mathrm{T},i}(t)$ = total intensity of solar radiation incident upon the external surface of wall i at time t (including direct and diffuse radiation from the sun and the sky, and reflected solar radiation from the ground and other surfaces)
$\alpha_{\mathrm{W},i,0}$ = solar absorptivity of the external surface of wall i

Determination of the total solar irradiation on a surface requires the knowledge about the direct normal and diffuse horizontal intensities incident upon the surface, and the geometric relation between the surface and the sky and the position of the sun at a particular time. The required year-round solar data may be obtained from a nearby meteorological station but some meteorological stations may only measure the global solar radiation. In this case, the global solar radiation would need to be discomposed into the direct and diffuse components through the use of an appropriate model.

Table 2.1 MoWiTT model constants (ASHRAE Transactions 100(1) 1994. © ASHRAE, www.ashrae.org.)

Wind direction	C_{t} (W m^{-2} K$^{-4/3}$)	a (W m^{-2} K (m s^{-1})^b)	b (–)
Windward	0.84	2.38	0.89
Leeward	0.84	2.86	0.617

Apart from solar radiation, the external surface will also exchange long wave radiation with the surroundings, such as outdoor air, the ground and nearby buildings, and the sky. A rigorous treatment taking into account all these factors is complicated and simplification is often made. The following summarises the derivation presented by McClellan & Pedersen (1997):

$$q_{W,i,0,\text{Rad}}(t) = \varepsilon_{W,i,0}\sigma\left\{F_{W,i,0,\text{A}}\left[T_o^4(t) - T_{W,i,0}^4(t)\right] + F_{W,i,0,\text{G}}\left[T_G^4(t) - T_{W,i,0}^4(t)\right]\right.$$
$$\left. + F_{W,i,0,\text{Sky}}\left[T_{\text{Sky}}^4(t) - T_{W,i,0}^4(t)\right]\right\}$$

where:

$q_{W,i,0,\text{Rad}}(t)$ = net long wave radiant gain of the external surface of wall i at time t
$\varepsilon_{W,i,0}$ = long wave emissivity of the external surface of wall i
σ = Stefan–Boltzmann constant = 5.67×10^{-8} W m^{-2} K^{-4}
$F_{W,i,0,\text{A}}$ = shape factor between the external surface of wall i and the outdoor air
$F_{W,i,0,\text{G}}$ = shape factor between the external surface of wall i and the ground
$F_{W,i,0,\text{Sky}}$ = shape factor between the external surface of wall i and the sky
$T_{\text{Sky}}(t)$ = the sky temperature at time t
$T_G(t)$ = the ground temperature at time t

Note that in this part of the calculation, the temperatures are absolute temperatures in Kelvin. Recognising that the sum of the three shape factors must equal one ($F_{W,i,0,\text{A}} + F_{W,i,0,\text{G}} + F_{W,i,0,\text{Sky}} = 1$), the following algebraic manipulations with the above equations can be performed:

$$q_{W,i,0,\text{Rad}}(t) = \varepsilon_{W,i,0}\sigma\left\{\left[F_{W,i,0,\text{A}}T_o^4(t) + F_{W,i,0,\text{G}}T_G^4(t) + F_{W,i,0,\text{Sky}}T_{\text{Sky}}^4(t)\right]\right.$$
$$\left. - \left[F_{W,i,0,\text{A}} + F_{W,i,0,\text{G}} + F_{W,i,0,\text{Sky}}\right]T_{W,i,0}^4(t)\right\}$$

$$q_{W,i,0,\text{Rad}}(t) = \varepsilon_{W,i,0}\sigma\left\{\left[F_{W,i,0,\text{A}}T_o^4(t) + F_{W,i,0,\text{G}}T_G^4(t) + F_{W,i,0,\text{Sky}}T_{\text{Sky}}^4(t)\right.\right.$$
$$\left. - \left(F_{W,i,0,\text{A}} + F_{W,i,0,\text{G}} + F_{W,i,0,\text{Sky}}\right)T_o^4(t)\right]$$
$$\left. + \left[T_o^4(t) - T_{W,i,0}^4(t)\right]\right\}$$

$$q_{W,i,0,\text{Rad}}(t) = \varepsilon_{W,i,0}\sigma\left\{\left[F_{W,i,0,\text{G}}\left(T_G^4(t) - T_o^4(t)\right) + F_{W,i,0,\text{Sky}}\left(T_{\text{Sky}}^4(t) - T_o^4(t)\right)\right.\right.$$
$$\left. + \left(T_o^4(t) - T_{W,i,0}^4(t)\right)\right\} \tag{2.48}$$

Combining the three components of heat transfer in Equations 2.45, 2.47 and 2.48:

$$q_{W,i,0}(t) = q_{W,i,0,\text{Conv}}(t) + q_{W,i,0,\text{Sol}}(t) + q_{W,i,0,\text{Rad}}(t)$$
$$= h_o\left(T_o(t) - T_{W,i,0}(t)\right) + \alpha_{W,i,0}I_{T,i}(t)$$
$$+ \varepsilon_{W,i,0}\sigma\left\{\left[F_{W,i,0,\text{G}}\left(T_G^4(t) - T_o^4(t)\right) + F_{W,i,0,\text{Sky}}\left(T_{\text{Sky}}^4(t) - T_o^4(t)\right)\right.\right.$$
$$+ \left(T_o^4(t) - T_{W,i,0}^4(t)\right)\right\}$$

$$q_{W,i,0}(t) = \alpha_{W,i,0} I_{T,i}(t) + h_o\left(T_o(t) - T_{W,i,0}(t)\right) + \varepsilon_{W,i,0}\sigma\left(T_o^4(t) - T_{W,i,0}^4(t)\right)$$
$$+ \varepsilon_{W,i,0}\sigma\left\{\left[F_{W,i,0,G}\left(T_G^4(t) - T_o^4(t)\right) + F_{W,i,0,Sky}\left(T_{Sky}^4(t) - T_o^4(t)\right)\right]\right\}$$

Defining a linearised coefficient of radiant heat transfer to outdoor air as:

$$h_{W,i,0,Rad} = \frac{\varepsilon_{W,i,0}\sigma\left(T_o^4 - T_{W,i,0}^4(t)\right)}{T_o - T_{W,i,0}} \tag{2.49}$$

and substituting Equation 2.49 into the preceding equation, we finally get:

$$q_{W,i,0}(t) = \alpha_{W,i,0} I_{T,i}(t) + \left(h_o + h_{W,i,0,Rad}\right)\left(T_o(t) - T_{W,i,0}(t)\right)$$
$$+ \varepsilon_{W,i,0}\sigma\left\{\left[F_{W,i,0,G}\left(T_G^4(t) - T_o^4(t)\right) + F_{W,i,0,Sky}\left(T_{Sky}^4(t) - T_o^4(t)\right)\right]\right\} \tag{2.50}$$

Note that through the algebraic manipulations summarised above, the long wave radiation exchange between the external surface of a wall or a roof and the surroundings can now be estimated with reference to the temperature difference between the outdoor air and the external surface of the wall or roof, plus a correction (the last term in Equation 2.50) that is dependent on the temperatures of the ground, the sky and the outdoor air. Note also that the shape factor between the external surface of the wall or roof with outdoor air ($F_{W,i,0,A}$) has been eliminated during the algebraic manipulations, and is ignored in further calculations. Thus, the shape factors between a vertical wall and the ground and between the wall and the sky are both taken as 0.5. For a tilted surface with a tilting angle Σ, the shape factors can be determined as follows:

$$F_{W,i,0,G} = \frac{1 - \cos\Sigma}{2} \tag{2.51a}$$

$$F_{W,i,0,Sky} = \frac{1 + \cos\Sigma}{2} \tag{2.51b}$$

Involving the fourth power of the absolute temperatures in Equation 2.50 significantly complicates the solution process. Therefore, the equation is often further simplified into a linear equation as follows:

$$q_{W,i,0}(t) = \alpha_{W,i,0} I_{T,i}(t) + \left(h_o + h_{W,i,0,Rad}\right)\left(T_o(t) - T_{W,i,0}(t)\right)$$
$$+ h_{WR,i,0,G}\left(T_G(t) - T_o(t)\right) + h_{WR,i,0,Sky}\left(T_{Sky}(t) - T_o(t)\right) \tag{2.52}$$

Similar to Equation 2.49, $h_{WR,i,0,G}$ and $h_{WR,i,0,Sky}$ are the radiant heat transfer coefficients defined as follows:

$$h_{WR,i,0,G} = \frac{\varepsilon_{W,i,0}\sigma F_{W,i,0,G}\left(T_G^4(t) - T_o^4(t)\right)}{T_G - T_o} \tag{2.53a}$$

$$h_{WR,i,0,Sky} = \frac{\varepsilon_{W,i,0}\sigma F_{W,i,0,Sky}\left(T_{Sky}^4(t) - T_o^4(t)\right)}{T_{Sky} - T_o} \qquad (2.53b)$$

The above two equations can be approximated, to a sufficiently high degree of accuracy, as:

$$h_{WR,i,0,G} = 4\varepsilon_{W,i,0}\sigma F_{W,i,0,G} T_{Avg,G,o}^3(t) \qquad (2.54a)$$

$$h_{WR,i,0,Sky} = 4\varepsilon_{W,i,0}\sigma F_{W,i,0,Sky} T_{Avg,Sky,o}^3(t) \qquad (2.54b)$$

where:

$$T_{Avg,G,o}(t) = \frac{T_G(t) + T_o(t)}{2} \qquad (2.55a)$$

$$T_{Avg,Sky,o}(t) = \frac{T_{Sky}(t) + T_o(t)}{2} \qquad (2.55b)$$

Equation 2.52 is often abbreviated as:

$$q_{W,i,0}(t) = \alpha_{W,i,0} I_{T,i}(t) + h_{W,i,0,RC}\left(T_o(t) - T_{W,i,0}(t)\right) + \varepsilon_{W,i,0}\Delta R(t) \qquad (2.56)$$

where $h_{W,i,0,RC}$ is a combined convective and radiant heat transfer coefficient for the wall or roof and $\Delta R(t)$ denotes the net long wave radiation exchange of the surface with the sky and the ground if the surface is 'black'.

Various approaches are available for estimating sky and ground temperatures. For a clear sky, which is the condition on which design cooling load calculations are based, the simplest method is to assume the sky temperature to be 6K below the outdoor dry-bulb temperature. Methods are also available for predicting the sky temperature under a cloudy sky (e.g. Kimura 1977; Clarke 2001). Ground temperature values can vary significantly, depending upon terrain, surrounding buildings (heat island effect); but general surface temperatures estimates can be based on a form of sol-air temperature (Clarke 2001). For conduction calculations involving sub-surface temperatures, local soil conditions and moisture content are dominant influences – see for example Kusuda & Achenbach (1965).

For vertical surfaces, the radiant heat loss to the sky compensates the heat gain from the ground. Consequently, the net long wave radiation gain is quite small and may be neglected (i.e., $\Delta R(t) \approx 0$). However, for horizontal surfaces, i.e. external surfaces of roofs, which exchange long wave radiation with the sky only, the heat loss to the sky could be rather significant, particularly during night-time under a clear sky.

The term at the left hand side of Equation 2.56 denotes the conduction heat flux that enters the wall or roof from its external surface. With the use of the concept of sol-air temperature ($T_{e,o}(t)$), the same heat transfer rate can be described by:

$$q_{W,i,0}(t) = h_{W,i,0,RC}\left(T_{e,o}(t) - T_{W,i,0}(t)\right) \qquad (2.57)$$

It follows that:

$$T_{e,o}(t) = T_o(t) + \frac{\alpha_{W,i,0} I_{T,i}(t)}{h_{W,i,0,RC}} + \frac{\varepsilon_{W,i,0} \Delta R(t)}{h_{W,i,0,RC}} \qquad (2.58)$$

It can be seen that the sol-air temperature, as defined above, will allow the heat transfer at the external surface of a wall or roof to be treated relatively easily, as the sol-air temperature can be used directly as a boundary condition in the calculation. However, the long wave radiation gain from the sky and the ground needs to be determined based on the surface temperature of the wall or roof (see Equation 2.52), and hence the sol-air temperature is also dependent on the surface temperature.

The surface temperature can be solved from Equation 2.52 as follows:

$$T_{W,i,0}(t) = \frac{\left(\begin{array}{c} -q_{W,i,0}(t) + \left(h_o + h_{W,i,0,Rad} - h_{WR,i,0,G} - h_{WR,i,0,Sky}\right) T_o(t) \\ + h_{WR,i,0,G} T_G(t) + h_{WR,i,0,Sky} T_{Sky}(t) + \alpha_{W,i,0} I_{T,i}(t) \end{array}\right)}{h_o + h_{W,i,0,Rad}} \qquad (2.59)$$

Equation 2.59, however, is an implicit equation, as the conduction heat flux at the surface and the radiant heat transfer coefficients are dependent on the surface temperature. The methods that can be used for relating the outer surface temperatures of building envelope elements to the heat transfer through the elements, and to the heat gain and cooling load of the space, are dependent on the methods used to model the heat transfer through the fabric, which will be discussed later.

Heat balance on internal surfaces of external walls and roofs

In studies on heat transfer in buildings, the simplifying assumption that often needs to be made is that the air inside a space is perfectly mixed such that the state of the air in the room can be represented by a single air state, to provide a basis for determining the heat transfers between the enclosing surfaces and the room air.

The heat transfer that takes place at the internal surface of an external wall or roof (denoted as wall i) includes the conduction heat transfer from within the wall or roof to the surface, the convective heat exchange between the surface and the indoor air, and the radiant heat exchanges of the surface with other enclosing surfaces of the room, and with the sources in the room that emit long wave radiation. Since the algebraic sum of all these heat fluxes at the surface must be zero, the heat balance equation for the surface can be expressed as follows:

$$q_{W,i,L}(t) = h_{i,i}\left(T_{W,i,L}(t) - T_i(t)\right) - q_{W,i,L,Sol}(t) - q_{W,i,L,Rad,W}(t) - q_{W,i,L,Rad,Int}(t) \qquad (2.60)$$

where:

$h_{i,i}$ = the convective heat transfer coefficient for the internal surface of wall i

$T_{W,i,L}(t)$ = the internal surface temperatures of wall i at time t

$T_i(t)$ = the indoor temperature in the room at time t

$q_{W,i,L,Sol}(t)$ = the radiant heat gain at the internal surface of wall i due to solar radiation transmitted through windows at time t

$q_{W,i,L,Rad,W}(t)$ = the net heat gain at the internal surface of wall i due to long wave radiation exchange with other enclosing surfaces of the room at time t

$q_{W,i,L,Rad,Int}(t)$ = the net gain heat gain at the internal surface of wall i due to long wave radiation emitted by internal heat sources in the room at time t

The convective heat transfer coefficient, h_i, for internal surfaces in buildings may vary by a factor of 5 to 6, depending on the mode of convection (natural or forced) and the direction of heat flow relative to the surface (e.g. horizontal, upward or downward). Extensive results can be found in the literature based on both experimental procedures and the use of computational fluid dynamics (Alamdari & Hammond 1983; Awbi 1998; Awbi & Hatton 1999). For values recommended for general design applications see also ASHRAE (2001). In indoor environments, the convective heat transfer from the internal surface of a wall or roof to the room air is often estimated together with the net gain or loss of energy at the surface due to long wave radiation heat exchange between the surface and other enclosing surfaces of the room, through the use of a combined convective and radiant heat transfer coefficient in the calculation.

Only the convective heat exchange with the room air, described by the first term at the right hand side of Equation 2.60, will instantaneously become the cooling load of the room. The other terms, however, will affect the surface temperature of the wall or roof, and thus the convective heat transfer from the surface to the room air.

The internal surface of wall i can be solved from the above as follows:

$$T_{W,i,L}(t) = \frac{q_{W,i,L}(t) + h_{i,i}T_i(t) + q_{W,i,L,Sol}(t) + q_{W,i,L,Rad,W}(t) + q_{W,i,L,Rad,Int}(t)}{h_{i,i}} \qquad (2.61)$$

Similar to the equation for the external surface temperature, the above equation is an implicit equation, as the conduction heat flux at the surface, the long wave radiation heat exchanges among the internal surfaces of the room, and the heat exchanges of the surface with the internal sources are all dependent on the internal surface temperature.

Radiation heat exchanges among the internal surfaces of the enclosure

The radiation heat exchanges among the enclosing surfaces of a room can be modelled by using the conventional radiation heat transfer network analysis method, based on the assumption that all the surfaces are plain grey surfaces. Figure 2.8 shows a network for a simple room. In this method, the net rate of radiation heat gain of each of the grey surfaces that are exchanging radiation with each other can be described by:

$$-q_{W,i,L,Rad,W}(t) = \frac{\varepsilon_{W,i,L}}{1 - \varepsilon_{W,i,L}}\left(E_{b,W,i,L}(t) - J_{W,i,L}(t)\right) \qquad (2.62)$$

where:

$q_{W,i,L,Rad,W}(t)$ = the rate of radiant heat absorbed by the surface per unit area of the surface (hence the negative sign)

$E_{b,W,i,L}(t)$ = the emissive power of a black body at temperature $T_{W,i,L}(t)$; note that $E_{b,W,i,L}(t) = \sigma T^4_{W,i,L}(t)$

$J_{W,i,L}(t)$ = the radiosity of the internal surface of wall i at time t

and, for each surface, a heat balance consideration allows the following equation to be established:

$$\left(\frac{\varepsilon_{W,i,L}}{1 - \varepsilon_{W,i,L}} \right)\left(E_{b,W,i,L}(t) - J_{W,i,L}(t)\right) + \sum_{j=1}^{N} F_{i,j}\left(J_{W,j,L}(t) - J_{W,i,L}(t)\right) = 0 \tag{2.63}$$

where $F_{i,j}$ is the radiation shape factor between surfaces i and j.

The equations for all the surfaces can be written in matrix form as follows:

$$\begin{bmatrix} a_{11} & a_{12} & a_{13} & \cdots & a_{1N} \\ a_{21} & a_{22} & a_{23} & \cdots & a_{2N} \\ \vdots & \vdots & \vdots & \vdots & \vdots \\ \vdots & \vdots & \vdots & \vdots & \vdots \\ a_{N1} & a_{N2} & a_{N3} & \cdots & a_{NN} \end{bmatrix} \begin{Bmatrix} J_{W,1,L}(t) \\ J_{W,2,L}(t) \\ \vdots \\ \vdots \\ J_{W,N,L}(t) \end{Bmatrix} = \begin{Bmatrix} C_1 \\ C_2 \\ \vdots \\ \vdots \\ C_N \end{Bmatrix} \tag{2.64}$$

where, if the N surfaces completely enclose the space:

$$a_{ii} = \delta_{ij}$$

$$a_{ij} = -(1 - \delta_{ij})(1 - \varepsilon_{W,i,L})F_{i,j}$$

where, δ_{ij} is the Kronecker delta (i.e., $\delta_{ij} = 1$ when $i = j$; otherwise 0).

$$C_i = \varepsilon_{W,i,L}\sigma T^4_{W,i,L}(t)$$

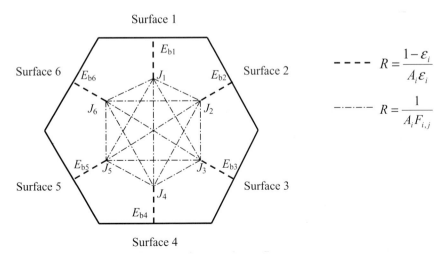

Fig. 2.8   Radiation thermal network for an enclosure with six surfaces.

Since the radiation shape factors can be determined based on the dimensions and geometric relations among the surfaces, and the surface emissivities are of known values, the radiosities for the surfaces can be solved from Equation 2.64 once the temperatures of the surfaces are known. The net radiation heat flux entering each surface can then be determined using Equation 2.62.

The above method can be implemented in a straightforward manner for modelling the long wave radiation exchanges among the enclosing surfaces of a room in using the explicit Finite Difference Method to model the building fabric heat transfer, as the temperatures at the internal surfaces at the current time step will all be known. If an implicit scheme is used instead, the process will become more complex, as the solution of the set of implicit equations would involve the solution of the non-linear radiation heat exchange equations through the use of an iterative calculation procedure. Likewise, an iterative method would need to be used to solve for the surface temperatures when the Response Factor Method or the Transfer Function Method is used.

Simplified treatment of the radiant heat exchanges among the enclosing surfaces of a room

Using the fundamental approach as outlined above to model the long wave radiation heat exchanges among the enclosing surfaces of a room imposes a substantial burden to the building heat transfer simulation calculations. Many methods, therefore, have been proposed to simplify the calculation.

Calculation of the shape factors can be substantially simplified by using the area-weighted method, as shown below, to yield approximate values for the shape factors:

$$F_{i,j} \approx \frac{A_i}{\sum\limits_{j=1}^{N} A_j} \qquad (2.65)$$

In calculating the sum of areas, the area of surface i, and the areas of any surfaces facing the same direction as surface i, are excluded. Use of the approximate shape factors thus determined, however, can lead to imbalance in the total radiant heat exchange among the surfaces.

Liesen & Pedersen (1997) examined several simplified methods for approximate calculation of the long wave radiation heat exchanges among the enclosing surfaces of a room. Among the methods included in their study, the mean radiant temperature with balance (MRT/Bal) method seems to be a good comprise between accuracy and simplicity (Walton 1980). In the MRT/Bal method, each surface is assumed to exchange long wave radiation with and only with a fictitious surface. The area, temperature and emissivity of this surface are estimated as follows:

$$A_{\text{MRT},i} = \sum\limits_{j=1}^{N} A_j \ (\text{for } j \neq i) \qquad (2.66)$$

$$\varepsilon_{\text{MRT},i} = \sum\limits_{j=1}^{N} \frac{A_j \varepsilon_j}{A_{\text{MRT},i}} \ (\text{for } j \neq i) \qquad (2.67)$$

$$T_{\mathrm{MRT},i} = \sum_{j=1}^{N} \frac{A_j \varepsilon_j T_j}{A_{\mathrm{MRT},i} \varepsilon_{\mathrm{MRT},i}} \ (\text{for } j \neq i) \tag{2.68}$$

In addition to the surface concerned (surface i), any surfaces facing the same direction as surface i are also excluded from the calculation of the area, emissivity and temperature of the fictitious surface.

The shape factor between a surface and the fictitious surface with which it is exchanging long wave radiation is determined from (Walton 1980):

$$F_{\mathrm{MRT},i} = \frac{1}{\dfrac{1-\varepsilon_i}{\varepsilon_i} + 1 + \dfrac{A_i(1-\varepsilon_{\mathrm{MRT},i})}{A_{\mathrm{MRT},i}\varepsilon_{\mathrm{MRT},i}}} \tag{2.69}$$

The net heat gain per unit area of a surface due to radiant heat exchange with other enclosing surfaces of the room can then be estimated by:

$$-q_{\mathrm{W},i,\mathrm{Rad,W}} = \sigma F_{\mathrm{MRT},i}\left(T_i^4 - T_{\mathrm{MRT},i}^4\right) \tag{2.70}$$

The above MRT method will lead to some net imbalance in the total radiant energy due to the approximations made in determining the shape factors and the mean radiant temperature, which needs to be corrected by distributing the imbalance to the surfaces. The balance heat flux to be adjusted for each surface can be determined by:

$$q_{\mathrm{Bal}} = \frac{\displaystyle\sum_{i=1}^{N} \sigma A_i F_{\mathrm{MRT},i}\left(T_{\mathrm{MRT},i}^4 - T_i^4\right)}{\displaystyle\sum_{i=1}^{N} A_i} \tag{2.71}$$

With this correction included:

$$-q_{\mathrm{W},i,\mathrm{Rad,W}} = \sigma F_{\mathrm{MRT},i}\left(T_i^4 - T_{\mathrm{MRT},i}^4\right) - q_{\mathrm{Bal}} \tag{2.72}$$

A further step of simplification can also be used to linearise the calculation by the following approximation:

$$4T_{\mathrm{Avg,MRT},i}^3\left(T_i - T_{\mathrm{MRT},i}\right) \approx T_i^4 - T_{\mathrm{MRT},i}^4 \tag{2.73}$$

where:

$$T_{\mathrm{Avg,MRT},i} = \frac{T_i - T_{\mathrm{MRT},i}}{2} \tag{2.74}$$

A radiant heat transfer coefficient can also be defined as follows to yield a linearised model:

$$-q_{W,i,\text{Rad},W} = h_{R,i}(T_i - T_{\text{MRT},i}) - q_{\text{Bal}} \qquad (2.75)$$

and, it follows that:

$$h_{R,i} = 4\sigma F_{\text{MRT},i} T_{\text{Ave,MRT},i}^3 \qquad (2.76)$$

Heat balance on window glazing

The radiation heat exchange with the sky and the ground at the external surface, and with the internal surfaces of other fabric elements at the internal surface, considered above for walls and roofs, equally apply to the window glazing. The method to be used to account for these heat exchanges is similar except that, due to the relatively thin glass thickness, a uniform temperature may be assumed for the internal and the external surfaces.

Assuming that the heat capacity of the glass is negligible, a heat balance on a single pane of window glass can be described by:

$$h_{RC}(T_o - T_{WG}) + \alpha_\theta I_D + \alpha_d I_d + \varepsilon_{WG}\Delta R + q_{WG,\text{Rad},W} + q_{WG,\text{Rad,Int}} = h_i(T_{WG} - T_i) \quad (2.77)$$

The glass temperature, therefore, can be solved from the above as follows:

$$T_{WG} = \frac{h_{RC}T_o + h_i T_i + \alpha_\theta I_D + \alpha_d I_d + \varepsilon_{WG}\Delta R + q_{WG,\text{Rad},W} + q_{WG,\text{Rad,Int}}}{h_i + h_{RC}} \qquad (2.78)$$

For windows with more than one pane of glass, the temperature of each pane of glass will need to be determined separately, taking into account the heat transfer between each pair of glass panes.

Treatment of solar radiation: diffuse and direct components of solar radiation

To estimate the incident irradiance on a surface at any orientation, estimations of diffuse and direct fractions of radiation on the horizontal plane are required since most solar data are presented as global measurements on the horizontal plane. From these values, an estimation of incident irradiance on a tilted surface may be made. For most building energy modelling purposes, hourly estimates of incident irradiance will be sufficient, and reasonably extensive data sets are available for most world sites from which direct and diffuse fractions may be derived (e.g. World Meteorological Organisation (1964–1985); *European Solar Radiation Atlas* 2000). Various correlations are available from which the estimation of diffuse and direct components of total radiation from the horizontal radiation only can be arrived at (e.g. Erbs *et al.* 1982, Reindl *et al.* 1990, Skartveit & Olseth 1987). These methods are generally based on correlating the hourly total horizontal radiation, $I_{T,h}$ with the hourly clearness index, k_c, defined as:

$$k_C = \frac{I_{T,h}}{I_E} \qquad (2.79)$$

where I_E is the extraterrestrial radiation – the total solar radiation incident on a horizontal surface of the earth if there were no atmosphere to be considered. The extraterrestrial radiation can be determined from (Duffie & Beckman 1991):

$$I_E = I_0\left[1 + 0.033\cos\left(360 n_{day}/365\right)\right]\cos(\theta_h) \tag{2.80}$$

in which I_0 is the solar constant (taken to be 1367 Wm^{-2}), n_{day} is the day number of the year ($1 \le n_{day} \le 365$) and θ_h is the angle of incidence of the solar beam with the horizontal plane.

Of the various empirical modelling approaches for splitting the global horizontal radiation into its direct and diffuse components, the following example recommended by Skartveit & Olseth (1987) is especially applicable to high northern latitudes. Models based on other global sites can be found in the literature (for instance several methods are compared in Duffie & Beckman 1991):

$$\frac{I_d}{I_{T,h}} = \begin{cases} 1.00 & \left(\text{for } k_C \le k_{C(0)}\right) \\ f(k_C) & \left(\text{for } k_{C(0)} \le k_C \le \alpha k_{C(1)}\right) \\ 1 - \alpha k_{C(1)}\left[1 - f\left(\alpha k_{C(1)}\right)/k_C\right] & \left(\text{for } k_C \ge \alpha k_{C(1)}\right) \end{cases} \tag{2.81}$$

where:

$$f(k_C) = 1 - (1-d)\left(a\sqrt{K} + (1-a)K^2\right)$$

and:

$$K = 0.5\left\{1 + \sin\left[\pi\,\frac{k_C - k_{C(0)}}{k_{C(1)} - k_{C(0)}} - 0.5\right]\right\}$$

The recommended parameters of the model are (Skartveit & Olseth 1987):

$$k_{C(0)} = 0.2;\ k_{C(1)} = 0.87 - 0.56\exp[-0.06(90 - \theta_h)]$$
$$\alpha = 1.09;\ a = 0.27;\ d = 0.15 + 0.43\exp[-0.06(90 - \theta_h)]$$

In Equation 2.81, I_d is the diffuse component of solar radiation. Thus the direct component on the horizontal plane may now be found by difference: $I_{D,h} = I_{T,h} - I_d$.

Treatment of solar radiation: estimation of irradiance on a surface

The total irradiance on a surface inclined at an angle of β to the horizontal can be obtained from the direct and diffuse components of solar radiation. For a surface whose angle of incidence with the solar beam is θ, the following may be used (Liu & Jordan 1963):

$$I_T = I_{D,h}\cos\theta + I_d\left(\frac{1 + \cos\beta}{2}\right) + \rho_g\left(I_{D,h} + I_d\right)\left(\frac{1 - \cos\beta}{2}\right) \tag{2.82}$$

where ρ_g is the ground reflectance (usually taken to be at 0.2 for normal conditions (Lui & Jordan 1963)).

The angle of incidence, θ, of direct radiation on a surface is calculated by geometric considerations. A thorough treatment can be found in Duffie & Beckman (1991):

$$\cos\theta = \sin(d)\sin(l)\cos(\beta) - \sin(d)\cos(l)\sin(\beta)\cos(\alpha) + \ldots$$
$$\ldots\cos(d)\cos(l)\cos(\beta)\cos(\omega) + \cos(d)\sin(l)\sin(\beta)\cos(\alpha)\cos(\omega) + \ldots \quad (2.83)$$
$$\ldots\cos(d)\sin(\beta)\sin(\alpha)\sin(\omega)$$

where d is the declination angle, the angular position of the sun with respect to the plane of the equator at solar noon, north positive ($-23.45° \leq d \leq 23.45°$), and may be estimated for any day of the year, n_{day} ($1 \leq n_{day} \leq 365$) from the following:

$$d = 23.45\sin\left(360(284 + n_{day})/365\right) \quad (2.84)$$

l is the latitude angle, the angular location of the site north or south of the equator, north positive ($-90° \leq l \leq 90°$).

α is the surface azimuth angle, the angle of projection on the horizontal plane of the normal to the surface with the local meridian (the local line of constant longitude) passing through the surface. The azimuth angle defines the orientation of the surface in the horizontal plane and is measured with respect to due south, east negative, west positive ($-180 \leq \alpha \leq 180$).

ω is the hour angle, the angular displacement of the sun east or west of the local meridian ($15°$ h from solar noon; east negative). The sun crosses the local meridian at solar noon. The calculation for hour angle requires solar time, t_{Sol}, rather than local clock time, t_{Clock}. Conversion from clock time requires a longitude correction to account for the difference between the standard longitude clock time and the clock time at the surface of interest. A further correction is required that takes into account perturbations in the earth's rotation, which affect the time the sun crosses the local meridian (Duffie & Beckman 1991), thus:

$$t_{Sol} = t_{Clock} + 4(L_0 - L_{loc}) + e \quad (2.85)$$

in which L_0 is the longitude of the standard meridian (degrees west), L_{loc} is the longitude of the location of interest (degrees west) and e is obtained from:

$$e = 9.87\sin\left[720(n_{day} - 81)/364\right] - 7.53\cos\left[360(n_{day} - 81)/364\right]$$
$$-1.5\sin\left[360(n_{day} - 81)/364\right]$$

Treatment of solar radiation: insolation and shadowing

In some applications a rigorous treatment of the distribution of solar radiation within a room space is needed. Examples can for instance be found in passive solar designs in which high glazing areas are combined with the selection of high thermal capacity construction materials for certain key surfaces to act, in effect, as collectors.

It is possible to develop a method of geometric projections of internal room surfaces onto window surfaces thereby pairing portions of the window surface with corresponding insolated internal surfaces (and, likewise, external shadowing obstruction surfaces).

Figure 2.9 illustrates the three possible projections onto a window plane from the internal surfaces of a room space for a given solar beam trajectory. Each surface is projected onto the plane containing a window in the direction of the solar beam, I_D, thus locating an overlapping patch on the window plane. The direct solar radiation attributable to this patch is then applied to the surface whose projection formed it.

It is possible to show (e.g. Mortensen 1989) that the geometric projection of any point, (x, y, z), in the room space is translated to a point on the window plane, (x_p, y_p), where:

$$x_p = x + z \cdot \left(\cos \theta_{x(0)} / \cos \theta_{z(0)} \right) \tag{2.86}$$

$$y_p = y + z \cdot \left[\tan^2 \theta_{z(0)} - \left(\cos \theta_{x(0)} / \cos \theta_{z(0)} \right)^2 \right]^{\frac{1}{2}} \tag{2.87}$$

(in which $\theta_{x(0)}$ and $\theta_{z(0)}$ are the angles of incidence on the planes $x = 0$ and $z = 0$ respectively).

In a similar manner, it is possible to project a window surface along the beam in order to establish overlaps with any external obstructions such as adjacent buildings. The overlapping patches then form shaded regions lying on the window surface that can be used to correct the solar beam trajectory entering the room space.

The result of this coordinate transformation yields receiving surfaces and shadowing surfaces as polygons that lie in the same plane. There are now two approaches for identifying overlaps between receiving and shadowing surfaces. The first, a discrete element approach, involves casting a mesh over the receiving surface and then testing each grid point for enclosure by the shadow polygon – a set of grid containment tests. The accuracy of this method depends on the mesh size used. An alternative method, based on homogenous coordinates, was first developed by Walton (1979) and has been used widely since (e.g. Athienitis & Stylianou 1991). In most circumstances this method has the advantage over the discrete element approach in that it is computationally efficient. Homogenous coordinates are used to determine the vertices of an overlap between a receiving surface and a shadow surface. A polygon containment test is then carried out to establish the extent of any overlap. A procedure for this is summarised as follows – further details can be found in Walton (1979).

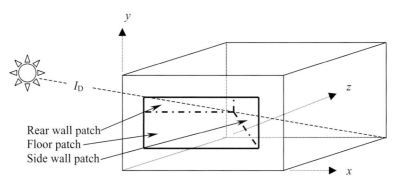

Fig. 2.9   Internal surface projections onto a window plane.

- Step 1: Transform the coordinates of the receiving surface and all shadowing surfaces into a single *x–y* plane.
- Step 2: Express all surfaces as homogenous coordinates and remove any vertices of shadowing polygons contained within other shadowing polygons to leave a pair of surfaces (one receiving and one shadowing).
- Step 3: Identify the vertices of the receiving polygon enclosed by the shadow polygon; then identify the vertices of the shadow polygon enclosed by the receiving polygon and locate the intercepts of the sides of the receiving polygon with the sides of the shadow polygon. Hence determine the vertices of the overlap and sort them into clockwise sequence.
- Step 4: Calculate the area of the overlap polygon. This then forms an insolation patch in situations where the receiving surface is an internal surface and the mating surface is a window or a shadow patch, in the case of the window surface being the receiving surface and an external obstruction forming the mating surface.

The above deals with first-received direct solar radiation at internal surfaces. Though the majority of this energy will be absorbed in the case of opaque room surfaces, a small amount will be reflected. Since most reflected solar energy will be absorbed by the second or third reflection, it is reasonable in most applications to apportion the first reflected component based on surface view factors as described earlier.

A more rigorous treatment of reflected surface radiation can be found in Clarke (2001). Also, the direct component of solar radiation only has been dealt with here. The treatment of the diffuse component is altogether more difficult due to the complex relationships of internal and external surface reflections. However, an acceptable compromise is discussed in the following section.

Simplified treatment of solar and other radiant heat gains

The above treatment of the heat gain due to solar transmission through windows is complicated, as it requires the determination of:

(1)   the parts of the internal surfaces that the transmitted direct solar radiation will fall upon, which vary with the time in the day and the day in the year;

(2)   the amount of energy absorbed and reflected by the surfaces in each incidence of the radiation, from the first surface it strikes after entering the space to other surfaces that the reflected radiation will be incident upon, until the solar radiation is completely dissipated.

Though task (2) above can be simplified by regarding the reflected radiation as entirely diffuse in nature, the problem remains complicated for practical applications. For simplicity, the assumption is often made that the direct component of the transmitted solar radiation would be distributed uniformly over the floor only, whereas the diffuse radiation would be uniformly distributed to all the enclosing surfaces of the room. According to this assumption, the radiant heat gain of an internal surface due to transmitted solar radiation is described by:

for the floor:

$$q_{W,i,L,Sol}(t) = \frac{\sum_j A_{W,j} \tau_{d,j} I_{j,d}(t)}{\sum_k A_{W,k}} + \frac{\sum_j SLF_j(t) \cdot A_{W,j} \tau_{D,j} I_{j,D}(t)}{A_{W,floor}} \tag{2.88}$$

and for other walls and the ceiling:

$$q_{W,i,L,Sol}(t) = \frac{\sum_j A_{W,j} \tau_{d,j} I_{j,d}(t)}{\sum_k A_{W,k}} \tag{2.89}$$

where:

$q_{W,i,L,Sol}(t)$ = rate of solar heat absorbed by internal surface of wall i at time t
$A_{W,k}$ = area of wall k ($k = 1, 2, \ldots, N$ where N = total number of enclosing surfaces of the room, including the floor and the windows)
$A_{W,floor}$ = area of the floor
$A_{W,j}$ = area of window j ($j \in k$)
$\tau_{d,j}$ = diffuse solar radiation transmissivity of window j
$\tau_{D,j}$ = direct solar radiation transmissivity of window j
$I_{j,d}(t)$ = intensity of diffuse radiation on a surface in the same direction as window j at time t
$I_{j,D}(t)$ = intensity of direct radiation on a surface in the same direction as window j at time t
$SLF_j(t)$ = the fraction of area of window j that has direct solar radiation falling upon it

Liesen & Pedersen (1997) found that only small difference would arise in making different assumptions on the distribution of the transmitted energy to the internal surfaces.

Many internal heat sources would also dissipate heat through both convection and radiation. Similar to transmitted diffuse solar radiation, the assumption is often made that the radiant heat emissions from internal sources would be distributed uniformly to all the enclosing surfaces of the room. Hence, the intensity of the radiation from the internal sources on each of the enclosing surfaces is described by:

$$q_{W,i,L,Rad,Int}(t) = \frac{\sum_j q_{j,Rad}(t)}{\sum_k A_{W,k}} \tag{2.90}$$

where:

$q_{W,i,L,Rad,Int}(t)$ = rate of radiation emission from internal sources that is absorbed by internal surface of wall i at time t
$q_{i,Rad}(t)$ = rate of radiation emission of internal source j at time t

It should be noted that the radiation emission from lighting is primarily short wave radiation whereas many other internal sources, such as office equipment, would emit primarily long wave radiation, but no distinction is made in the wavelength of the emitted radiation in this simplified treatment.

Zone heat balance

Consider a room of internal volume V, which is enclosed by N building fabric elements and contains M internal heat sources. The net sensible heat gain of the air in the room at time t and the corresponding effect on the room air temperature, assuming the indoor air is perfectly mixed, can be expressed as follows:

$$\rho_a V c_a \frac{dT_i(t)}{dt} = \sum_{i=1}^{N} A_{W,i} h_{i,i} \left(T_{W,i,L}(t) - T_i(t) \right) + m_{Inf}(t) c_a \left(T_o(t) - T_i(t) \right)$$

$$+ \sum_{j=1}^{M} q_{j,Conv}(t) - Q_{Sys,Sen}(t) \tag{2.91}$$

where:

ρ_a = density of the indoor air
c_a = specific heat capacity of air
$T_i(t)$ = the indoor air temperature at time t
$T_o(t)$ = the outdoor air temperature at time t
$A_{W,i}$ = the surface temperatures of the ith element at the room side at time t
$h_{i,i}$ = the convective heat transfer coefficient for the surface of the ith element
$T_{W,i,L}(t)$ = the surface temperature of the ith element at the room side at time t
$m_{Inf}(t)$ = the rate of infiltration of outdoor air into the room at time t
$q_{j,Conv}(t)$ = the rate of convective heat gain from internal source j
$Q_{Sys,Sen}(t)$ = the rate of sensible heat extraction from the room by the air-conditioning system at time t

With the assumption that during the air-conditioned periods, the indoor temperature will be maintained steadily at the set point level, the time derivative of the indoor temperature at the left hand side of Equation 2.91 will become zero. The equation can, therefore, be re-arranged to allow the cooling load that the air-conditioning system should provide to maintain the steady indoor temperature to be determined, as shown below:

$$Q_{Sys,Sen}(t) = \sum_{i=1}^{N} A_{W,i} h_{i,i} \left(T_{W,i,L}(t) - T_i(t) \right) + m_{Inf}(t) c_a \left(T_o(t) - T_i(t) \right) + \sum_{j=1}^{M} q_{j,Conv}(t) \tag{2.92}$$

Equation 2.92 links the heat transfer through each fabric element to the indoor air temperature whereas the fabric elements are linked by the equations that describe the long wave radiation heat exchange among the internal surfaces of the elements, such as Equations 2.63 and 2.64.

By setting the rate of sensible heat extraction of the air-conditioning system to zero, Equation 2.91 can also be used in the Finite Difference Method to predict the indoor temperature during the non-air-conditioned period. For the analytical methods that involve hourly calculations, the heat capacity of the volume of air in the room may be considered to be negligible, which allows the indoor temperature to be solved from Equation 2.92 as follows:

$$T_i(t) = \frac{\sum_{i=1}^{N} A_{W,i} h_{i,i} T_{W,i,L}(t) + m_{\text{Inf}}(t) c_a T_o(t) + \sum_{j=1}^{M} q_{j,\text{Conv}}(t)}{\sum_{i=1}^{N} A_{W,i} h_{i,i} + m_{\text{Inf}}(t) c_a} \tag{2.93}$$

2.3 Synthesis of heat transfer methods

Implementation of the Response Factor Method in a room heat balance

Recall that, in the Response Factor Method, the conduction heat fluxes at the two surfaces of a wall or roof (denoted as wall i) are related to the surface temperatures (Section 1.3) by:

$$q_{W,i,0}(n) = \sum_{j=0}^{\infty} X_i(j) T_{W,i,0}(n-j) - \sum_{j=0}^{\infty} Y_i(j) T_{W,i,L}(n-j) \tag{2.94}$$

$$q_{W,i,L}(n) = \sum_{j=0}^{\infty} Y_i(j) T_{W,i,0}(n-j) - \sum_{j=0}^{\infty} Z_i(j) T_{W,i,L}(n-j) \tag{2.95}$$

Taking the current temperature terms out of the summation processes yields:

$$q_{W,i,0}(n) = X_i(0) T_{W,i,0}(n) - Y_i(0) T_{W,i,L}(n) + \sum_{j=1}^{\infty} X_i(j) T_{W,i,0}(n-j) - \sum_{j=1}^{\infty} Y_i(j) T_{W,i,L}(n-j)$$

$$q_{W,i,L}(n) = Y_i(0) T_{W,i,0}(n) - Z_i(0) T_{W,i,L}(n) + \sum_{j=1}^{\infty} Y_i(j) T_{W,i,0}(n-j) - \sum_{j=1}^{\infty} Z_i(j) T_{W,i,L}(n-j)$$

The summation terms in the above equations are all based on past values of the temperatures, and hence are of known values. Expressing these equations as:

$$q_{W,i,0}(n) = X_i(0) T_{W,i,0}(n) - Y_i(0) T_{W,i,L}(n) + S_{\text{Past},W,i,0}(n) \tag{2.96}$$

$$q_{W,i,L}(n) = Y_i(0) T_{W,i,0}(n) - Z_i(0) T_{W,i,L}(n) + S_{\text{Past},W,i,L}(n) \tag{2.97}$$

substituting Equation 2.96 into Equation 2.59, and Equation 2.97 into Equation 2.61, and replacing t by n:

$$T_{W,i,0}(n) = \frac{\left(\begin{array}{c} Y_i(0)T_{W,i,L}(n) - S_{\text{Past},W,i,0}(n) + \left(h_o + h_{W,i,0,\text{Rad}} - h_{\text{WR},i,0,G} - h_{\text{WR},i,0,\text{Sky}}\right)T_o(n) \\ + h_{\text{WR},i,0,G}T_G(n) + h_{\text{WR},i,0,\text{Sky}}T_{\text{Sky}}(n) + \alpha_{W,i,0}I_{T,i}(n) \end{array}\right)}{X_i(0) + h_o + h_{W,i,0,\text{Rad}}}$$

$$(2.98)$$

$$T_{W,i,L}(n) = \frac{\left(\begin{array}{c} Y_i(0)T_{W,i,0}(n) + S_{\text{Past},W,i,L}(n) + h_{i,i}T_i(n) + q_{W,i,L,\text{Sol}}(n) \\ + q_{W,i,L,\text{Rad},W}(n) + q_{W,i,L,\text{Rad},\text{Int}}(n) \end{array}\right)}{Z_i(0) + h_{i,i}}$$

$$(2.99)$$

For each fabric element, two equations similar to Equations 2.98 and 2.99 can be derived. All these equations are coupled by the relations that describe the long wave radiant heat exchanges among the internal surfaces, and the whole set of equations needs to be solved simultaneously for each time step. An iterative method, such as successive back substitution, is often employed for this purpose. Having determined the internal surface temperatures for all fabric elements, the sensible cooling load can be determined using Equation 2.92. For predicting the indoor temperature during the non-air-conditioned periods, Equation 2.93 should be used instead, and solved together with the surface temperature equations.

Implementation of the Transfer Function Method in a room heat balance

Application of the Heat Balance Method in conjunction with the Transfer Function Method for modelling the heat transfer through the fabric elements is similar to the above method for the Response Factor Method. In the Transfer Function Method, the conduction heat transfer through the walls and roofs is modelled by:

$$q_{W,i,0}(n) = \sum_{j=0}^{N_E} E_i(j)T_{W,i,0}(n-j) - \sum_{j=1}^{N_F} F_i(j)T_{W,i,L}(n-j) - \sum_{j=1}^{N_q} H_i(j)q_{W,i,0}(n-j) \quad (2.100)$$

$$q_{W,i,L}(n) = \sum_{j=0}^{N_F} F_i(j)T_{W,i,0}(n-j) - \sum_{j=1}^{N_G} G_i(j)T_{W,i,L}(n-j) - \sum_{j=1}^{N_q} H_i(j)q_{W,i,L}(n-j) \quad (2.101)$$

Comparing the above equations with Equations 2.94 and 2.95 shows that with the summation terms up to the last time step revised to be as follows:

$$S_{\text{Past},W,i,0}(n) = \sum_{j=1}^{N_E} E_i(j)T_{W,i,0}(n-j) - \sum_{j=1}^{N_F} F_i(j)T_{W,i,L}(n-j) - \sum_{j=1}^{N_q} H_i(j)q_{W,i,0}(n-j) \quad (2.102)$$

$$S_{\text{Past},W,i,L}(n) = \sum_{j=1}^{N_F} F_i(j)T_{W,i,0}(n-j) - \sum_{j=1}^{N_G} G_i(j)T_{W,i,L}(n-j) - \sum_{j=1}^{N_q} H_i(j)q_{W,i,L}(n-j) \quad (2.103)$$

the following expressions for the internal and external surface temperatures can be derived:

$$T_{\mathrm{W},i,0}(n) = \frac{\left(\begin{array}{c} F_i(0)T_{\mathrm{W},i,\mathrm{L}}(n) - S_{\mathrm{Past},\mathrm{W},i,0}(n) + \left(h_\mathrm{o} + h_{\mathrm{W},i,0,\mathrm{Rad}} - h_{\mathrm{WR},i,0,\mathrm{G}} - h_{\mathrm{WR},i,0,\mathrm{Sky}}\right)T_\mathrm{o}(n) \\ + h_{\mathrm{WR},i,0,\mathrm{G}}T_\mathrm{G}(n) + h_{\mathrm{WR},i,0,\mathrm{Sky}}T_{\mathrm{Sky}}(n) + \alpha_{\mathrm{W},i,0}I_{\mathrm{T},i}(n) \end{array}\right)}{E_i(0) + h_\mathrm{o} + h_{\mathrm{W},i,0,\mathrm{Rad}}}$$

(2.104)

$$T_{\mathrm{W},i,\mathrm{L}}(n) = \frac{\left(\begin{array}{c} F_i(0)T_{\mathrm{W},i,0}(n) + S_{\mathrm{Past},\mathrm{W},i,\mathrm{L}}(n) + h_{\mathrm{i},i}T_\mathrm{i}(n) + q_{\mathrm{W},i,\mathrm{L},\mathrm{Sol}}(n) \\ + q_{\mathrm{W},i,\mathrm{L},\mathrm{Rad},\mathrm{W}}(n) + q_{\mathrm{W},i,\mathrm{L},\mathrm{Rad},\mathrm{Int}}(n) \end{array}\right)}{G_i(0) + h_{\mathrm{i},i}}$$

(2.105)

Similar to when the Response Factor Method is used, the temperatures at the wall and roof surfaces need to be solved by an iterative procedure before the sensible cooling load of a space can be determined using Equation 2.92.

The Transfer Function Method for cooling load simulation

The integration of the Transfer Function Method for modelling heat transfer through walls and roofs with the detailed Heat Balance Method, as described above, yields a rigorous analytical method for building energy simulation. The Transfer Function Method described in the ASHRAE *Handbook* (ASHRAE 2001) and in other relevant textbooks for cooling load prediction is a simplified version of this rigorous method. In this Transfer Function Method, the conduction transfer function for a wall or roof is based on the sol-air temperature that represents the combined effect of outdoor air temperature and radiant heat gain of the external surface. For the internal surface, the assumption is made that both the convective and radiant heat transfers are related to the indoor air temperature, which is maintained at a constant level (T_i = constant). These simplifications in the heat transfer calculations require the thermal resistances to convective heat transfer at the two surfaces of a wall or roof to be included in the calculation of the coefficients for the conduction transfer function, and the use of a combined convective and radiant heat transfer coefficient for the indoor surface. The conduction transfer function for modelling the heat gain from the internal surface of a wall, therefore, becomes:

$$q_{\mathrm{W},i,\mathrm{L}}(n) = \sum_{j=0}^{N_b} b_i(j)T_{\mathrm{e},\mathrm{o}}(n-j) - \sum_{j=0}^{N_c} c_i(j)T_\mathrm{i} - \sum_{j=1}^{N_q} d_i(j)q_{\mathrm{W},i,\mathrm{L}}(n-j)$$

(2.106)

Pre-calculated conduction transfer function coefficients for a set of wall and roof constructions are given in the *Handbook*, from which appropriate coefficients can be selected for use in cooling load simulations without the need to go through the tedious procedures of evaluating the conduction transfer function coefficients. However, if the U-value of the wall for which coefficients are given differs from that of the actual wall, the coefficients for the sol-air temperature ($b_i(0 \ldots N_b)$) and the indoor temperature ($c_i(0 \ldots N_C)$) should be adjusted by multiplying each of the coefficients by the ratio of the U-values of the actual wall and the given wall ($U_{\mathrm{actual}}/U_{\mathrm{given}}$).

The Transfer Function Method goes a step further by using a room transfer function as shown below for modelling the conversion of the heat gains of a space into the sensible cooling load of the space (Q_{Sen}):

$$Q_{Sen}(n) = \sum_{j=0}^{N_v} v_j q(n-j) - \sum_{j=1}^{N_w} w_j Q_{sen}(n-j) \qquad (2.107)$$

Note that determination of the cooling load components due to the heat gains from transmitted solar radiation, the conduction heat gains from walls and roofs, and the heat gains from lights and appliances requires different sets of room weighting factors v_j and w_j. The values of the weighting factors can be determined using a room model development based on the detailed Heat Balance Method (Cumali 1979; Kerrisk *et al.* 1981), and by exciting the room model with a unit pulse of each of the several types of heat gains, according to a predetermined relative distribution of the energy among the surfaces. In this process, a fixed value (typically zero) would be assumed for the sol-air temperature, and the initial temperatures of all components would be assumed to be uniform (also at zero temperature).

The sensible cooling loads at different time steps in response to the unit pulse heat gain excitation (denoted as $d(j)$ for $j = 0, 1, 2, \ldots$) represent the coefficients in the z-transfer function of the room as follows:

$$K(z) = d(0) + d(1)z^{-1} + d(2)z^{-2} + \ldots \ldots \qquad (2.108)$$

For avoiding a large number of steps in the summation process, and thus the need to use a large set of room transfer function coefficients, the transfer function is expressed as the division of two polynomials of z as follows:

$$K(z) = \frac{v_0 + v_1 z^{-1} + v_2 z^{-2}}{1 + w_1 z^{-1} + w_2 z^{-2}} = d(0) + d(1)z^{-1} + d(2)z^{-2} + \ldots \ldots \qquad (2.109)$$

The five unknown weighting factors v_0, v_1, v_2, w_1 and w_2, can then be solved from the following five equations, the first four of which are derived from equating the coefficients in the polynomial of z^{-j} obtained by multiplying out the above equation, and the last being the result of a heat balance consideration (Sowell 1988):

$$v_0 = d(0) \qquad (2.110a)$$

$$v_1 = d(1) + w_1 d(0) \qquad (2.110b)$$

$$v_2 = d(2) + w_1 d(1) + w_2 d(0) \qquad (2.110c)$$

$$0 = d(3) + w_1 d(2) + w_2 d(1) \qquad (2.110d)$$

$$\sum_{j=0}^{\infty} d(j) = \frac{v_0 + v_1 + v_2}{1 + w_1 + w_2} \qquad (2.110e)$$

From the way in which the weighting factors are derived, it can be seen that, strictly speaking, they would be applicable only to the specific room construction based upon which the room model was established for obtaining the weighting factors.

Pre-calculated weighting factors for rooms of different characteristics are given in the *Handbook of Fundamentals* such that the weighting factors for one of the rooms that is closest in thermal characteristics to the specific room for which the design cooling load is to be calculated can be selected for use for this purpose.

Note that the room weighting factors given in the ASHRAE method are 'normalised', meaning that all heat gains that enter the room will eventually become the cooling load of the room (Sowell 1988). Use of the 'normalised' weighting factors, therefore, can lead to over-estimation of the cooling load, as some of the heat gains can be transmitted out of the room. The values of the $v_{0,1,2}$ can be 'un-normalised' by the factor F_c, which can be estimated from (ASHRAE 1993):

$$F_C = 1 - 0.0116K_\theta \qquad (2.111)$$

where K_θ can be calculated from the areas and U-values of the fabric components, and the length of the exterior walls of the room (L), as follows:

$$K_\theta = \frac{1}{L}(A_W U_W + A_R U_R + A_G U_G + A_P U_P) \qquad (2.112)$$

where A_W, A_R, A_G and A_P are, respectively, the areas of external wall, roof, window and partition enclosing the room, and U_W, U_R, U_G and U_P are the corresponding U-values. Note also that in the Transfer Function Method, the heat gains used are the room sensible heat gains, which include both the radiant and convective heat gains, and the weighting factors are determined based on a pre-determined split between the radiant and convective components of the sensible heat gains. The weighting factors, therefore, may need to be corrected before being used for the cooling load calculation for a specific room. The following method (McQuiston & Spitler 1992) can be used for correcting the room weighting factors for different radiant/convective splits:

$$v_{0,C} = \frac{r}{r_o}v_0 + \left(1 - \frac{r}{r_o}\right) \qquad (2.113a)$$

$$v_{1,C} = \frac{r}{r_o}v_1 + \left(1 - \frac{r}{r_o}\right)w_1 \qquad (2.113b)$$

$$v_{2,C} = \frac{r}{r_o}v_2 + \left(1 - \frac{r}{r_o}\right)w_2 \qquad (2.113c)$$

where v_0, v_1 and v_2 are the uncorrected weighting factors for the sensible gain that comprises a radiant faction of r_o; and $v_{0,C}$, $v_{1,C}$ and $v_{2,C}$ are the corrected weighting factors for a radiation faction r.

The Radiant Time Series (RTS) Method for design cooling load estimation

More recently, ASHRAE has recommended an alternative design cooling load calculation method that is described here briefly (ASHRAE 2001). The new method is called the Radiant Time Series (RTS) Method, and was introduced as the standard cooling load estimation method, replacing all the previous methods, such as the Cooling Load Temperature Difference/Solar Cooling Load/Cooling Load Factor (CLTD/SCL/CLF) Method and the

Total Equivalent Temperature Difference/Time Averaging (TETD/TA) Method. Similar to the older methods, the RTS Method is based on the Response Factor Method (RFM) and the Transfer Function Method (TFM), which are detailed methods for hour-by-hour building cooling energy simulation.

In the RFM, the heat flux at the internal surface of an external wall or roof may be expressed as a function of the current and past values of the sol-air temperature and the indoor air temperature as shown in the following equation:

$$q_1(n) = \sum_{j=0}^{\infty} Y(j)T_{e,o}(n-j) - \sum_{j=0}^{\infty} Z(j)T_i(n-j) \tag{2.114}$$

where:

$q_1(n)$ = heat flux at the internal surface of an external wall or roof at time step n
$T_{e,o}(i)$ = sol-air temperature at time step i
$T_i(i)$ = the indoor temperature at time step i
$Y(j)$ = response factor of the wall or roof at time step j due to a triangular unit pulse temperature excitation applied to the external surface at time step 0
$Z(j)$ = response factor of the wall or roof at time step j due to a triangular unit pulse temperature excitation applied to the internal surface at time step 0

Since, instead of the temperatures at the two surfaces, the sol-air temperature and indoor air temperature are used in Equation 2.114, the response factors should have included the effects of the convective heat transfer coefficient at the two surfaces. For the internal surface, a combined heat transfer coefficient that relates both the convective heat flow and the net long wave radiation heat gain at the internal surface to the indoor air temperature should be used in the derivation of the response factors. The extent to which the summation process in Equation 2.114 should go back in time depends on the thermal capacitance of the wall being modelled.

The RTS method takes advantage of the assumptions made in design cooling load calculations that:

(1)   the external weather conditions would repeat in a 24-hour cycle;
(2)   the indoor condition would be maintained at a steady level throughout the day.

On the basis of assumption (1) above, a set of 24 periodic response factors can be derived from the conventional response factors as follows:

$$T_{p,k} = \sum_{j=0}^{\infty} Y(24j + k) \ (\text{for } k = 0, 1, 2, \ldots, 23) \tag{2.115}$$

$$Z_{p,k} = \sum_{j=0}^{\infty} Z(24j + k) \ (\text{for } k = 0, 1, 2, \ldots, 23) \tag{2.116}$$

Thus, the heat flux at the internal surface of the wall or roof at hour h in the design day can be determined from:

$$q_1(h) = \sum_{j=0}^{23} Y_{p,j} T_{e,o}(h-j) - \sum_{j=0}^{23} Z_{p,j} T_i(h-j) \tag{2.117}$$

Noting that $\sum_{j=0}^{23} Y_{p,j}$ equals $\sum_{j=0}^{23} Z_{p,j}$ and, based on the second assumption, Equation 2.117 can be simplified to:

$$q_1(h) = \sum_{j=0}^{23} Y_{p,j} \left(T_{e,o}(h-j) - T_i(h-j) \right) \tag{2.118}$$

The above equation represents the method used in the RTS method for determining the convective heat gain from opaque envelope elements.

Similarly, in the RTS method, the room sensible cooling load at time step n ($Q_{Sen}(n)$) is estimated from the heat gains at the same time step and over the past 23 hours, based on a set of 24 periodic zone response factors (r_j) as follows:

$$Q_{Sen}(n) = \sum_{j=0}^{23} r_j q(h-j) \tag{2.119}$$

The periodic zone response factors can be determined based on the custom room weighting factors in the TFM. In the TFM, the cooling load at a particular time step is expressed as a weighted sum of the heat gain at the current time step, and of the heat gains and cooling loads in the previous time steps, as shown in Equation 2.107, which is shown once again below:

$$Q_{Sen}(n) = \sum_{j=0}^{M} v_j q(n-j) - \sum_{j=1}^{N} w_j Q_{Sen}(n-j)$$

where v_j and w_j are the room weighting factors.

By moving the past cooling load terms at the right hand side of the above equation to the left hand side, the equations for the cooling loads at the 24 hours in the design day, which are assumed to repeat day after day, can be written as:

$$
\begin{bmatrix}
1 & 0 & 0 & \cdots & & w_2 & w_1 \\
w_1 & 1 & 0 & 0 & \cdots & & w_2 \\
w_2 & w_1 & 1 & 0 & 0 & \cdots & \\
& & & \cdots & & & \\
& & & w_2 & w_1 & 1
\end{bmatrix}
\begin{Bmatrix}
Q_0 \\ Q_1 \\ Q_2 \\ \cdots \\ Q_{23}
\end{Bmatrix}
=
\begin{bmatrix}
v_0 & 0 & 0 & \cdots & & v_2 & v_1 \\
v_1 & v_0 & 0 & 0 & \cdots & & v_2 \\
v_2 & v_1 & v_0 & 0 & 0 & \cdots & \\
& & & \cdots & & & \\
& & & v_2 & v_1 & v_0
\end{bmatrix}
\begin{Bmatrix}
q_0 \\ q_1 \\ q_2 \\ \cdots \\ q_{23}
\end{Bmatrix}
$$

which is abbreviated as:

$$\{Q\} = [w]^{-1}[v]\{q\} \tag{2.120}$$

Likewise, Equation 2.119, when applied to the 24 hours in the design day, can be written as:

$$
\begin{Bmatrix} Q_0 \\ Q_1 \\ Q_2 \\ \cdots \\ Q_{23} \end{Bmatrix}
=
\begin{bmatrix}
r_0 & r_{23} & r_{22} & \cdots & & r_2 & r_1 \\
r_1 & r_0 & r_{23} & r_{22} & \cdots & & r_2 \\
r_2 & r_1 & r_0 & r_{23} & r_{22} & \cdots & \\
& & & \cdots & & & \\
& & & & r_2 & r_1 & r_0
\end{bmatrix}
\begin{Bmatrix} q_0 \\ q_1 \\ q_2 \\ \cdots \\ q_{23} \end{Bmatrix}
$$

which is abbreviated as:

$$\{Q\} = [r]\{q\} \tag{2.121}$$

The values of the periodic zone response factors can, therefore, be solved from Equations 2.120 and 2.121 as:

$$[r] = [w]^{-1}[v] \tag{2.122}$$

Note that, in the RTS method, the periodic zone response factors apply only to the radiant portion of the heat gains, and are, therefore, also called the radiant time factors. The convective component of the heat gains will instantaneously contribute to the cooling load and no radiant time factors need to be applied. However, in the TFM, the heat gains used are the room sensible heat gains and the weighting factors are typically determined based on a pre-determined split between the radiant and convective components of the heat gains. The weighting factors, therefore, should be corrected before being used for determination of the radiant time factors, through the use of Equation 2.113a–c.

Two different sets of radiant time factors are required; one for heat gains due to transmitted direct solar radiation and the other for heat gains due to transmitted diffuse solar gains and other non-solar heat gains, such as radiant heat gains from lighting and equipment. They have to be treated differently because, for the solar heat gain, the assumption was made that the transmitted direct solar radiation would be absorbed solely by the floor, whereas the diffuse radiation and other radiant heat from the internal sources were assumed to be uniformly distributed to all internal surfaces in the room.

Apart from the above major differences between the RTS method and the TFM, as described above, similar methods are used for the determination of other heat gains, such as transmitted solar gain from windows and heat gains from lighting, equipment, occupants, infiltration, etc.

Implementation of the Finite Difference Method in a room heat balance

In the preceding discussions on the use of the forward difference or explicit method, the boundary node temperatures (denoted as node 1 and node N) were related to the adjacent internal node temperature and the ambient condition at the respective sides of its two surfaces, quantified using the sol-air temperature for the external side and the environmental temperature at the internal side, as follows:

$$T_1^{n+1} = 2Fo\left(BiT_{e,1}^n - \left(1 + Bi - \frac{1}{2Fo}\right)T_1^n + T_2^n \right) \tag{2.123}$$

$$T_N^{n+1} = 2Fo\left(T_{N-1}^n - \left(1 + Bi - \frac{1}{2Fo}\right)T_N^n + BiT_{e,N}^n \right) \tag{2.124}$$

The sol-air temperature at the last time step $(T_{e,1}^n)$ can be determined using Equation 2.58 based on the surface temperature at the last time step (T_1^n). Likewise, the environmental temperature at the indoor side $(T_{e,N}^n)$ can be determined based on the internal surface temperature (T_N^n), and those of other enclosing surfaces of the room at the last time step (for determining $q_{N,\mathrm{Rad,W}}^n$), as follows:

$$T_{e,N}^n = T_i^n + \frac{q_{N,\mathrm{Sol}}^n + q_{N,\mathrm{Rad,W}}^n + q_{N,\mathrm{Rad,Int}}^n}{h_i} \tag{2.125}$$

The heat gains due to the shares of the solar radiation transmitted through windows and the radiation emitted by the internal sources, which are imposed onto the internal surface of the wall $(q_{N,\mathrm{Sol}}^n + q_{N,\mathrm{Rad,Int}}^n)$, are dependent on the pre-determined way of splitting these heat gains. The net long wave radiation heat gain of the internal surface due to radiation heat exchanges with other enclosing surfaces can also be determined according to the concurrent surface temperatures of these surfaces, using the chosen radiation heat transfer model for the room.

When the backward or the Crank–Nicholson schemes are employed, due to their dependence on the unknown surface temperatures, the sol-air temperature and the environmental temperature would have to be solved together with the surface and internal temperatures of all the envelope elements using an iterative solution scheme. This may start from using the surface temperatures determined for the last time step to determine the sol-air and environmental temperatures, followed by prediction of the temperatures at the surfaces and the internal node temperatures of each envelope element. The surface temperatures obtained can then be used to update the estimates of the sol-air temperature and the environmental temperatures for each fabric element. The calculation procedures of estimating the envelope element temperatures and the sol-air and environmental temperature can then be repeated until a converged set of surface temperatures is obtained.

To reduce the complexity of the calculations, the simplified method for treating radiant heat exchange, as outlined above, may be applied.

Implementing the Lumped Capacitance Method in a room heat balance

The simplest interpretation of a lumped capacitance room model involves combining the higher thermal capacity (HTC) building elements, typically consisting of external wall and floor elements, together into an equivalent capacitance and adding a capacitance for room air. The lower thermal capacity (LTC) elements, typically consisting of windows, internal partitions and external roof constructions are then treated as a purely resistive path and are combined with ventilation. The result is a second-order room space model. In this interpretation, a separate treatment of inter-surface radiation is not carried out and an allowance is therefore made for radiant heat exchange by adjusting values of the thermal resistances.

A method for calculating the resistance weightings has been offered by Lorenz & Masy (1982) and a general interpretation of the method has been proposed by Crabb *et al.* (1987). Though the method is not suitable in applications involving significant radiant exchanges and is known to be unsuited to situations involving high thermal capacity room elements, it enjoys a simple structure in that a single room space can be described by as few as five parameters. Figure 2.10 gives an interpretation based on the methods described in the previous sections.

The high thermal capacity path is expressed as a single lump equivalent of the various contributions from these elements. For illustration, suppose this path consists of an external wall element and a floor element, then the inner and outer thermal resistances would first be determined using the method prescribed in Section 1.4. Then parameters for an equivalent single lump 'T section' can be arrived at as follows:

$$r_{i,HTC} = r_{i,F} \cdot r_{i,W} / (r_{i,F} + r_{i,W}) \tag{2.126a}$$

$$r_{o,HTC} = (r_{i,F} + r_{o,F}) \times (r_{i,W} + r_{o,W}) / (r_{i,F} + r_{o,F} + r_{i,W} + r_{o,W}) - r_{i,HTC} \tag{2.126b}$$

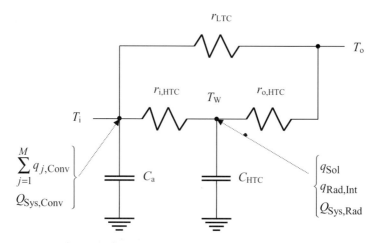

Fig. 2.10   Second-order lumped capacity room space model.

$$C_{HTC} = C_F + C_W \tag{2.126c}$$

(where, for the purpose of this illustration, $r_{i,F}$, $r_{i,W}$ and $r_{o,F}$, $r_{o,W}$ are the inner and outer thermal resistances respectively for the floor and roof elements, and C_F, C_W are the corresponding overall thermal capacities).

The room space heat balance equations for this case will therefore be as follows, in which q_{Sol} and $q_{Rad,Int}$ are respectively the total sources of solar radiation and internal equipment radiation entering the space, and $Q_{Sys,Conv}$ and $Q_{Sys,Rad}$ are respectively the convective and radiant energy exchanges with the space due to plant.

For the air node:

$$\rho_a V c_a \frac{dT_i}{dt} = (T_W - T_i)/r_{i,HTC} + (T_o - T_i)/r_{LTC} + \sum_{j=1}^{M} q_{j,Conv} \pm Q_{Sys,Conv} \tag{2.127}$$

and, for the structure node:

$$C_{HTC} \frac{dT_W}{dt} = (T_i - T_W)/r_{i,HTC} + (T_o - T_W)/r_{o,HTC} + q_{Sol} + q_{Rad,Int} \pm Q_{Sys,Rad} \tag{2.128}$$

Note that the sign for $Q_{Sys,Conv}$ (and $Q_{Sys,Rad}$ for surface exchanges) can be either negative or positive depending on whether the plant is respectively extracting heat (i.e. summer cooling mode) or adding heat (i.e. winter heating mode).

Expressing each construction element by a second-order function can arrive at a more rigorous lumped capacity formulation as detailed in Section 1.4. This method releases surface temperatures for each construction element enabling inter-surface radiant exchanges to be dealt with. In the following, this method is applied to a building space based, for illustration, on a space consisting of one external wall with a window and one internal wall, floor and ceiling resulting in a 9th-order room space model (Fig. 2.11). In this case the internal elements are treated adiabatically (i.e. the conditions in the adjoining room spaces are assumed to be identical to the subject space). It should be clear from this illustration that other room space geometries could be easily dealt with using this method.

The heat balance equations for this case are given as follows.

For the room air heat balance an equation similar to Equation 2.91 may be used with adjustments to the surface convection terms:

$$\rho_a V c_a \frac{dT_i}{dt} = \sum_{i=1}^{N} A_{W,i} (T_{W,i,L} - T_i)/x_{1,W,i} \cdot r_{W,i} + \dots$$

$$\dots A_{WG} (T_i - T_o)/(1/h_i + 1/h_o) + m_{inf} c_a (T_o - T_i) + \sum_{j=1}^{M} q_{j,Conv} \pm Q_{Sys,Conv} \tag{2.129}$$

(in which $x_{1,W,i}$ is the resistance share for the inner zone of the second order i^{th} element description and $r_{W,i}$ is the overall thermal resistance of the ith element (see Section 1.4)).

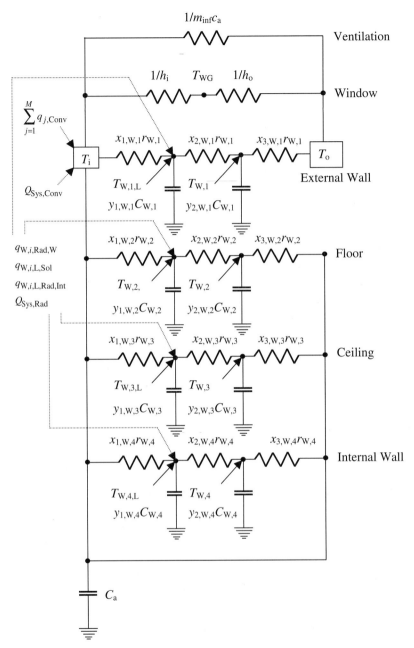

Fig. 2.11 Ninth-order room model based on second-order lumped capacitance elements.

For the inner layer of the ith construction element:

$$y_{1,W,i} \cdot C_{W,i} \cdot \frac{\mathrm{d}T_{W,i,L}}{\mathrm{d}t} = A_{W,i}\left[(T_i - T_{W,i,L})/x_{1,W,i} \cdot r_{W,i} - (T_{W,i,L} - T_{W,i})/x_{2,W,i} \cdot r_{W,i}\right] - \cdots$$

$$\cdots q_{W,i,\mathrm{Rad},W} + q_{W,i,L,\mathrm{Sol}} + q_{W,i,L,\mathrm{Rad},\mathrm{Int}} \pm Q_{\mathrm{Sys},\mathrm{Rad}}$$

$$(2.130)$$

For the outer layer of the external wall element (element 1) and the ith internal element:

$$y_{2,W,1} \cdot C_{W,1} \cdot \frac{dT_{W,1}}{dt} = A_{W,1}\left[\left(T_{W,1,L} - T_{W,1}\right)/x_{2,W,1} \cdot r_{W,1} - \left(T_{W,1} - T_o\right)/x_{3,W,1} \cdot r_{W,1}\right]$$

(2.131)

and:

$$y_{2,W,i} \cdot C_{W,i} \cdot \frac{dT_{W,i}}{dt} = A_{W,i}\left[\left(T_{W,i,L} - T_{W,i}\right)/x_{2,W,i} \cdot r_{W,i} - \left(T_{W,i} - T_i\right)/x_{3,W,i} \cdot r_{W,i}\right]$$

(2.132)

in which $y_{1,W,i}$ and $y_{2,W,i}$ are the capacitance shares for the inner and outer layers of the element; $C_{W,i}$ is the corresponding overall thermal capacity of the element; $x_{2,W,i}$ and $x_{3,W,i}$ are the resistance shares of the middle and outer zones of the element (see Section 1.4).

Note that, in Equation 2.131, a simplification is made in that T_o is used in place of the external surface temperature, which takes into account the radiant and convective heat balance at the external surface of the element. This is acceptable for the sort of short-term simulations often carried out using simplified lumped capacitance methods so long as the overall element resistance includes a surface radiation component, but for a more rigorous interpretation, T_o should be replaced by $T_{W,1,0}$, which can be determined from Equation 2.59.

2.4 Latent loads and room moisture content balance

The discussions so far have been limited to the sensible heat transfers that take place in a room. In modelling the total heat transfer in a room, the effects of various moisture sources on the moisture content of the room air should also be determined. Unlike the sensible heat gains, the latent heat gains of a room will instantaneously become a cooling load component. The sources of moisture include the occupants, certain materials/processes and equipment or appliances that would generate moisture (such as cooking, foodstuffs, liquid water exposed to the air, etc.) and moisture carried by the outdoor air into the room through infiltration or the ventilation system. The hygroscopic building fabric elements, and the furniture, would also exchange moisture with the indoor air. Furthermore, the air-conditioning system may add or remove moisture from the indoor air through provision of humidification or dehumidification, which may either be under active control or be processes associated with the provision of cooling.

Consider, once again, a room of internal volume V, which is enclosed by N building fabric elements and contains M internal moisture sources. The net moisture gain of the air in the room at time t and the corresponding effect on the room air moisture content, assuming the indoor air is perfectly mixed, can be expressed as follows:

$$\rho_a V \frac{dw_i(t)}{dt} = \sum_{i=1}^{N} A_{w,i} h_{d,i}\left(w_{w,i,L}(t) - w_i(t)\right) + m_{Inf}(t)\left(w_o(t) - w_i(t)\right)$$

$$+ \sum_{j=1}^{M} M_j(t) - M_{Sys}(t) \tag{2.133}$$

where:

$w_i(t)$ = the indoor air moisture content at time t

$w_o(t)$ = the outdoor air moisture content at time t

$A_{W,i}$ = the surface temperatures of the ith element at the room side at time t

$h_{d,i}$ = the convective mass transfer coefficient for the surface of the ith element

$w_{W,i,L}(t)$ = the surface moisture content of the ith element at the room side at time t

$m_{Inf}(t)$ = the rate of infiltration of outdoor air into the room at time t

$M_j(t)$ = the rate of moisture gain from internal source j

$M_{Sys}(t)$ = the rate of moisture extraction from the room by the air-conditioning system at time t

The dehumidification rate for maintaining the indoor moisture content steadily at a preset level can be solved from the above, by setting the time derivative of the indoor moisture content to zero, as follows:

$$M_{Sys}(t) = \sum_{i=1}^{N} A_{W,i} h_{d,i}\left(w_{W,i,L}(t) - w_i(t)\right) + m_{Inf}(t)\left(w_o(t) - w_i(t)\right) + \sum_{j=1}^{M} M_j(t) \tag{2.134}$$

The latent load on the air-conditioning system can be determined by multiplying the latent heat of vaporisation (h_{fg}) by the moisture extraction rate above.

With the assumption that the moisture capacitance of the room air is negligible, the indoor air moisture content can be related to the source terms as follows:

$$w_i(t) = \frac{\sum_{i=1}^{N} A_{W,i} h_{d,i} w_{W,i,L}(t) + m_{Inf}(t) w_o(t) + \sum_{j=1}^{M} M_j(t) - M_{Sys}(t)}{\sum_{i=1}^{N} A_{W,i} h_{d,i} + m_{Inf}(t)} \tag{2.135}$$

If only passive control over the indoor air moisture content is provided, the above equation can be used to estimate the indoor moisture content at different times in the air-conditioned period, as a balance of the moisture transport into and generated within the room with the moisture removal rate due to dehumidification of the system and ex-filtration. During the non-air-conditioned period, the dehumidification rate of the air-conditioning system ($M_{Sys}(t)$) can be set to zero.

References

Alamdari, F. & Hammond, G.P. (1983) Improved data correlations for buoyancy-driven convection in rooms. *Building Services Engineering Research & Technology* 4(3) pp.106–12.

ASHRAE (1993) *Handbook of Fundamentals*. American Society of Heating, Refrigerating and Air Conditioning Engineers, Inc., Atlanta, GA.

ASHRAE (2001) *Handbook of Fundamentals*. American Society of Heating, Refrigerating and Air Conditioning Engineers, Inc., Atlanta, GA.

Athienitis, A.K. & Stylianou, M. (1991) Method and global relationship for estimation of transmitted solar energy distribution in passive solar rooms. *Energy Sources* 13 pp.319–36.

Awbi, H.B. (1998) Calculation of convective heat transfer coefficients of room surfaces for natural convection. *Energy & Buildings* 28 pp.219–27.

Awbi H.B. & Hatton, A. (1999) Natural convection from heated room surfaces. *Energy & Buildings* 30 pp.233–44.

Clarke, J.A. (2001) *Energy Simulation in Building Design*. Butterworth-Heinemann, Oxford.

Crabb, J.A., Murdoch, N. & Penman, J.M. (1987) A simplified thermal response model. *Building Services Engineering Research & Technology* 8(1) pp.13–19.

Cumali, Z. (1979) *Interim Report: Passive Solar Calculation Methods – Report for US ERDA, Contract no. EM-78–C-01–5221*. Cumali Associates, Oakland, GA.

Duffie, J.A. & Beckman, W.A. (1991) *Solar Engineering of Thermal Processes*. John Wiley, New York.

Erbs, D.G., Klein, S.A. & Duffie, J.A. (1982) Estimation of the diffuse radiation fraction for hourly, daily and monthly-average global radiation. *Solar Energy* 28(4) pp.293–302.

European Solar Radiation Atlas (2000). École des Mines de Paris, Paris.

Kerrisk, J., Schnurr, J.E., Moore, J.E. & Hunn, B.D. (1981) The custom weighting-factor method for thermal load calculations in the DOE-2 computer program. *ASHRAE Transactions* 87(2) pp.569–84.

Kimura, K. (1977) *Scientific Basis of Air Conditioning*. Applied Science, London.

Kusuda, T. & Achenbach, P.R. (1965) Earth temperatures and thermal diffusivity at selected stations in the United States. *ASHRAE Transactions* 71(1) pp.61–75.

Liesen, R.J. & Pedersen, C.O. (1997) An evaluation of inside surface heat balance models for cooling load calculations. *ASHRAE Transactions* 103(2) pp.485–502.

Liu, B.Y.H. & Jordan, R.C. (1963) The long term average performance of flat plate solar energy collectors. *Solar Energy* 7(2) pp.53–73.

Lorenz, F. & Masy, G. (1982) *Méthode d'évaluation de l'économie d'énergie apportée par l'intermittence de chauffage dans les bâtiments: Traitement par différences finies d'un modèle à deux constantes de temps (Report No. GM820130–01)*. Université de Liège, Liège.

Loveday, D.L. & Taki, A.H. (1996) Convective heat transfer coefficients at a plane surface on a full-scale building façade. *Inernational Journal Heat Mass Transfer* 39(8) pp.1729–42

McLellan, T.M. & Pedersen, C.O. (1997) The sensitivity of cooling load calculations to outside heat balance models. *ASHRAE Transactions* 103(2) pp.469–84.

McQuiston, F.C. & Spitler, J.D. (1992) *Cooling and Heating Load Calculation Manual*. American Society of Heating, Refrigerating and Air Conditioning Engineers, Inc., Atlanta, GA.

Mortensen, M.E. (1989) *Computer Graphics*. Heinemann Newnes, Oxford.

Reindl, D.T., Beckman, W.A. & Duffie, J.A. (1990) Diffuse fraction correlations. *Solar Energy* 45(1) pp.1–7.

Skartveit, A. & Olseth, J.A. (1987) A model for the diffuse fraction of hourly global radiation. *Solar Energy* 38(4) pp.271–4.

Sowell, E.F. (1988) Cross-check and modification of the DOE-2 program for calculation of zone weighting factors. *ASHRAE Transactions* 94(2) pp.737–53.

Walton, G.N. (1979) The application of homogenous coordinates to shadowing calculations. *ASHRAE Transactions* 85(1) pp.174–80.

Walton, G.N. (1980) A new algorithm for radiant interchange in room load calculations. *ASHRAE Transactions* 86(2) pp.190–208.

World Meteorological Organisation (1964–1985) *Solar Radiation and Radiation Balance Data (The World Network) – Annual Data Compilations 1964–1985*. World Radiation Data Centre, Leningrad.

Yazdanian, M. & Klems, J.H. (1994) Measurement of the exterior convection film coefficient for windows in low rise buildings. *ASHRAE Transactions* 100(1).

Chapter 3

Mass Transfer, Air Movement and Ventilation

The previous chapters have dealt with heat transfer in building elements and room spaces through the mechanisms of conduction, convection and radiation. Throughout, uniformity within the structural elements of the building and perfect mixing within the air capacity of the room space has been taken for granted. In this chapter, two departures from these key assumptions are explored. The first of these deals with the transport of moisture within building elements and the second deals with the essentially unmixed behaviour of the air in the room space.

It is useful to note at this point that complete closure of the thermal environment of a space cannot be made until mass transfer and air movement are properly described. Few of the existing models that will be explored in later chapters deal with these additional features in any detail, mainly due to historical limitations in computer power. Nonetheless, the seamless 'marriage' of the principles given in Chapter 2 with those of the present chapter forms the ultimate goal of a deterministic model of the thermal environment of a space.

3.1 Heat and mass transfer in building elements

Heat and mass transfer in building elements is of particular interest in applications where interstitial condensation is prevalent. This can occur in cold climates, especially in badly ventilated spaces, where high rates of internal vapour production occur (e.g. kitchens and bathrooms) resulting in a driving vapour flow from the internal space through the fabric, and in air conditioned building spaces in tropical and sub-tropical climates during summer conditions when the driving moisture flow is liable to be from outside to inside. Most building construction elements are porous and thus provide pathways for vapour-rich air to pass through the material together with sites at which condensed vapour can collect as liquid water. With diurnal swings in internal and external boundary conditions, it is feasible, and likely in some cases, that this flow will be a bi-directional resulting in the building fabric absorbing moisture at certain times of the day and desorbing it back to the space at a later time in the cycle.

If it is assumed that the vapour pressure of air is the dominant driving potential for vapour moisture transfer then transfer through material is taken to be always in a pendular state in which liquid water can transfer between regions of material only after it has been evaporated, and 'filtration' flow arising from chains of condensed liquid moving through the pores of material is, therefore, neglected. With this assumption, a two-equation version of Huang's three-equation model (which included filtration flow – Huang 1979) can be expressed as the following differential permeability-based model:

$$\left(\rho_1 \frac{\partial \varepsilon_1}{\partial p_v}\right) \frac{\partial p_v}{\partial t} + \left(\rho_1 \frac{\partial \varepsilon_1}{\partial T}\right) \frac{\partial T}{\partial t} = \mu_m \frac{\partial^2 p_v}{\partial x^2} + \frac{\partial \mu_m}{\partial x} \frac{\partial p_v}{\partial x} \tag{3.1}$$

$$\left(-h_{fg}\rho_1 \frac{\partial \varepsilon_1}{\partial p_v}\right) \frac{\partial p_v}{\partial t} + \left((\rho c_p)_B - h_{fg}\rho_1 \frac{\partial \varepsilon_1}{\partial T}\right) \frac{\partial T}{\partial t} = \phi_B \frac{\partial^2 T}{\partial x^2} + \frac{\partial \phi_B}{\partial x} \frac{\partial T}{\partial x} \tag{3.2}$$

where:

ρ = the density (kg m^{-3})
ρ_1 = the liquid (water) density (kg m^{-3})
ε_1 = the volume fraction of liquid (water) in porous medium
p_v = the partial pressure of water vapour (N m^{-2})
T = the temperature (K)
μ_m = the differential permeability (s)
x = the spatial dimension in the direction of transfer (m)
h_{fg} = the latent heat of vapourisation (of water) (kJ kg^{-1})
ϕ = the thermal conductivity of porous medium (W m^{-1}K^{-1})
c_p = the specific heat capacity (kJ kg^{-1} K^{-1})
(subscript B refers to the bulk porous material).

Equation 3.1 describes mass conservation of liquid water (the mass of water vapour in the pores is assumed to be negligible due to the low vapour density compared with that of water). Equation 3.2 describes heat energy conservation. The local volume fraction of liquid in porous material (ε_1) is governed by the vapour pressure (p_v) and temperature (T) through the material sorption isotherm. The sorption isotherm of a material relates the local volume fraction of liquid in material to the local relative humidity of air at constant temperature. Relationships for a variety of construction materials can be found in the literature (e.g. Hansen 1996).

Partial discretisation and solution of equations for heat and mass transfer

In many applications, reducing Equations 3.1 and 3.2 to a pair of ordinary differential equations (ODEs) by discretising the spatial terms only can have advantages over full discretisation. First, the equations can be reduced to a form that is consistent with any other equations present in the model (for instance, if plant were to be included – see Chapter 5). Second, the equations can then be solved using one of a large number of 'standard' ODE solvers, some of which are especially suited to problems involving mixed dynamics – so called 'stiff' problems (see Section 5.4).

The spatial terms, for example, the vapour pressure terms on the right hand side of Equation 3.1, can be dealt with as follows:

$$\frac{\partial^2 p_v}{\partial x^2} \cong \frac{p_v(n+1) - 2p_v(n) + p_v(n-1)}{\Delta x^2}$$

$$\frac{\partial p_v}{\partial x} \cong \frac{p_v(n+1) - p_v(n-1)}{2\Delta x}$$

where $P_v(n+1), p_v(n-1)$ are the vapour pressures evaluated at the next and previous nodes in space respectively and $p_v(n)$ is the vapour pressure at the current node. Furthermore, if these difference equations are evaluated as a weighted sum of the finite difference approximations at the current (m) next time $(m+1)$ time steps:

$$\frac{\partial^2 p_v}{\partial x^2} \cong \frac{\lambda}{\Delta x^2} (\Delta p_v(n+1) - 2\Delta p_v(n) + \Delta p_v(n-1)) + \dots$$

$$\dots \frac{1}{\Delta x^2} (p_v(n+1,m) - 2p_v(n,m) + p_v(n-1,m)) \tag{3.3}$$

$$\frac{\partial p_v}{\partial x} \cong \frac{\lambda}{2\Delta x} (\Delta p_v(n+1) - \Delta p_v(n-1)) + \frac{1}{2\Delta x} (p_v(n+1,m) - p_v(n-1,m)) \tag{3.4}$$

where λ is a weighting factor and:

$$\Delta p_v(n-1) = p_v(n-1,m+1) - p_v(n-1,m)$$
$$\Delta p_v(n) = p_v(n,m+1) - p_v(n,m)$$
$$\Delta p_v(n+1) = p_v(n+1,m+1) - p_v(n+1,m)$$

It is evident that explicit differencing arises when $\lambda = 0$ and an implicit scheme, requiring an evaluation of variables at a future time step, arises when $\lambda = 1$. The implicit scheme requires an iterative procedure for solution of the resulting partially discretised equations with a corresponding increase in computational cost. The advantage, however, is that this scheme is unconditionally stable.

The above time-differenced forms of the vapour pressure variable can be expressed in derivative form as follows:

$$\Delta p_v(n-1) = \Delta t \frac{d}{dt} p_v(n-1) = \Delta t \dot{p}_v(n-1) \tag{3.5}$$

$$\Delta p_v(n) = \Delta t \frac{d}{dt} p_v(n) = \Delta t \dot{p}_v(n) \tag{3.6}$$

$$\Delta p_v(n+1) = \Delta t \frac{d}{dt} p_v(n+1) = \Delta t \dot{p}_v(n+1) \tag{3.7}$$

Thus by substituting Equations 3.3–3.7 into Equation 3.1 together with similar expressions for the temperature spatial derivatives in Equation 3.2, and re-arranging, leads to:

$$-a_{pi}\dot{p}_v(n-1) + b_{pi}\dot{p}_v(n) - c_{pi}\dot{p}_v(n+1)$$
$$= d_{pi}(p_v(n+1,m) - 2p_v(n,m) + p_v(n-1,m))$$
$$+ e_{pi}(p_v(n+1,m) - p_v(n-1,m)) - f_{pi}\dot{T}(n) \tag{3.8}$$

$$-a_{Ti}\dot{T}(n-1)+b_{Ti}\dot{T}_v(n)-c_{Ti}\dot{T}_v(n+1)$$
$$= d_{Ti}(T(n+1,m)-2T(n,m)+T(n-1,m))+e_{Ti}(T(n+1,m)-T(n-1,m))-f_{Ti}\dot{p}_v(n) \quad (3.9)$$

where:

$$
\begin{cases}
a_{pi} = \lambda\Delta t\left(\dfrac{\mu_m}{\Delta x^2} - \dfrac{1}{2\Delta x}\times\dfrac{\partial\mu_m}{\partial x}\right) \\[2ex]
b_{pi} = \rho_1\dfrac{\partial\varepsilon_1}{\partial p_v} + \lambda\Delta t\dfrac{2\mu_m}{\Delta x^2} \\[2ex]
c_{pi} = \lambda\Delta t\left(\dfrac{\mu_m}{\Delta x^2} + \dfrac{1}{2\Delta x}\times\dfrac{\partial\mu_m}{\partial x}\right)
\end{cases},\qquad
\begin{cases}
d_{pi} = \dfrac{\mu_m}{\Delta x^2} \\[2ex]
e_{pi} = \dfrac{1}{2\Delta x}\times\dfrac{\partial\mu_m}{\partial x} \\[2ex]
f_{pi} = \rho_1\dfrac{\partial\varepsilon_1}{\partial T}
\end{cases}
$$

and:

$$
\begin{cases}
a_{Ti} = \lambda\Delta t\left(\dfrac{\phi_B}{\Delta x^2} - \dfrac{1}{2\Delta x}\times\dfrac{\partial\phi_B}{\partial x}\right) \\[2ex]
b_{Ti} = (\rho c_p)_B - h_{fg}\rho_1\dfrac{\partial\varepsilon_1}{\partial T} + \lambda\Delta t\dfrac{2\phi_B}{\Delta x^2} \\[2ex]
c_{Ti} = \lambda\Delta t\left(\dfrac{\phi_B}{\Delta x^2} + \dfrac{1}{2\Delta x}\times\dfrac{\partial\phi_B}{\partial x}\right)
\end{cases},\qquad
\begin{cases}
d_{Ti} = \dfrac{\phi_B}{\Delta x^2} \\[2ex]
e_{Ti} = \dfrac{1}{2\Delta x}\times\dfrac{\partial\phi_B}{\partial x} \\[2ex]
f_{Ti} = h_{fg}\rho_1\dfrac{\partial\varepsilon_1}{\partial p_v}
\end{cases}
$$

Whereas Equations 3.8 and 3.9 can be used to model heat and mass transfer in the internal nodes of an homogenous material layer, alternative forms of these equations are required for boundary nodes and for interface nodes between adjacent layers of differing materials as found in traditional multi-leaf construction.

For the two boundary nodes at the inside and outside of the construction element, the layer thickness can be taken to be half of that forming inner layers (see Section 2.1). Just two nodes therefore participate in this boundary half-layer, the current node and the surface node, defined by the boundary air condition, giving the following for the outside surface boundary half-layer:

$$a_{pb}\dot{p}_v(N-1)-b_{pb}\dot{p}_{va} = c_{pb}(p_{va}(m)-p_v(N-1,m))+d_{pb}-e_{pb}\dot{T}(N-1) \quad (3.10)$$

$$a_{Tb}\dot{T}(N-1)-b_{Tb}\dot{T}_a = c_{Tb}(T_a(m)-T(N-1,m))+d_{Tb}-e_{Tb}\dot{p}_v(N-1) \quad (3.11)$$

where:

$$
\begin{cases}
a_{pb} = \rho_1\dfrac{\partial\varepsilon_1}{\partial p_v} + \dfrac{2\mu_m\lambda\Delta t}{\Delta x^2} \\[2ex]
b_{pb} = \dfrac{2\mu_m\lambda\Delta t}{\Delta x^2}
\end{cases},\qquad
\begin{cases}
c_{pb} = \dfrac{2\mu_m}{\Delta x^2} \\[2ex]
d_{pb} = \dfrac{2m_v}{\Delta x} \\[2ex]
e_{pb} = \rho_1\dfrac{\partial\varepsilon_1}{\partial T}
\end{cases}
$$

and:

$$\begin{cases} a_{Tb} = \left(\rho c_p\right)_B - h_{fg}\rho_1 \dfrac{\partial \varepsilon_1}{\partial T} + \dfrac{2\phi_B \lambda \Delta t}{\Delta x^2} \\[2ex] b_{Tb} = \dfrac{2\phi_B \lambda \Delta t}{\Delta x^2} \end{cases} \quad , \quad \begin{cases} c_{Tb} = \dfrac{2\phi_B}{\Delta x^2} \\[2ex] d_{Tb} = \dfrac{2q}{\Delta x} \\[2ex] e_{Tb} = -h_{fg}\rho_1 \dfrac{\partial \varepsilon_1}{\partial p_v} \end{cases}$$

In the above, N refers to the total number of nodes defining the construction element, m_v and q refer to the moisture flux ($kg\,m^{-2}\,s^{-1}$) and heat flux ($W\,m^{-2}$) entering the surface. The additional subscript 'a' refers to the air state at the surface (a pair of equations similar to 3.10 and 3.11 can also be defined for the inside surface boundary half-layer).

In the case of composite (multi-layer) elements, for nodes that lie at the junctions of adjacent layers of material, changes in material properties and Δx require to be accounted for. In the following n lies at the interface; $n-1$ and $n+1$ lie in the alternative materials either side of the interface (see Fig. 2.5):

$$-a_{pc}\dot{p}_v(n-1) + b_{pc}\dot{p}_v(n) - c_{pc}\dot{p}_v(n+1)$$
$$= d_{pc}\left(p_v(n,m) - p_v(n-1,m)\right) + e_{pc}\left(p_v(n+1,m) - p_v(n-1,m)\right) - f_{pc}\dot{T}(n) \qquad (3.12)$$

$$-a_{Tc}\dot{T}(n-1) + b_{Tc}\dot{T}_v(n) - c_{Tc}\dot{T}_v(n+1)$$
$$= -d_{Tc}\left(T(n,m) - T(n-1,m)\right) + e_{Tc}\left(T(n+1,m) - T(n,m)\right) - f_{Tc}\dot{p}_v(n) \qquad (3.13)$$

where:

$$\begin{cases} a_{pc} = \lambda\Delta t\,\dfrac{\mu_m(n-1)}{\Delta x(n-1)} \\[2ex] b_{pc} = \rho_1\left(\dfrac{\partial \varepsilon_1}{\partial p_v}\right)_C + \lambda\Delta t\left(\dfrac{\mu_m(n-1)}{\Delta x(n-1)} + \dfrac{\mu_m(n+1)}{\Delta x(n+1)}\right) \\[2ex] c_{pc} = \lambda\Delta t\,\dfrac{\mu_m(n+1)}{\Delta x(n+1)} \end{cases} , \quad \begin{cases} d_{pc} = \dfrac{\mu_m(n-1)}{\Delta x(n-1)} \\[2ex] e_{pc} = \dfrac{\mu_m(n+1)}{\Delta x(n+1)} \\[2ex] f_{pc} = \rho_1\left(\dfrac{\partial \varepsilon_1}{\partial T}\right)_C \end{cases}$$

and:

$$\begin{cases} a_{Ti} = \lambda\Delta t\,\dfrac{\phi_B(n-1)}{\Delta x(n-1)} \\[2ex] b_{Ti} = \left(\rho c_p\right)_C - \rho_1\left(h_{fg}\dfrac{\partial \varepsilon_1}{\partial T}\right)_C + \lambda\Delta t\left(\dfrac{\phi_B(n-1)}{\Delta x(n-1)} + \dfrac{\phi_B(n+1)}{\Delta x(n+1)}\right) \\[2ex] c_{Ti} = \lambda\Delta t\,\dfrac{\phi_B(n+1)}{\Delta x(n+1)} \end{cases} , \quad \begin{cases} d_{Ti} = \dfrac{\phi_B(n-1)}{\Delta x(n-1)} \\[2ex] e_{Ti} = \dfrac{\phi_B(n+1)}{\Delta x(n+1)} \\[2ex] f_{Ti} = \rho_1\left(h_{fg}\dfrac{\partial \varepsilon_1}{\partial p_v}\right)_C \end{cases}$$

and:

$$\left(\frac{\partial \varepsilon_1}{\partial p_v}\right)_C = \frac{\Delta x(n-1)}{2} \times \frac{\partial \varepsilon_1(n-1)}{\partial p_v(n)} + \frac{\Delta x(n+1)}{2} \times \frac{\partial \varepsilon_1(n+1)}{\partial p_v(n)}$$

$$\left(\frac{\partial \varepsilon_1}{\partial T}\right)_C = \frac{\Delta x(n-1)}{2} \times \frac{\partial \varepsilon_1(n-1)}{\partial T(n)} + \frac{\Delta x(n+1)}{2} \times \frac{\partial \varepsilon_1(n+1)}{\partial T(n)}$$

$$\left(h_{fg}\frac{\partial \varepsilon_1}{\partial p_v}\right)_C = h_{fg}(n-1) \times \frac{\Delta x(n-1)}{2} \times \frac{\partial \varepsilon_1(n-1)}{\partial p_v(n)} + h_{fg}(n+1) \times \frac{\Delta x(n+1)}{2} \times \frac{\partial \varepsilon_1(n+1)}{\partial p_v(n)}$$

$$\left(h_{fg}\frac{\partial \varepsilon_1}{\partial T}\right)_C = h_{fg}(n-1) \times \frac{\Delta x(n-1)}{2} \times \frac{\partial \varepsilon_1(n-1)}{\partial T(n)} + h_{fg}(n+1) \times \frac{\Delta x(n+1)}{2} \times \frac{\partial \varepsilon_1(n+1)}{\partial T(n)}$$

$$\left(\rho c_p\right)_C = \frac{\Delta x(n-1)}{2}\left(\rho c_p\right)_{B(n-1)} + \frac{\Delta x(n+1)}{2}\left(\rho c_p\right)_{B(n+1)}$$

Equations 3.8–3.13 form a set of ordinary differential equations that can now be solved using one of a number of standard algorithms. Methods for determining the differential permeability, μ_m, can be found in the literature (e.g. Galbraith *et al.* 1998).

Despite neglecting the influence of filtration flow, the model can be shown to give good results. Figure 3.1 gives results of the model when applied to the drying of a sample of 13.2 mm (thick) gypsum plasterboard compared with the experimental results of Thomas & Burch (1990).

3.2 Air movement

For building ventilation problems, including fluid flow in ducts and pipes, a static coordinate system prevails through which a moving flow pattern is established. Thus the conservation form of the equations of continuity, momentum and energy is applicable. Equally, when dealing with this class of fluid flow problem, low flow velocities are encountered for which

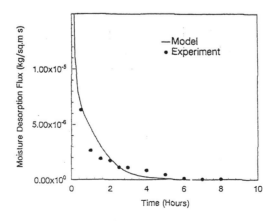

Fig. 3.1 Differential permeability model results compared with experimental data.

incompressible flow and negligible viscous dissipation may be assumed. These considerations simplify the particular equation set for building flow applications.

In fluid flow problems, the independent variables of time and space apply. For the latter, a Cartesian coordinate system may be used. Consider the control volume of Fig. 3.2.

For some variable, θ, dependent on space and time so that:

$$\theta = \theta(x, y, z, t)$$

then from the differential chain rule:

$$d\theta = \frac{\partial \theta}{\partial x} dx + \frac{\partial \theta}{\partial y} dy + \frac{\partial \theta}{\partial z} dz + \frac{\partial \theta}{\partial t} dt$$

and hence:

$$\frac{\partial \theta}{\partial t} = \theta_x \frac{\partial \theta}{\partial x} + \theta_y \frac{\partial \theta}{\partial y} + \theta_z \frac{\partial \theta}{\partial z} + \frac{\partial \theta}{\partial t}$$

in which:

$$\theta_x = \frac{dx}{dt}; \theta_y = \frac{dy}{dt}; \theta_z = \frac{dz}{dt}$$

Alternatively:

$$\frac{D\theta}{Dt} = \frac{\partial \theta}{\partial t} + \mathbf{\Phi} \cdot \nabla \theta \tag{3.14}$$

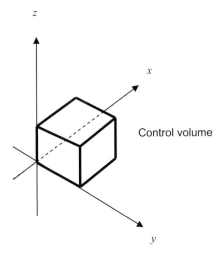

Fig. 3.2 Cartesian coordinate system.

where:

∇	is the vector differential operator
$\dfrac{D\theta}{Dt}$	is the substantive derivative
$\Phi \cdot \nabla\theta$	is the convective derivative
$\dfrac{\partial\theta}{\partial t}$	is the local derivative

Conservation equations for mass, momentum and energy

The governing equations, simplified as applicable to typical building flow problems, are as follows.

Continuity

For incompressible flow, density $\rho =$ constant and mass continuity about the control volume reduces to:

$$\frac{\partial u_x}{\partial x} + \frac{\partial u_y}{\partial y} + \frac{\partial u_z}{\partial z} = \nabla \cdot \mathbf{u} = 0 \tag{3.15}$$

where u_x, u_y, u_z are the x,y,z components of velocity respectively and $\nabla\cdot$ is the vector divergence operator.

Momentum

The momentum equations, sometimes referred to as the Navier–Stokes equations, expanded for all three velocity components are as follows:

$$\rho\left[\frac{\partial u_x}{\partial t} + u_x\frac{\partial u_x}{\partial x} + u_y\frac{\partial u_x}{\partial y} + u_z\frac{\partial u_x}{\partial z}\right] = S_{u_x} - \frac{\partial p}{\partial x} + \mu\left[\frac{\partial^2 u_x}{\partial x^2} + \frac{\partial^2 u_x}{\partial y^2} + \frac{\partial^2 u_x}{\partial z^2}\right] \tag{3.16}$$

$$\rho\left[\frac{\partial u_y}{\partial t} + u_x\frac{\partial u_y}{\partial x} + u_y\frac{\partial u_y}{\partial y} + u_z\frac{\partial u_y}{\partial z}\right] = S_{u_y} - \frac{\partial p}{\partial y} + \mu\left[\frac{\partial^2 u_y}{\partial x^2} + \frac{\partial^2 u_y}{\partial y^2} + \frac{\partial^2 u_y}{\partial z^2}\right] \tag{3.17}$$

$$\rho\left[\frac{\partial u_z}{\partial t} + u_x\frac{\partial u_z}{\partial x} + u_y\frac{\partial u_z}{\partial y} + u_z\frac{\partial u_z}{\partial z}\right] = S_{u_z} - \frac{\partial p}{\partial z} + \mu\left[\frac{\partial^2 u_z}{\partial x^2} + \frac{\partial^2 u_z}{\partial y^2} + \frac{\partial^2 u_z}{\partial z^2}\right] \tag{3.18}$$

where p is pressure; $S_{u_x}, S_{u_y}, S_{u_z}$ are momentum source terms in the x,y,z directions respectively; μ and ρ are the viscosity and density of air respectively.

Comparing Equations 3.16–3.18 with the general form of Equation 3.14, a more compact form of the momentum equations emerges:

$$\rho\frac{Du_x}{Dt} = S_{u_x} - \frac{\partial p}{\partial x} + \mu\nabla^2 u_x \tag{3.19}$$

$$\rho \frac{Du_y}{Dt} = S_{u_y} - \frac{\partial p}{\partial y} + \mu \nabla^2 u_y \qquad (3.20)$$

$$\rho \frac{Du_z}{Dt} = S_{u_z} - \frac{\partial p}{\partial z} + \mu \nabla^2 u_z \qquad (3.21)$$

Energy

Low velocity flows imply negligible viscous dissipation and, since the specific heat capacity and thermal conductivity of air, c_p and ϕ, may be assumed constant through the flow domains of relevance to building ventilation problems, then temperature, T, can be viewed as a conserved property. Accordingly, the energy equation can be expressed as follows:

$$\rho c_p \left[\frac{\partial T}{\partial t} + \frac{\partial (u_x T)}{\partial x} + \frac{\partial (u_y T)}{\partial y} + \frac{\partial (u_z T)}{\partial z} \right] = S_T + \phi \left[\frac{\partial^2 T}{\partial x^2} + \frac{\partial^2 T}{\partial y^2} + \frac{\partial^2 T}{\partial z^2} \right] \qquad (3.22)$$

or, in compact form:

$$\rho c_p \frac{\partial T}{\partial t} + \rho c_p \nabla \cdot \mathbf{u} T = S_T + \phi \nabla^2 T \qquad (3.23)$$

in which S_T is a volumetric thermal energy source term.

In many building ventilation applications, buoyancy influences within room space flow patterns may be significant and it is appropriate to include a gravity term in the vertical component: $g(\rho - \rho_\infty)$ (in which ρ now becomes the local air density and ρ_∞ is the constant effective mean air density throughout the calculation domain; g is acceleration due to gravity). According to the Boussinesq approximation, instantaneous temperature-dependent fluid properties (density, viscosity, etc) are assumed constant throughout the calculation domain in the above with the exception of this local density. Thus nominally, there are five independent variables (u_x, u_y, u_z, p, T) and five equations have been defined. The local density adds a further dependent variable for which a sixth equation requires to be sought. This can be satisfied through the introduction of the Equation of State:

$$\rho = p/RT \qquad (3.24)$$

in which R is the gas constant.

The above particular form of the conservation equations is appropriate to most forms of building flow applications, including indoor ventilation and air movement around building structures, and fluid flow in pipes and ducts applicable to building services systems. There is one additional consideration, however, which applies to certain building ventilation problems – that of chemical species. Typically, this crops up in indoor air quality analyses and is also applicable to the treatment of moisture migration studies in room space environments. To deal with this, an additional equation is needed for each candidate species. The general form of this additional equation is as follows for a chemical species with mass fraction, m, in the air/species mixture:

$$\frac{\partial}{\partial t}(\rho m) + \frac{\partial}{\partial x}(\rho m u_x) + \frac{\partial}{\partial y}(\rho m u_y) + \frac{\partial}{\partial z}(\rho m u_z) = S_m + \Gamma_m\left[\frac{\partial^2 m}{\partial x^2} + \frac{\partial^2 m}{\partial y^2} + \frac{\partial^2 m}{\partial z^2}\right] \qquad (3.25)$$

or, in the more compact vector operator form:

$$\frac{\partial m}{\partial t} + \nabla \cdot \mathbf{u}\rho m = S_m + \Gamma_m \nabla^2 m \qquad (3.26)$$

in which S_m is a volume rate of chemical species generation and Γ_m is the species diffusion co-efficient.

In most cases, species concentrations in indoor building air mixtures will be weak and the bulk density will be approximated to that of air (and, hence can be assumed constant through-out the calculation domain except for local buoyancy considerations). For richer mixtures, additional equations may be needed to describe mixture density.

It should be very evident now that all equations forming the conservation are structurally similar and this leads to some helpful rationale when dealing with solution of the equations, as will be seen a little later.

Turbulence modelling

Turbulence fluid flow is characterised by randomly fluctuating rotational eddies whose size and frequency vary considerably. The largest eddies are the same size as the flow domain whereas the smaller high frequency eddies are determined by viscous forces. It is also pos-sible to have steady recirculating cells in fluid flow, particularly in the wake of a flow disturb-ance, that are not part of the fluctuating pattern and are thus not turbulence.

Approaches to dealing with turbulent motion in fluid flow more or less group into three categories:

- direct numerical simulation (DNS)
- Reynolds-averaged Navier–Stokes (RANS) approaches
- large eddy simulation (LES).

Direct numerical simulation

Large eddy motion extracts kinetic energy from the mean flow and is highly anisotropic. Where these larger eddies are spectrally widely separated from the smaller eddies, the small-er eddies lose directional dependence and hence become almost isotropic. For the Newtonian flow assumption, the momentum equations (Equations 3.16–3.18) are theoretically appli-cable to both laminar and turbulent flow conditions and thus may be solved directly – that is, direct numerical simulation. However, in order to accurately construct the turbulent eddies at the smallest scale, the grid size resolution (see later; *Solution Schemes*) will need to be corre-spondingly small. This can be estimated from the Kolmogorov lengthscale, η (see for example Tennekes & Lumley 1972):

$$\eta = \left[\frac{v^3}{\varepsilon}\right]^{\frac{1}{4}} \qquad (3.27)$$

where υ is the kinematic viscosity (μ/ρ) and ε is the energy dissipation rate per unit mass of fluid ($m^2 s^{-3}$). ε in turn can be estimated from:

$$\varepsilon \cong \frac{\bar{u}_x^3}{L} \tag{3.28}$$

where $\bar{u}_x$ is the mean velocity in the streamwise direction (assumed here to be x) and L is the size of the flow domain (i.e. the maximum lengthscale). Thus combining Equations 3.27 and 3.28, the minimum relative turbulent lengthscale can be estimated from:

$$\frac{\eta}{L} = Re^{-\frac{3}{4}} \tag{3.29}$$

For example, air flowing at $50 \, \text{mm} \, s^{-1}$ along a 3 m wall at 20°C ($\mu = 1.846 \times 10^{-5} \, \text{kg} \, m^{-1} \, s^{-1}$) will have a minimum lengthscale of 3 mm. If this wall formed part of a $3 \, \text{m} \times 3 \, \text{m} \times 3 \, \text{m}$ room then the flow domain would require to be described by 10^9 grid cells. Thus calculations involved in this small problem for a low Reynolds number would suggest a problem size well beyond the scope of all but the most powerful computers. Problems with higher Reynolds numbes become more intractable and it is generally accepted that this approach as a general turbulence modelling method may never prove viable even with the most optimistic predictions of computer hardware development progress. Thus the DNS approach is restricted to small-scale test cases for verifying, or calibrating the parameters of, other turbulence modelling approaches (see for example Murakami 1998; Sarkar & So 1997; Apsley & Leschziner 1998).

Simplified and Reynolds-averaged approaches

The simplest approach to turbulence modelling is to treat the fluid as a pseudo-fluid with increased 'effective' viscosity, μ_{eff}, which can be obtained for instance from:

$$\mu_{eff} = \frac{\rho u_{typ} L}{Re_{eff}} \tag{3.30}$$

where μ_{typ} is the typical flow field velocity and L in this case is the cross-stream length. Re_{eff} is chosen to prompt an increased viscosity beyond the molecular viscosity. Since for many typical flow problems it can be stated with some confidence that $Re \leq 100$ will not be turbulent, then putting Re_{eff} equal to say 50 will force μ_{eff} to a higher value than would be the case with the normal molecular viscosity property. This value is then treated as a constant in the flow modelling problem; a 'constant effective viscosity'. This approach may be suitable in some instances for predicting the main features of a turbulent flow but is not suitable for predicting local details. It has been used for low-velocity internal room air modelling by Whittle 1987.

For most practical problems a constant effective viscosity approach gives insufficient information about local characteristics of the flow field. In most cases, the time-averaged turbulence characteristics are of primary interest permitting a turbulence model to be used for practical problem solving – one or a set of additional equations which when solved along with the governing equations of continuity, energy and momentum closely simulate the behaviour of the real turbulent flow. In turbulent flows, the instantaneous velocity, u, may be considered to consist of a mean component, $\bar{u}$, and a fluctuating component, u', thus $u = \bar{u} + u'$. For a practical interpretation of turbulent fluid motion, the time-averaged velocities in the spatial coordinate system are required and, by definition, the time-averaged fluctuating components of these will be zero. Consider the non-linear velocity components on the left hand side of the x-component momentum equation (Equation 3.16):

$$u_x^2 = (\bar{u}_x + u'_x)^2 = \bar{u}_x^2 + 2\bar{u}_x u'_x + u'_x$$

The time-averaged u_x^2 will be:

$$\overline{\bar{u}_x^2} = \overline{u_x^2} = \bar{u}_x^2 + \overline{u'^2_x}$$

Similarly, other time-averaged velocity component products give:

$$\overline{u_y u_x} = \bar{u}_y \bar{u}_x + \overline{u'_y u'_x}$$

$$\overline{u_z u_x} = \bar{u}_z \bar{u}_x + \overline{u'_z u'_x}$$

Similar expressions arise in the case of the y and z component momentum equations (Equations 3.17 and 3.18). Thus expressing Equations 3.16–3.18 in their time-averaged forms with velocities adopting their mean and fluctuating components leads to the following:

$$\rho \left[\frac{D\bar{u}_x}{Dt} - \left(\frac{\partial \overline{u'^2_x}}{\partial x} + \frac{\partial \overline{u'_y u'_x}}{\partial y} + \frac{\partial \overline{u'_z u'_x}}{\partial z} \right) \right] = S_{u_x} - \frac{\partial \bar{p}}{dx} + \mu \nabla^2 \bar{u}_x \tag{3.31}$$

$$\rho \left[\frac{D\bar{u}_y}{Dt} - \left(\frac{\partial \overline{u'_x u'_y}}{\partial x} + \frac{\partial \overline{u'^2_z}}{\partial y} + \frac{\partial \overline{u'_z u'_y}}{\partial z} \right) \right] = S_{u_y} - \frac{\partial \bar{p}}{dy} + \mu \nabla^2 \bar{u}_y \tag{3.32}$$

$$\rho \left[\frac{D\bar{u}_z}{Dt} - \left(\frac{\partial \overline{u'_x u'_z}}{\partial x} + \frac{\partial \overline{u'_y u'_z}}{\partial y} + \frac{\partial \overline{u'^2_z}}{\partial z} \right) \right] = S_{u_z} - \frac{\partial \bar{p}}{dz} + \mu \nabla^2 \bar{u}_z \tag{3.33}$$

The additional terms that appear on the left hand side of the original equations are termed 'apparent stresses' or 'Reynolds stresses'. A similar analysis leads to the following form of the original instantaneous form of energy equation (Equation 3.23):

$$\frac{\partial \rho T}{\partial t} + \nabla \cdot \overline{\rho \mathbf{u} T} - \left[\frac{\partial \overline{\rho u'_x T'}}{\partial x} + \frac{\partial \overline{\rho u'_y T'}}{\partial y} + \frac{\partial \overline{\rho u'_z T'}}{\partial z} \right] = \frac{S_T}{c_p} + \frac{\phi}{c_p} \nabla^2 T \tag{3.34}$$

where the additional terms on the left hand side $\left(\frac{\partial \overline{\rho u'_x T'}}{\partial x} , \text{ etc} \right)$ are turbulent heat fluxes.

Methods using this time-averaged form of the momentum equations and energy equations tend to be referred to as 'Reynolds-averaged Navier–Stokes' (RANS) methods.

In one of the first attempts to deal with turbulent flow quantitatively, Boussinesq (1877) introduced a virtual viscosity termed the 'turbulent viscosity', μ_t, defined here for example with respect to the $-\rho u'_x u'_y$ Reynolds stress term:

$$-\overline{\rho u'_x u'_y} = \mu_t \frac{\partial u_x}{\partial y} \tag{3.35}$$

Similarly, for turbulent heat fluxes:

$$-\overline{\rho u'_x T'} = \Gamma_t \frac{\partial T}{\partial x} \tag{3.36}$$

μ_t is not a property of the fluid like the molecular dynamic viscosity, but varies from point to point in the flow stream according to the turbulent characteristic at each point. In relation to the turbulent heat flux, Γ_t is the turbulent thermal diffusivity and is related to μ_t (i.e. $\Gamma_t = \mu_t / \sigma_t$ in which σ_t is the turbulent Prandtl number).

Arising from one of the earliest turbulence models, which is still used to good purpose today in a number of applications, is Prandtl's mixing length hypothesis (Prandtl 1925):

$$\mu_t = \rho L_m^2 \left| \frac{\partial u_x}{\partial y} \right| \tag{3.37}$$

(u_x is taken to be the streamwise velocity in the x direction in this instance).

The lengthscale, L_m, requires to be determined. Typical values of the ratio of streamwise mixing lengthscale to the width of the turbulent flow region vary from about 0.07 for a plane mixing layer to 0.125 for a spreading jet in stagnant surroundings (Launder & Spalding 1972). This is an effective method with low computational cost for applications such as boundary layer problems, fully developed flow in ducts and pipes and applications involving isothermal flows without turbulence diffusion. However, by far the most common collection of RANS approaches for a wide variety of applications in use today are the so called two-equation methods. Though there are various forms, the most popular and enduring two-equation turbulence model describes turbulence kinetic energy (k) and its dissipation rate (ε); the so called $k-\varepsilon$ model.

Multiplication of the instantaneous momentum equations (Equations 3.16–3.18 and their Reynolds stress forms Equations 3.31–3.33) by the corresponding fluctuating velocity components, addition of the terms and subtraction of the resulting two equations leads to a general form for the k equation. Though there are many variations and refinements to the way in which the k equation can be expressed, the following is a common interpretation for general high Reynolds number applications, and includes a term for z-component buoyancy production:

$$\rho \left[\frac{\partial k}{\partial t} + \frac{\partial k \bar{u}_x}{\partial x} + \frac{\partial k \bar{u}_y}{\partial y} + \frac{\partial k \bar{u}_z}{\partial z} \right] = \frac{\partial}{\partial x} \left(\frac{\mu_t}{\sigma_k} \frac{\partial k}{\partial x} \right) + \frac{\partial}{\partial y} \left(\frac{\mu_t}{\sigma_k} \frac{\partial k}{\partial y} \right) + \frac{\partial}{\partial z} \left(\frac{\mu_t}{\sigma_k} \frac{\partial k}{\partial z} \right) + \dots$$

$$\dots \mu_t \left(\frac{\partial \bar{u}_x}{\partial y} + \frac{\partial \bar{u}_y}{\partial x} \right) \frac{\partial \bar{u}_x}{\partial y} + \frac{\mu_t \beta g}{\sigma_t} \frac{\partial T}{\partial z} - \rho \varepsilon \tag{3.38}$$

or, in the more compact vector form:

$$\rho\left[\frac{\partial k}{\partial t}+\nabla\cdot k\overline{\mathbf{u}}\right]=\nabla\cdot\left(\frac{\mu_t}{\sigma_k}\nabla k\right)+\mu_t\left(\frac{\partial\overline{u}_x}{\partial y}+\frac{\partial\overline{u}_y}{\partial x}\right)\frac{\partial\overline{u}_x}{\partial y}+\frac{\mu_t\beta g}{\sigma_t}\frac{\partial T}{\partial z}-\rho\varepsilon \tag{3.39}$$

In words, the terms which balance with the rate of change of k plus transport by convection are, to the right of the equals sign from left to right, transport of k by diffusion, the volume rate of production of k by shear, the volumetric production of k by gravity and the rate of destruction of k by dissipative forces.

The rate of turbulence dissipation is described by a second equation that is of similar form to that of k:

$$\rho\left[\frac{\partial\varepsilon}{\partial t}+\nabla\cdot\varepsilon\overline{\mathbf{u}}\right]=\nabla\cdot\left(\frac{\mu_t}{\sigma_\varepsilon}\nabla\varepsilon\right)+\mu_t C_1\frac{\varepsilon}{k}\left(\frac{\partial\overline{u}_x}{\partial y}+\frac{\partial\overline{u}_y}{\partial x}\right)\frac{\partial\overline{u}_x}{\partial y}+\frac{\mu_t\beta g c_3}{\sigma_t}\frac{\partial T}{\partial z}-\rho C_2\varepsilon \tag{3.40}$$

The turbulent viscosity in the $k-\varepsilon$ model is given by:

$$\mu_t = \rho C_\mu \frac{k^2}{\varepsilon} \tag{3.41}$$

The constants in the standard form of the $k-\varepsilon$ turbulence model have for many years been based on those proposed by Launder & Spalding 1973 (see Table 3.1).

Whereas the instantaneous conservation equations, Equations 3.15–3.23, are closed (i.e. independent variables are u_x,u_y,u_z,T,p,ρ and there are five conservation equations plus an equation of state), the introduction of the six different Reynolds stresses in the time-averaged form of the momentum equations requires the model to close through the turbulence modelling method. The two-equation turbulence model together with the equation for turbulent viscosity lead to μ_t and the overall model is now closed by substituting an extended form of the Boussinesq relationship for each of the Reynolds stresses in the momentum equations. In Cartesian–tensor notation (i.e. i and $j = 1,2,3$ corresponding to the x,y,z coordinates of the Cartesian system; a repeated subscript in a term implies summation across all three coordinates); this is as follows:

$$-\rho\overline{u_i'u_j'} = \mu_t\left(\frac{\partial\overline{u}_i}{\partial x_j}+\frac{\partial\overline{u}_j}{\partial x_i}\right)-\frac{2}{3}\rho k\delta_{ij} \tag{3.42}$$

(in which δ_{ij} is the Kronecker delta, i.e. $\delta_{ij=1}$ when $i=j$; otherwise 0).

The above form of the $k-\varepsilon$ model is applicable to the higher Reynolds number modelling applications and requires a high calculation grid resolution (see later; *Solution Schemes*) in

Table 3.1 Standard $k-\varepsilon$ model constants (Launder & Spalding 1973).

C_1	C_2	C_3	C_μ	σ_k	σ_ε	σ_t
1.44	1.92	1.0	0.09	1.0	1.314	0.9

the near wall and, even then, results are not always satisfactory in this region because viscous stresses tend to dominate over turbulent Reynolds stresses. To deal with this, the universal log-law of the wall is applied (Schlichting 1979) in which the mean velocity, $\bar{u}_p$, at a point in the surface boundary layer, y_p, (assuming for this illustration that the streamwise boundary layer is in the y direction) is given by the following and, correspondingly, k and ε in the near wall region are obtained (Launder & Spalding 1973):

$$\frac{\bar{u}_p}{u_\tau} = \frac{1}{\kappa} \ln(Ey_p) \tag{3.43a}$$

$$k = \frac{u_\tau^2}{\sqrt{C_u}} \tag{3.43b}$$

$$\varepsilon = \frac{u_\tau^3}{\kappa y} \tag{3.43c}$$

where u_τ is the friction velocity. The constants, κ and E, may be taken to be 0.41 and 9.8 respectively for smooth walls (alternative values for rough walls may be found in Schlichting 1979). For convective heat flux through the boundary layer, a similar analysis is possible (e.g. Jayatillaka 1969) but this method tends to underestimate convective heat transfer in conditions applicable to building internal surfaces. Hence an empirical expression is preferable for the calculation of surface heat transfer coefficients – see for example Alamdari & Hammond (1983) and Awbi & Hatton (1999).

While the general form of the k–ε model described above works well for high Reynolds number applications, it is not valid in near wall regions when the Reynolds number is low. Such conditions apply frequently in internal building space ventilation modelling. There are two main alternatives within the RANS group of turbulence models for dealing with this:

- use an adapted low Reynolds number (low-Re) k–ε model
- use a two-layer turbulence model

In the low-Re k–ε case, the general idea is to apply wall damping to ensure that viscous stresses dominate over turbulent Reynolds stresses when the Reynolds number is low. This involves the following modifications to the general k–ε model (Equations 3.39–3.41):

$$k: \quad \rho\left[\frac{\partial k}{\partial t} + \nabla \cdot k\bar{\mathbf{u}}\right] = \nabla \cdot \left(\left(\mu + \frac{\mu_t}{\sigma_k}\right)\nabla k\right) + \mu_t\left(\frac{\partial \bar{u}_x}{\partial y} + \frac{\partial \bar{u}_y}{\partial x}\right)\frac{\partial \bar{u}_x}{\partial y} + \frac{\mu_t\beta g}{\sigma_t}\frac{\partial T}{\partial z} - \rho\varepsilon \tag{3.44}$$

$$\varepsilon: \quad \rho\left[\frac{\partial \varepsilon}{\partial t} + \nabla \cdot \varepsilon\bar{\mathbf{u}}\right] = \nabla \cdot \left(\left(\mu + \frac{\mu_t}{\sigma_\varepsilon}\right)\nabla \varepsilon\right) + \frac{\mu_t}{k}\varepsilon C_1 f_1\left(\frac{\partial \bar{u}_x}{\partial y} + \frac{\partial \bar{u}_y}{\partial x}\right)\frac{\partial \bar{u}_x}{\partial y} + \dots$$

$$\dots \frac{\mu_t\beta g C_3}{\sigma_t}\frac{\partial T}{\partial z} - C_2 f_2 \rho\varepsilon \tag{3.45}$$

and:

$$\mu_t = \rho C_\mu f_\mu \frac{k^2}{\varepsilon} \tag{3.46}$$

Several prescriptions are available for the wall damping functions, f_1, f_2, f_μ, perhaps the most common of which being those recommended by Lam & Bremhorst (1981):

$$f_\mu = \left[1 - \exp(-0.0165\,\mathrm{Re}_y)\right]^2 (1 + 20.5/\mathrm{Re}_t) \tag{3.47a}$$

$$f_1 = \left(1 + 0.05/f_\mu\right)^3 \tag{3.47b}$$

$$f_2 = 1 - \exp(-\mathrm{Re}_t^2) \tag{3.47c}$$

In the above, Re_y is defined as $\rho k^{\frac{1}{2}} y / \mu$ (y is the y-coordinate location) and Re_t is the turbulence Reynolds number ($= \rho k^2 / \varepsilon$). This approach to turbulence modelling has been successfully applied to the calculation of room surface convection coefficients for natural convection by Awbi (1998), with the conclusion that improved accuracy over experimental data might be expected when compared with the standard log-wall function form of the $k{-}\varepsilon$ model.

One problem with the low-Re $k{-}\varepsilon$ model is that it requires high grid resolution (see *Solution Schemes*) near the wall due to the steep rate of decline in ε towards the wall and this in turn leads to high computational costs. An alternative is to use a two-layer $k{-}\varepsilon$ model in which the standard high-Re $k{-}\varepsilon$ model is used to deal with the main flow field away from the wall surface, and in the near wall region where viscous stresses dominate a one-equation model using a length-scale is substituted. Various strategies for expressing two-layer $k{-}\varepsilon$ models have been reported in the literature; see for example Norris & Reynolds (1975) and Xu & Chen (2001a, 2001b). For example, in the inner flow region near the wall, the turbulent momentum equations and standard k equation are used but the ε differential equation and standard turbulence viscosity equation are replaced with the following (Norris & Reynolds 1975):

$$\varepsilon = \frac{k^{\frac{3}{2}}}{L}\left(1 + \frac{C_\varepsilon \mu}{\rho L k^{\frac{1}{2}}}\right) \tag{3.48}$$

where L is a lengthscale that is proportional to the disturbance from the wall, y:

$$L = C_D \kappa y \tag{3.49}$$

Values for the new constants are $C_\varepsilon = 13.2$ and $C_D = 6.41$ (Norris & Reynolds 1975) whereas values for the other constants are as defined earlier. The turbulent viscosity with this method is then given by the following:

$$\mu_t = f_d C_\mu \rho k^{\frac{1}{2}} L \tag{3.50}$$

in which f_d is a damping function ($= 1 - \exp(-0.0198\mathrm{Re}_t)$) (Norris & Reynolds 1975).

As well as improving computational efficiency over the low-Re $k{-}\varepsilon$ model, the two-layer model is also found to improve the prediction of flow separation and related suction pressure prediction when used in bluff body applications (such as wind flow over buildings) when compared with the standard $k{-}\varepsilon$ approach – see Zhou & Stathopoulos (1997).

In some applications such as impingement flow and separating flows at high Reynolds number, the Boussinesq relationship fails to account adequately for the anisotropic nature of the individual Reynolds stresses. It is possible to derive exact equations for the solution of the Reynolds stresses – the so-called Reynolds Stress Model (RSM) approach to solving turbulence flow (Launder *et al.* 1975). The six individual Reynolds stresses expressed in kinematic form $(u'^2_x, u'^2_y, u'^2_z, u'_x u'_y, u'_x u'_z, u'_y u'_z)$ are described by six partial differential equations (PDEs) in which the convective stress derivative is balanced with transport by diffusion, pressure-strain interaction, rotation and rates of production and dissipation. k is obtained by this method by addition of the three normal stress components but a seventh PDE is required for the rate of turbulence dissipation to effect model closure. Whereas this approach is the most general of all the RANS methods, there are five additional PDEs to solve over the conventional two-equation models and, thus, the computational budget is high. There are also fewer validations of this method reported in the literature for practical applications compared with the widely used two-equation methods.

A computationally cheaper alternative is to calculate the six Reynolds stresses using a set of six simultaneous algebraic equations – an Algebraic Stress Model (ASM) (Demuren & Rodi 1984). Like RSM, this method is better at dealing with the anisotropic nature of the Reynolds stresses in some applications than is the case with the Boussinesq expression but, again, this method is not as well validated for practical applications as is the case with the two-equation methods. Also, the development of an ASM in its simplest form requires the neglect or simplification of convection and diffusion terms, which is not suited to applications involving heat transfer.

Another relatively recent approach that has been successfully applied to high-Re applications is the use of Renormalisation Group (RNG) methods (see Yakhot & Orszag 1986). Here, RNG theory is used to develop a technique for predicting the effect on turbulence large scales of the turbulence small scales appearing in the flow field. The net result involves alternative values of the model constants $(\sigma_\varepsilon, \sigma_k, C_1, C_2, C_3, C_\mu)_1$ and an additional source term in the ε equation of the standard $k-\varepsilon$ model. In principle this method can be used to predict turbulence conditions in both low-Re and high-Re applications through the use of built-in model parameter corrections. The model has proved successful for high-Re separated flow prediction such as occurs in shearing flows over buildings.

Large eddy simulation

The $k-\varepsilon$ turbulence model is by far the most commonly encountered method for practical problem-solving. However, it is well known that it fails to predict accurately flow separation and reattachment in impingement flow conditions over bluff bodies. Specifically, turbulent intensity and suction pressure conditions over right-angled surfaces causing flow separation are not well accounted for. This failing applies to problems involving wind flow around buildings and is well documented in the literature (see for example Zhou & Stathopoulos 1997, Murakami 1998, Kawamoto 1997, Oliveira & Younis 2000). These failings have prompted interest in a new approach to turbulence modelling based on segregating the treatment of the turbulent flow into large scale and sub-grid scale (SGS). These techniques generally fall into a category of turbulence modelling called large eddy simulation (LES), based on the pioneering work of Deardorff (1970).

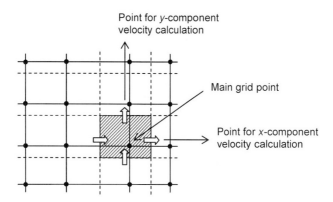

Fig. 3.5   Staggered grid layout.

Throughout this section dealing with discretisation, it has been assumed that the grid pattern throughout the domain will be regular. In most cases this assumption is inappropriate since those domain regions in the vicinity of solid boundaries will require grid refinement. Thus a flexibility to create an irregular grid pattern will be needed for most practical problems. The general discretised equations are applicable to regions of both coarse and refined grid densities, but there is a need to deal with the translation of variables and parameters at the interfaces between different grid densities. Details of a method of handling these translations can be found in Patankar (1980).

Grid strategies

The grid itself can be cast in one of two possible ways. The first of these is the structured grid (Fig. 3.6(a)) in which grid is, or close to, orthogonal throughout the domain. The second is the unstructured grid in which grid nodes can be placed more or less anywhere and the mesh connections made to suit (Fig. 3.6(b)).

In the structured grid case, all the neighbour cells are located and the connections between them are, thus, clearly defined. In the unstructured case, the additional information is required to define grid cell connectivity and this places further demands on computer memory and, correspondingly, computational overhead. However, the unstructured grid is easier to map than the structured grid. Techniques such as fine mesh embedding can be used with a structured grid strategy to improve problem definition in regions such as boundary layers, as is evident in Fig. 3.6(a). Several techniques are available for grid spacing in these regions – such as the use of a geometric sequence (Fig. 3.7) in which f is a grid spacing factor ($f > 1$ in this case).

Solution of the discretised equation set

The discretised equation set can be solved using a line-by-line method incorporating the tri-diagonal matrix algorithm (as described in Section 2.1). For three-dimensional problems, successive planes are solved throughout one of the dimensions (e.g. the x-y plane may be solved along the z-direction in successive sweeps).

The main complication in the solution of these equations lies in the fact that the pressure field is unknown. Hence most methods for processing the discretised equations start with a

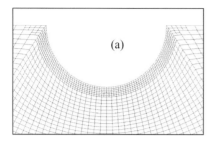 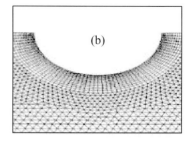

Fig. 3.6   Example grid formations. (a) Structured. (b) Unstructured.

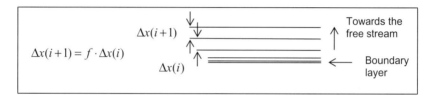

Fig. 3.7   Geometric sequence.

guessed pressure velocity and pressure field from which an iterative procedure can be defined to force a satisfaction of the continuity equation. These methods work on the basis of setting pressure and velocity results as follows:

$$p = p* - P'$$
(3.82a)

$$u_x = u_x^* - u_x'$$
(3.82b)

$$u_y = u_y^* - u_y'$$
(3.82c)

$$u_z = u_z^* - u_z'$$
(3.82d)

where the * superscript refers to guessed values and the prime superscript denotes corrections to guessed values. It is possible to derive an equation for pressure correction from the continuity equation and, in turn, a set of equations may be derived which express required corrections to the velocity components as a function of the corresponding pressure correction. Details can be found in Patankar (1980).

A number of solution procedures have evolved over the years, such as the explicit marker and cell (MAC) method (Harlow & Welch 1965). Later methods such as the simultaneous variable adjustment (SIVA) method and its later variant SIMPLE (Caretto *et al.* 1972) adopt the implicit procedure essential for pressure field correction. In summary, the SIMPLE algorithm works as follows (Caretto *et al.* 1972).

- STEP 1: Set the guessed pressure field values, $P*$.
- STEP 2: Solve the discretised momentum equations to give $\mu_x^*, \mu_y^*, \mu_z^*$.

- STEP 3: Solve a pressure correction equation to give p'.
- STEP 4: Calculate p from Equation 3.72a.
- STEP 5: Calculate μ_x, μ_y, μ_z from a set of velocity corrections equations.
- STEP 6: Calculate the remaining field variables (turbulence energy and dissipation, T (and m if present) from their respective discretised equations.
- STEP 7: Set the pressure calculated at STEP 4 equal to the new guessed pressure and return to STEP 2 until satisfactory convergence is achieved.

Often, numerical instability will arise in those regions of the domain where a sharp increase or decrease in any one of the dependent variables is experienced, such as in near-wall regions. It is therefore helpful to limit the maximum change in the dependent variable between iterations and this can be accomplished through under-relaxation by restricting the value of the dependent variable as follows:

$$\theta_P = \theta_P' + f_u \left(\theta_P'' - \theta_P' \right) \tag{3.83}$$

in which θ_P'', θ_P' are current and previous grid-point values of the dependent variable respectively and f_u is an under-relaxation factor.

After each iteration, a set of residuals is calculated:

$$R = \sum a_{nb} \theta_{nb} + b - a_P \theta_P \tag{3.84}$$

Thus convergence is deemed to have been achieved (STEP 7) when $|R|$ for all grid points is less than a specified tolerance.

For problems involving transient calculations an additional loop in the SIMPLE algorithm is needed that would involve a time 'marching' function and the computational costs become correspondingly higher.

A refinement to the SIMPLE algorithm is provided in an updated SIMPLER algorithm that incorporates an improved procedure for calculating the pressure correction, which in turn improves the convergence characteristics of the SIMPLE algorithm (see Patankar 1979).

3.3 Network ventilation models

Because the governing CFD equations are applicable at microscopic scale they are applicable for any type of flow prediction problem where a description of the characteristics of the flow domain are needed. Though flexibility exists to customise calculations to specific problems (i.e. supression of dynamic calculations, reduction to two- or even one-dimensionality, choice of simplified turbulence modelling assumptions), the computational cost of using CFD is usually high and often very high. It is possible to represent room space convection using intermediate models with reduced computational cost, such as those that split the space into a number of small zones about which energy and continuity balances can be applied (for example, see Musy *et al.* 2001). These types of models are generally called zonal models. In many building flow problems the bulk transfer of air between spaces and groups of spaces is of more interest than the precise characteristics of its distribution within a single space.

Examples include the analysis of natural ventilation rates, transfer of outdoor pollutants, and peak summertime temperature calculations in naturally ventilated spaces. In these cases it is frequently sufficient to understand conditions at macroscopic scale, which leads to the possibility of an alternative class of air movement model that is simpler and, therefore, computationally cheaper than is the case with CFD. This type of model has come to be known as a network ventilation model.

Basic equations

The essential precept for this type of model is that, in a uniform building zone or space, ventilation mass continuity holds within the zone boundary and, therefore, for all zone leakage paths, l:

$$\sum_l m_l = 0 \tag{3.85}$$

where m_l is the mass flow rate through a zone leakage path. A zone in this context is either a single building space or a cluster of connected like spaces around which a boundary can be defined and about which there are a number of air leakage paths. It follows that for a building of B zones, the above also applies to the entire building.

Based on the Bernoulli equation and applying the convention that all flows entering a zone are positive and all flows leaving are negative then the pressure difference across any leakage path in a zone can be summarised as:

$$\Delta P_{j,i} = P_{Tj} - P_{Ti} + \Delta P_s \text{ for an inter-zone path} \tag{3.86}$$

and

$$\Delta P_{w,i} = P_w - P_{Ti} + \Delta P_s \text{ for an external path} \tag{3.87}$$

where $\Delta P_{j,i}$ is the total pressure difference across a given leakage path separating the jth zone and the ith zone; $\Delta P_{w,i}$ is the total pressure difference due to wind pressure acting on a leakage path in an external element of the ith zone; P_{Tj}, P_{Ti} are the total pressures at the leakage path connections to zone j and zone i respectively; P_w is the total pressure acting on the external surface due to wind and ΔP_s is the pressure difference due to elevation ('stack' effect). In a complete building simulation, the air flow equations will be solved in tandem with the energy equations from which the zone temperature will be know; thus ΔP_s can be modelled using:

$$\Delta P_S = 352.98 \cdot g \cdot h_S (1/T_o - 1/T_i) \tag{3.88}$$

in which h_s is the difference in height of the leakage path, T_o is the air temperature in either the connecting zone or the external air temperature as appropriate (K) and T_i is the air temperature in the ith zone (K).

The wind pressure acting on an external building surface (P_w) can be found from:

$$P_w = C_p \cdot \rho \cdot u_w^2 / 2 \tag{3.89}$$

where u_w is the wind velocity, ρ is the air density based on local conditions and C_p is the wind pressure coefficient, which depends on the prevailing wind direction, building surface geometry and local topography. Methods for estimating typical UK wind velocities and wind pressure coefficients can be found in British Standard BS6399–2 (1997).

Modelling leakage and source paths

When large openings form a zone envelope leakage path, such as an openable window ventilation source, the orifice equation is applicable:

$$m = C \cdot A \cdot \rho\sqrt{2\Delta p/\rho} \qquad\qquad (3.90)$$

where C is the flow coefficient and A is the area of the opening.

Many leakage paths in building envelope details are small crack-like openings for which it is found that a pressure difference exponent, n, of between 0.5 and 1.0 (typically 0.65) is applicable and the following power-law model can be used:

$$m = \rho \cdot C \cdot \Delta p^n \qquad\qquad (3.91)$$

The flow constant, C, and exponent, n, can be determined from fan pressurisation tests and extensive databases exist for a range of construction details (for example AIVC 1994 and ASHRAE 2001). Alternatively, a convenient empirical method can be used for crack-like power-law model constants (Clarke 2001):

$$n = [1 + \exp(-W/2)]/2 \qquad\qquad (3.92)$$

$$C = 9.7 \times L(0.0092)^n \qquad\qquad (3.93)$$

in which W is the crack width and L is the crack length.

Baker *et al.* (1987) point out that a quadratic model can more accurate than a power-law model for the commonly encountered infiltration cracks:

$$\Delta P = a \cdot m + b \cdot m^2 \quad (m, \Delta P > 0) \quad \text{or} \quad \Delta P = a \cdot m - b \cdot m^2 \quad (m, \Delta P < 0) \qquad (3.94)$$

where a and b are constants (Baker *et al.* 1987). Equation 3.94 can then be solved for m by solving the quadratic in the usual way.

The other major category of ventilation source paths is due to mechanical ventilation, for which modelling methods are considered in Chapter 4.

Solution scheme for a network ventilation model

A solution to the equations describing all flow paths in a building of B zones requires that the continuity balance of Equation 3.85 be satisfied and a recursive method involving zone pressure and flows is needed. Perhaps the most common method used for the solution of these equations is based on a Newton–Raphson procedure as described by Walton (1984). According to this method successive values of room pressure for each room are estimated using the following:

$$\mathbf{P^+ = P^- - D} \tag{3.95}$$

where $\mathbf{P^+}$ is a B-element vector of updated room pressures, $\mathbf{P^-}$ is a B-element vector of previous room pressure estimates and $\mathbf{D}$ is a B-element vector of room pressure corrections.

$\mathbf{D}$ is updated at each iteration from the following:

$$\mathbf{J \times D = F} \tag{3.96}$$

where $\mathbf{J}$ is a $B \times B$ Jacobian matrix of derivatives and $\mathbf{F}$ is a B-element vector of room flow balances whose elements are calculated from:

$$\mathbf{F} = \begin{bmatrix} \sum m_1 \\ \sum m_2 \\ \cdot \\ \cdot \\ \cdot \\ \sum m_B \end{bmatrix} \tag{3.97}$$

(where $\sum m_1$ is the flow balance (i.e. Equation 3.85) for zone 1, $\sum m_2$ is the flow balance for zone 2, etc).

Diagonal elements of the Jacobian are calculated from the following, for the ith zone which has l flow paths:

$$J_{i,i} = \sum_l \frac{\partial m_l}{\partial P_i} \tag{3.98}$$

and off-diagonal elements for all openings l between subject zone i and connecting j:

$$J_{j,i} = \sum_l \frac{\partial m_l}{\partial P_j} \tag{3.99}$$

The derivatives are easily expressed. For example, for a power-law flow path between subject zone i and connecting zone j:

$$\frac{\partial m_i}{\partial P_i} = -n \cdot C \cdot \Delta P_{j,i}^{n-1} = -n \cdot m_i / \Delta P_{j,i} \tag{3.100a}$$

similarly:

$$\frac{\partial m_i}{\partial P_j} = n \cdot m_i / \Delta P_{j,i} \tag{3.100b}$$

Solution is deemed to have occurred when all elements of the $\mathbf{F}$ matrix are zero or within a small error tolerance of zero. Solution of the matrix relationship of Equation 3.96 requires the inverse of $\mathbf{J}$, which will require to be carried out using a method such as Gaussian elimination.

In practice, convergence can be accelerated if necessary using a modified updating equation:

$$\mathbf{P}^+ = \mathbf{P}^- - f \cdot \mathbf{D} \tag{3.101}$$

where f is a relaxation coefficient. A typical value for f of 0.75 has been suggested (Axley 1988).

References

AIVC (1994) *AIVC Technical Note 44: An analysis and data summary of the AIVC numerical database.* Air Infiltration and Ventilation Centre, Warwick.

Alamdari, F. & Hammond, G.P. (1983) Improved data correlations for buoyancy-driven convection in rooms. *Building Services Engineering Research & Technology* 4(3) pp.106–12.

Apsley, D.D. & Leschziner, M.A. (1998) A new low-Reynolds-number non-linear two-equation turbulence model for complex flows. *International Journal Heat and Fluid Flow* 19 pp.209–22.

ASHRAE (2001) *ASHRAE Handbook: Fundamentals.* American Society of Heating, Refrigerating and Air Conditioning Engineers, Atlanta, GA.

Awbi, H.B. (1998) Calculation of convective heat transfer coefficients of room surfaces for natural convection. *Energy and Buildings* 28 pp.219–27.

Awbi, H.B. & Hatton, A. (1999) Natural convection from heated room surfaces. *Energy and Buildings* 30 pp.233–44.

Axley, J.W. (1988) *Progress Towards a General Analytical Method for Predicting Indoor Air Pollution in Buildings: Indoor Air Quality Modelling Phase III Report: NBSIR 88–3814.* National Institute of Standards and Technology, Gaithersburg, MA.

Baker, P.K., Sharples, S. & Ward, I.C. (1987) Air flow through cracks. *Building and Environment* 22(4) pp.293–304.

Boussinesq, J. (1877) Essai sur la théorie des eaux courantes. *Mémoires Présentés par divers savants à l'Académie des Sciences de l'Institut de France* XXIII 46, Paris.

British Standard BS6399–2 (1997) *Loading for Buildings: Code of Practice for Wind Loads.* British Standards Institute, London.

Caretto, L.S., Gosman, A.D., Patankar, S.V. & Spalding, D.B. (1972) Two calculation procedures for steady three dimensional flows with recirculation. *Proceedings of 3rd International Conference, Numerical Methods in Fluid Dynamics, Paris.* Vol. II, p.60.

Ciofalo, M. (1996) Large-eddy simulations of turbulent flow with heat transfer in simple and complex geometries using Harwell-Flow3D. *Applied Mathematical Modelling* 20 pp.262–71.

Clarke, J.A. (2001) *Energy Simulation in Building Design.* Butterworth-Heinemann, Oxford.

Deardorff, J.W. (1970) A numerical study of three dimensional turbulent channel flow at large Reynolds numbers. *Journal of Fluid Mechanics* 41 pp.453–80.

Demuren, A.O. & Rodi, W. (1984) Calculation of turbulence-driven secondary motion in non-circular ducts. *Journal of Fluid Mechanics* 140 pp.189–222.

Galbraith, G.H., McLean, R.C. & Guo, J. (1998) Moisture permeability data: Mathematical representation. *Building Services Engineering Research & Technology* 19(1) pp.31–6.

Gan, G. (1995) Evaluation of room air distribution systems using computational fluid dynamics. *Energy and Buildings* 23 pp.83–93.

Hansen, K.K. (1996) *Sorption isotherms: A catalog and database. Water Transmission Through Building Materials and Systems: Mechanisms and Measurements – ASTMS STP 1039* (eds H.R.Trechsel & M.Blomberg). American Society for Testing of Materials, Philadelphia.

Harlow, F.H. & Welch, J.E. (1965) Numerical calculation of time-dependent viscous incompressible flow of fluid with free surface. *Journal of the Physics of Fluids* 8 p2821.

Havet, M., Blay, D. & Veyrat, O. (1994) Analysis of thermal comfort in large enclosures heated by radiant panels. *Proceedings of the International Conference On Air Distribution in Rooms: Roomvent '94.*

Homes, M.J., Lam, J.K-W., Ruddick, K.G. & Whittle, G.E. (1990) Computation of conduction, convection and radiation in the perimeter zone of an office. *Proceedings of the International Conference on Air Distribution in Rooms: Roomvent '90.* Oslo.

Huang, C.L.D. (1979) Multi-phase moisture transfer in porous media subject to temperature gradient. *International Journal of Heat Mass Transfer* 22(9) pp.1295–307.

Jayatillaka, C.V.L. (1969) The influence of Prandtl number and surface roughness on the resistance of the laminar sub-layer to momentum and heat transfer. *Progress in Heat and Mass Transfer* Vol I. Pergamon, London.

Jiang, Y. & Chen, Q. (2001) Study of natural ventilation in buildings by large eddy simulation. *Journal of Wind Engineering and Industrial Aerodynamics* 89 pp.1155–78.

Jiang, Y. & Chen, Q. (2002) Effect of fluctuating wind directions on cross flow natural ventilation in buildings from large eddy simulation. *Building and Environment* 37 pp.379–86.

Kawamoto, S. (1997) Improved turbulence models for estimating wind loading. *Journal of Wind Engineering and Industrial Aerodynamics* 67 & 68 pp.589–99.

Lam, C.K.G. & Bremhorst, K.A. (1981) Modified form of the k–ε model for predicting wall turbulence. *Transactions of the ASME, Journal of Fluids Engineering* 103 pp.456–60.

Launder, B.E., Reece, G.J. & Rodi, W. (1975) Progress in the development of a Reynolds-stress closure. *Journal of Fluid Mechanics* 68(3) pp.537–66.

Launder, B.E. & Spalding, D.B. (1972) *Lectures in Mathematical Models of Turbulence.* Academic Press, London.

Launder, B.E. & Spalding, D.B. (1973) The numerical computation of turbulent flows. *Computer Methods in Applied Mechanics and Engineering* 89.

Murakami, S. (1998) Overview of turbulence models applied in CWE–1997. *Journal of Wind Engineering and Industrial Aerodynamics* 74–76 pp.1–24

Musy, M., Wurtz, E., Winklemann, F. & Allard, F. (2001) Generation of a zonal model to simulate natural convection in a room with a radiative/convective heater. *Building and Environment* 36 pp.589-96.

Negrão, C.O.R. (1998) Integration of computational fluid dynamics with building thermal and mass flow simulation. *Energy and Buildings* 27 pp.155–65.

Norris, L.H. & Reynolds, W.C. (1975) Report No. FM-10. Stanford University Department of Mechanical Engineering, Stanford, CA.

Oliveira, P.J. & Younis, B.A. (2000) On the prediction of turbulent flows around full-scale buildings. *Journal of Wind Engineering and Industrial Aerodynamics* 86 pp.203–20.

Patankar, S.V. (1979) A calculation procedure for two-dimensional elliptic situations. *Numerical Heat Transfer* 2.

Patankar, S.V. (1980) *Numerical Heat Transfer and Fluid Flow.* Taylor & Francis, London.

Prandtl, L. (1925) Über die ausgebildete Turbulenz. *Zeitschrift für Angewandte Mathematik und Mechanik* 5 pp.136–9.

Sarkar, A. & So, R.M.C. (1997) A critical evaluation of near-wall two-equation models against direct numerical simulation data. *International Journal of Heat and Fluid Flow* 18 pp.197–208.

Schlichting, H. (1979) *Boundary Layer Theory.* McGraw-Hill, New York.

Tennekes, H. & Lumley, J.L. (1972) *A First Course in Turbulence.* MIT Press, Cambridge, MA.

Thomas, W.C. & Burch, D.M. (1990) Experimental validation of a mathematical model for predicting water vapour sorption at interior building surfaces. *ASHRAE Transactions* 96(1).

Tutar, M. & Oguz, G. (2002) Large eddy simulation of wind flow around parallel buildings with varying configurations. *Fluid Dynamics Research* 31(5–6) pp.289–315.

Walton, G.N. (1984) A computer algorithm for predicting infiltration and inter-room airflows. *ASHRAE Transactions* 90(1B) pp.601–10.

Whittle, G.E. (1987) *BSRIA Technical Note TN 2/87: Numerical Air Flow Modelling.* Building Services Research and Information Association, Bracknell.

Xu, W. & Chen, Q. (2001a) A two-layer turbulence model for simulating indoor airflow, part I: Model development. *Energy and Buildings* 33 pp.613–25.

Xu, W. & Chen, Q. (2001b) A two-layer turbulence model for simulating indoor airflow, part II: Applications. *Energy and Buildings* 33 pp.627–39.

Yakhot, V. & Orszag, S.A. (1986) Renormalisation group methods in turbulence. *Journal on Scientific Computing* 1(1) pp.1–51.

Zhou, Y. & Stathopoulos, Y. (1997) A new technique for the numerical simulation of wind flow around buildings. *Journal of Wind Engineering and Industrial Aerodynamics* 72 pp.137–47.

Chapter 4
Steady-State Plant Modelling

Building energy simulation is becoming an indispensable part of the design process for new buildings and heating, ventilating and air-conditioning (HVAC) systems. When facilitated by an appropriate simulation tool, a designer can predict building and system performance to ensure that the building and system being designed will meet all the regulatory and functional requirements, and to compare performance and conduct financial appraisals of alternative designs to guide selection. Methods for modelling the thermal behaviours of buildings are covered in Chapters 1 and 2. These chapters emphasise that it is crucial to model the dynamic responses of the building fabric in predicting cooling loads in buildings. However, as HVAC plants can respond to changes in operating conditions much more quickly than the building fabric, for the purpose of evaluating year-round performance of HVAC systems in buildings (e.g. annual air-conditioning energy use), the use of steady-state plant models will generally suffice. In this chapter, methods for modelling steady-state performance of HVAC systems and equipment are introduced. Modelling of dynamic performance of systems regulated by control systems is the subject of the next chapter.

4.1 Model formulations for plant

Modelling the performance of an HVAC system involves setting up mathematical models for various system components (e.g. the equipment and ducting/piping networks) and linking the component models together to form a system model. Input to the HVAC system model includes the loads on the system, which may be the predictions of a detailed building thermal model, and the boundary conditions under which the system operates (e.g. weather conditions). The output of the system model will be predictions of system parameters of interest, such as the thermodynamic states (e.g. temperature, humidity and pressure) and flow rates of the working fluids, and the energy input to the system.

Mathematical models for system components, each of which may also be regarded as a system, can be derived based on the physical laws that govern the processes taking place in the system and the state of the working substances within and crossing the system boundary. The building thermal models described in preceding chapters and the cooling coil model that will be described later are examples of this type of model. There are merits and limitations of using system models derived based on fundamental principles. Since the predictions of such models bear physical meanings, it is easier to detect modelling errors, input errors (e.g. unrealistic boundary conditions) or problems with the numerical solution processes (e.g. divergence in iterative calculations) by examining if the values of the model outputs are reasonable or physically possible, which is especially important to debugging of the model during the development stage. The models will also be more generally applicable to similar components, which is a return on the great efforts invested in their derivation. However, for complex equipment (e.g. a chiller), models developed using this approach can be rather complicated and often require input of a large set of parameters for description of the physical characteristics of the system and its components. This, in turn, can limit their applications, as the required information can be difficult to obtain.

Incomplete knowledge about the characteristics of system components is a barrier to the development and use of mathematical models that are based on fundamental principles. This is quite often the case for proprietary equipment where only performance data pertaining to a set of operating conditions are available from the manufacturer, but details of the equipment, including the constituent materials, configurations and dimensions of internal parts, operating conditions within the equipment etc., are incomplete. Hence, even if one can derive a rigorous and detailed model, there remain difficulties in finding appropriate values for some of the required input parameters for the model, rendering the model not practically usable.

An alternative way of setting up mathematical models for systems is to establish equations to numerically relate the output of the system of interest (e.g. cooling capacity or energy demand) to the influential input (e.g. operating conditions) while the system itself is regarded as a black box. The required output is the dependent variable(s) and the input the independent variable(s) of the mathematical function, which is to be derived by curve-fitting the function to a set of known output/input data. The resultant equation can then be taken as the system model. Equation fitting is particularly useful in modelling the steady-state performance of systems. Obviously, application of such a model is subject to restrictions, e.g. it cannot reflect the effects of any factors the influence of which was omitted in the model derivation; and the predictions will be unreliable when the model is applied to situations outside the set of conditions based upon which it was derived.

Curve-fit models

The objective of equation fitting is to derive a mathematical expression from a set of known operating data of a system or equipment, obtained either from measurement or information given by the manufacturer, as a model of the system or equipment. The model can then be used to predict the output of the system based on known input conditions (e.g. values of system parameters) that arise while the system is in operation. In most situations, it is impracticable or simply not necessary to have an excessively complicated model. First, derivation of such a model can be very time and effort consuming, rendering it not justifiable. Second, if a complex functional form is used, data fluctuations due to random disturbances (e.g. errors of measurement or calculations) may cause problems to the solution process. In general, complexity of the model should be minimised as far as possible, so long as the model can provide an accurate enough description of the physical system. Verification, however, is necessary to see if the mathematical expression is indeed a good model of the system.

Whether or not a mathematical expression is a good representation of the sample data (the best fit) can be determined by examining the total deviation of the model predictions from known 'true' values (measured data or data from a reliable source, e.g. the manufacturer). To determine the degree of complexity required, the level of complexity of the model can be increased step by step (e.g. adding more terms to the equation, one at a time) and, in each step, the deviation is compared with those of the less complex ones. If the deviation is within acceptable limits whilst the reduction in deviation brought about by an increase in complexity is minimal, there is no need to further increase the complexity.

The first step of equation fitting is to postulate an appropriate mathematical function. There are, however, no general guidelines as to what equation form should be used for a particular system and this often needs to be judged by the model developer. A preliminary visual inspection of how the sample data appear when plotted out graphically is usually very helpful; by inspecting the data points, one can make a reasonable judgement as to what type of mathematical function would be appropriate for the sample data under concern. Furthermore, selection of independent variables should, as far as possible, be based on physical understanding of the system and no variables that are known to have significant influence to the dependent variable should be omitted.

After a mathematical form for the model has been selected, the unknown coefficients in the model equation can be evaluated by regression techniques in which attempt is made to minimise the total error between the model predictions and the sample data. The least squared error method, which aims to minimise the sum of squared errors, is the most commonly adopted.

Linear regression

The simplest mathematical form of a model is a straight line, as depicted by Equation 4.1. A linear model is sometimes used even if the system it models is known to have a non-linear output/input relation, provided it can provide acceptable predictions when application is restricted only to a limited range of operating conditions within which the output/input relation of the system is close to linear:

$$y = ax + b \tag{4.1}$$

Although the number of output/input data pairs required for defining a straight line is two, based upon which the intercept (b) and the slope (a) of the line Equation 4.1 can be uniquely defined. The accuracy of the linear model is dependent on the accuracy of the two data pairs used for its derivation. The uncertainties in the predictions of the model can be reduced if a larger set of output/input data are available. As would be expected, not all the data points will lie exactly on the straight line that passes through any two data points even if the system is known to have a linear output/input relation. The task will then be finding a straight line (i.e. determination of the values of the slope and intercept (a and b)) that best fits the data, in the sense that the sum of squared deviations between the model predictions and the known data points is the smallest.

If N data pairs are available, a and b that would lead to the least sum of squared deviations can be evaluated using Equations 4.2 and 4.3:

$$a = \frac{N\sum_{i=1}^{N} x_i y_i - \left(\sum_{i=1}^{N} x_i\right)\left(\sum_{i=1}^{N} y_i\right)}{N\sum_{i=1}^{N}(x_i^2) - \left(\sum_{i=1}^{N} x_i\right)^2} \tag{4.2}$$

$$b = \frac{1}{N}\left(\sum_{i=1}^{N} y_i\right) - \frac{a}{N}\left(\sum_{i=1}^{N} x_i\right) = \bar{y} - a\bar{x} \tag{4.3}$$

where $\bar{x}$ and $\bar{y}$ are the mean values of x and y respectively.

Note that a regression model may not necessarily provide a meaningful representation of the sample data. The Student's t test can be used to examine if variations in the independent variable do have a significant influence on the variations in the dependent variable in the first place so as to justify if the use of a linear model to relate the two variables would be meaningful. The coefficient of determination, r^2, which is a measure of the proportion of the total variation in the dependent variable (y) explained by the regression model, is a simple indicator of the goodness of fit of the model.

Nowadays, performing linear regression analysis and the associated tests does not require special software, as the required calculation tools are available as standard functions in spreadsheet programmes, such as Microsoft Excel (the function LINEST).

Polynomials

When the sample data exhibit a non-linear relation between the dependent and independent variables, whilst no specific mathematical form is known to be the most appropriate, a convenient mathematical form that can be used is a polynomial. The general form of a polynomial of order m for one dependent and one independent variable is as shown in Equation 4.4, in which there are $m+1$ unknown coefficients (a_i; for $i=0, 1, \ldots, m$):

$$y = a_0 + a_1 x + a_2 x^2 + \ldots + a_m x^m \tag{4.4}$$

Therefore, at least $m+1$ pairs of sample data (x_i, y_i) must be available in order that all the unknown coefficients can be evaluated. If there are exactly $m+1$ data pairs, the a's can be evaluated by substituting the $m+1$ (x_i, y_i) values in turn into Equation 4.4 thereby yielding $m+1$ equations with $m+1$ unknowns, and thus the a coefficients can be evaluated by solving these equations simultaneously. Conventional methods for solving system of linear equations like the Gaussian elimination, Gauss–Seidel method etc. can be used to evaluate the coefficients. A standard function (MINVERSE) is also available within Microsoft Excel that can be used to solve a system of linear equations.

If more than $m+1$ data pairs are available, the least square error method can be used to find the unknown coefficients that best fit the sample data. A convenient way to solve for the model coefficients is to regard the polynomial (with x being the independent variable) as a multiple linear equation with the unknown coefficients treated as the independent variables (while the x terms, including all its higher order terms, will be of known values and are the coefficients of the unknowns). Once again, the standard Microsoft Excel function LINEST can be used for this purpose. Many statistics programs, e.g. SPSS+, include specific modules for multiple linear regression analysis, which may also be used.

Where a large number of data are available, one can improve the accuracy of the regression equation by choosing a polynomial with a high order. In theory, it is possible to have a polynomial that fits every point in the sample if the order of the polynomial is exactly one less than the total number of sample data. A high-order polynomial, however, is not always desirable. Besides being more complicated, the large number of associated inflexion points would cause numerical problems (e.g. instability or even divergence) while the set of model equations, formed by coupling the high-order polynomial (representing one component model) with equations for other components, is solved numerically in the simulation process. A high order also presents problems to the process of evaluation of the coefficients. Since the data have to be raised to a high power in the calculation and hence the value of the coefficient would be extremely small or large, accuracy of the coefficients so evaluated would be low because rounding-off errors that arise during the calculation process would become significant.

The polynomial regression method can be extended and applied to derive models for systems that involve more than one influencing variable. In this case, one dependent variable is expressed as a function of a number of independent variables. For the case of two independent variables, where each independent variable can appear up to the second order, the expression, which is a bi-quadratic equation, will be as follows:

$$z(x,y) = \left(a_0 + a_1 x + a_2 x^2\right) \cdot \left(b_0 + b_1 y + b_2 y^2\right)$$
$$z = c_0 + c_1 x + c_2 x^2 + c_3 y + c_4 y^2 + c_5 xy + c_6 x^2 y + c_7 xy^2 + c_8 x^2 y^2 \tag{4.5}$$

Similar to a single independent variable polynomial (Equation 4.4), Equation 4.5 can be treated as a multiple linear equation in the process of solving for the unknown coefficients.

Non-linear regression

A polynomial is a convenient mathematical function that can be used to formulate mathematical models for system components. There are, however, situations where other non-linear functions would be more appropriate compared to a polynomial, e.g. when a model can be derived based on theoretical considerations but the model coefficients need to be evaluated from a set of measured equipment performance data.

One common type of equation that often arises in system simulation is the exponential equation as shown below:

$$y = a \cdot e^{bx} \tag{4.6}$$

Equation 4.6 often arises as the solution of a simple dynamic system. The coefficient a is related to the initial state of the system whereas the coefficient b is related to the time constant of the system. When a mathematical function similar to the above is known to be an appropriate equation for representing a system component while the coefficients a and b need to be determined based on measured values of x and y, a non-linear regression technique should be adopted.

Box 4.1 shows the derivation of a method for evaluating the coefficients in Equation 4.6. For more complex non-linear models, evaluation of the model coefficients may need more sophisticated methods, e.g. modules in mathematical analysis programs like MATLAB.

Box 4.1 Least-square-error non-linear curve-fit method

For a simple exponential function $y = a \cdot e^{bx}$, following the standard least-square-error pro-cedure, the deviation between the model prediction and the sample data, e, is expressed as:

$$e = a \cdot e^{bx} - y$$

For convenience, each side of the above equation is divided by y, and e/y is denoted as E as follows:

$$E = \frac{a \cdot e^{bx}}{y} - 1$$

Hence, the sum of squared deviations, S, is:

$$S = \sum E^2 = \sum \left(\frac{a \cdot e}{y} - 1 \right)^2$$

Differentiating S with respective to a and b in turn and setting each derivative to zero, we obtain,

$$a \sum g^2 - \sum g = 0 \tag{a}$$

$$a \sum (x \cdot g^2) - \sum (x \cdot g) = 0 \tag{b}$$

where

$$g = \frac{e^{bx}}{y} \tag{c}$$

It can be seen that Equations a and b are two non-linear equations with unknown coef-ficients a and b. The method that can be used to solve this set of equation for a and b is as described below:

i)   express a in terms of g from Equation a

$$a = \frac{\sum g}{\sum g^2} \tag{d}$$

ii)   substituting Equation d into Equation b, we obtain a non-linear function of b (the exponent term in Equation c), denoted as $F(b)$, as follows:

$$F(b) = 0 = \left(\sum g \right) \cdot \left(\sum x \cdot g^2 \right) - \left(\sum g^2 \right) \cdot \left(\sum x \cdot g \right) \tag{e}$$

Box 4.1 *Continued*

A numerical method has to be used to solve for b from the non-linear Equation e. Here, the Newton–Ralphson method is used as explained below:

$$b_{j+1} = b_j - \frac{F(b)}{F'(b)} \tag{f}$$

where

$$F'(b) = \frac{\mathrm{d}F(b)}{\mathrm{d}b} = 2(\textstyle\sum g) \cdot (\sum x^2 g^2) - (\sum x \cdot g^2) \cdot (\sum x \cdot g) - (\sum g^2) \cdot (\sum x^2 \cdot g)$$

By repeatedly using Equation f to update the estimate of b until a further iteration step does not produce a significantly different value yields b. When b is evaluated, a can be found by substituting b into Equation c and then into Equation d.

4.2 Mathematical models of air-conditioning equipment using equation fitting

In this section, curve-fit models for several types of air-conditioning system components are described, as application examples of the above-discussed equation fitting technique.

Chillers

Chillers are typically the dominant electricity consumer in buildings with central air-conditioning. Therefore, models that can accurately predict operating energy use of chillers are essential components of a building energy simulation program to enable it to predict annual energy use in air-conditioned buildings. The chiller models described below (Yik & Lam 1998) were derived based on the assumption that the following system variables will be kept at their respective design values at all times, which is approximately true for most chiller plants:

* The chilled water flow rate through the evaporator
* The flow rate of condenser air or water (for air- and water-cooled chillers respectively) through the condenser
* The supply chilled water temperature

The resultant model, therefore, cannot predict changes in chiller performance due to variations in these parameters. For instance, if the chilled water supply temperature set point is reset upward under part-load conditions to reduce chiller energy consumption, the models will not apply.

Based on the above assumption, and discounting all parameters that can be assumed to be invariant (e.g. all the design characteristics and conditions of electricity supply), the variables that can significantly influence the power demand of a chiller (W_{CHR}) will be the cooling load on the chiller (Q_{CHR}) and the temperature of the cooling medium supplied to its condenser (T_{CIN}). A mathematical model that relates the power demand of a chiller to the influencing variables, in the form of Equation 4.7, can be developed to allow electricity consumption of a chiller plant to be predicted by computer simulation:

$$W_{CHR} = f(Q_{CHR}, T_{CIN}) \tag{4.7}$$

Many chiller manufacturers provide simplified part-load characteristic curves for their water-cooled chillers, which are rated according to the ARI Standard conditions (ARI 1992). This type of chiller part-load curve relates the chiller power demand simply to the load (i.e. in mathematical form, $W_{CHR} = f(Q_{CHR})$). The power inputs at various load levels were rated on the basis of the assumption that the inlet condenser water temperature would drop linearly with the load, from 29.4°C at full-load down to 15.6°C at zero load. This corresponds to a drop in the condenser water inlet temperature by 1.4 K per 10% reduction in load (called condenser water relief).

The chiller power and coefficient of performance (COP) curves for a 1,760 kW chiller with the effect of condenser water relief included are shown in Fig. 4.1 and compared with performance curves for the same chiller under different constant condenser water inlet temperatures. Note that the curves shown are based on the original design condenser inlet temperature of 32°C (instead of 29.4°C) from which condenser water temperature relief will start to take effect when the load on the chiller is reduced.

Although the chiller model can be simplified by assuming a certain condenser water relief rate, as in the case of the ARI Standard rating conditions, using such chiller models could lead to erroneous energy use predictions when applied to a chiller plant with multiple chiller units. For instance, in a day where the weather condition is relatively mild, the cooling load can be catered for by just a few chillers in the plant, with other chillers shut down. However, although the chillers that remain operating may be loaded quite close to their full capacity, the condenser water inlet temperature may be significantly lower than the temperature assumed in the simplified model. The electricity consumption of the chiller will be overestimated in this case. On the other hand, a chiller may be loaded up to just a small fraction of its full capacity, but the condenser water inlet temperature may stay at a high level. This will occur in the evening of a hot summer day when the majority of the air-handling equipment has been shut down and only one chiller is kept running. In this case, the chiller power will be underestimated.

Furthermore, if a chiller model with condenser water relief are used to determine an optimised chiller sequencing control strategy as depicted by Fig.4.2 (the curves shown included the effect of condenser water temperature relief), sequencing on and off chillers according to the switching-over points so determined will not result in an annual energy consumption as would be predicted on the basis of such models, but will be higher. The underestimation could be over 10% (Austin 1991). To avoid such problems, chiller performance models that can separately account for the effect of variations in the temperature of the cooling medium at the condenser inlet should be used.

As air-cooled and water-cooled chillers with centrifugal, screw and reciprocating compressors may have significantly different part-load characteristics, a different model has to be established for each of these types of chillers. All the chiller models developed are in the form

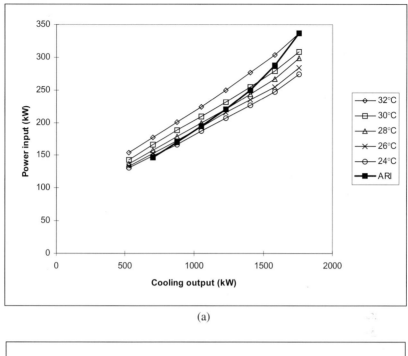

(a)

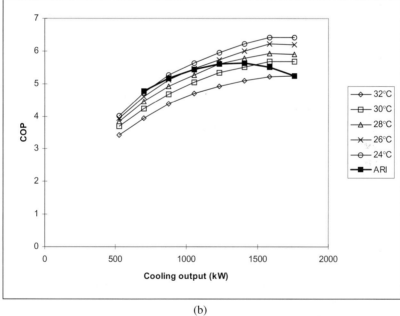

(b)

Fig. 4.1 The performance of a water-cooled chiller. (a) Power input. (b) Coefficient of performance (COP).

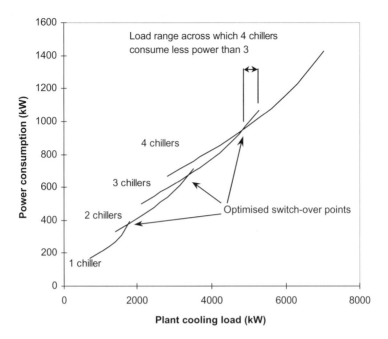

Fig. 4.2   Switching-over points for optimised chiller sequencing control.

of a bi-quadratic equation similar to Equation 4.5. This model form was chosen because it can account for the non-linear dependence of chiller power on the cooling load and the cooling medium temperature, and yet is relatively simple.

Instead of relating the actual value of the chiller power demand to the value of the corresponding cooling load, the model developed relates the normalised power demand (w_{CHR}) to the normalised load (q_{CHR}) of the chiller. These normalised values, each ranging between 0 and 1, can be evaluated by dividing the part-load power and the load by their respective rated full-load values for a given chiller. The cooling medium inlet temperatures in °C were used in the model without any conversion. The chiller model equation therefore becomes:

$$w_{CHR} = c_0 + c_1 q_{CHR} + c_2 q_{CHR}^2 + c_3 T_{CIN} + c_4 T_{CIN}^2 + c_5 q_{CHR} T_{CIN}$$
$$+ c_6 q_{CHR}^2 T_{CIN} + c_7 q_{CHR} T_{CIN}^2 + c_8 q_{CHR}^2 T_{CIN}^2 \qquad (4.8)$$

Performance data for air- and water-cooled centrifugal, screw and reciprocating chillers were collected from manufacturers as the basis for developing the chiller models. Table 4.1 summarises the types and capacities of chillers for which performance data have been obtained. All chiller performance data collected are pertaining to chillers using refrigerant R134a, and supplying chilled water at 7°C. The performance data for water-cooled chillers cover the range of condenser water inlet temperature from 32°C, the condition for the rated full output, to 24°C. The data for air-cooled chillers cover the range of condenser air inlet temperature from 23°C to 35°C, the ratings of the chillers based on 35°C.

Visual inspection of the graphs of the performance data revealed that for each type of chiller, the normalised performance of the chillers of different makes and unit sizes are

Table 4.1   Sources of chiller performance data and range of chillers covered (all with refrigerant R134a)

a) Water-cooled chillers

Type	Capacities of chillers (kW)[1]	Data source
Centrifugal	1759 & 3517	5 chiller models from 3 manufacturers
Screw	211, 352, 387 & 1055	4 chiller models from 2 manufacturers
Reciprocating	352 & 791	2 chiller models from 1 manufacturer

[1] Rated capacity at 32°C condenser water inlet temperature and 7°C chilled water leaving temperature.

b) Air-cooled chillers

Type	Capacities of chillers (kW)[2]	Data source
Centrifugal	703, 1055 & 1230	3 chiller models from 1 manufacturer
Screw	387, 457 & 739	3 chiller models from 1 manufacturer
Reciprocating	352 & 528	2 chiller models from 2 manufacturers

[2] Rated capacity at 35°C condenser air inlet temperature and 7°C chilled water leaving temperature.

very similar to each other (see Fig. 4.3 for water-cooled centrifugal chillers). Therefore, instead of developing a model for each size and make of chiller, one generic chiller model for each type of chiller was developed on the basis of the performance data of all chillers of the same type.

On the basis of the chiller performance data, the model coefficients (c_0 to c_8) were evaluated by using a regression analysis program. The regression method used assures that, for the given mathematical form of the model, the resultant set of coefficients would lead to the least sum of squared deviations between the model predictions and the chiller performance data. Table 4.2 summarises the values of the chiller model coefficients obtained for air- and water-cooled centrifugal, reciprocating and screw chillers.

Figure 4.3 shows the performance curve predicted by the model for water-cooled centrifugal chillers, together with the performance data obtained from the manufacturers. For this

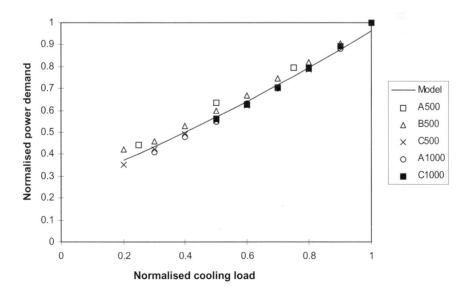

Fig. 4.3   Normalised performances of chillers.

Table 4.2 Coefficients in chiller models (Equation 4.8)

Model coefficient	Water cooled			Air cooled		
	Centrifugal	Screw	Reciprocating	Centrifugal	Screw	Reciprocating
c_0	2.365104E-01	-7.047963E-01	4.935148E-02	2.176147E-01	9.338920E-02	3.180071E-01
c_1	5.541562E-01	1.356044E+00	-1.874186E-01	3.970316E-02	3.604203E-01	-3.507387E-01
c_2	-2.740942E-01	1.971014E-01	5.688225E-01	4.466515E-01	9.226552E-02	4.613133E-01
c_3	-7.579433E-03	6.228124E-02	-9.706300E-04	-3.194077E-03	-2.914894E-03	-1.497708E-02
c_4	2.576668E-04	-7.257236E-04	2.560496E-05	1.353370E-04	8.828761E-05	2.561969E-04
c_5	-1.010772E-02	-7.754372E-02	2.906335E-02	9.742481E-03	6.493452E-04	4.304826E-02
c_6	2.600815E-02	-1.013358E-02	-7.051698E-03	-9.199176E-03	3.963624E-04	-1.474649E-02
c_7	2.958254E-04	1.038016E-03	-2.066242E-04	-1.260595E-04	1.662222E-05	-3.886984E-04
c_8	-3.759683E-04	6.371024E-04	8.163767E-05	3.080731E-04	6.363468E-04	2.155489E-04

model, the deviation in the predicted power demand from the manufacturers' data was within 7% of the full-load power. For the water-cooled screw chiller model, the deviation was within 5%, and for the reciprocating chiller model 1%. For the air-cooled chiller models for centrifugal, screw and reciprocating chillers, the deviations were within 1%, 2% and 6% respectively.

For reciprocating chillers comprising multiple cylinders and compressor units, capacity control is performed by unloading cylinders and/or sequencing on and off the compressors. Part-load performance of such chillers, therefore, will exhibit a 'saw-tooth' form rather than a smooth curve (Fig. 4.4). The performance data obtained, however, are limited to power demand of chillers at switching-over points between different numbers of operating compressors in the chillers. The models developed therefore are smoothed curves passing through these known performance data points (Fig. 4.4).

The chiller performance data used in the model development incurred the constraint that the models will yield the rated power consumption ($w_{CHR} = 1$) for a chiller, while being loaded to its full capacity ($q_{CHR} = 1$), only when the condenser water inlet temperature is at 32°C for a water-cooled chiller, or at an ambient temperature of 35°C for an air-cooled chiller. Although these temperatures represent the most widely adopted design conditions for indirect seawater cooled and air-cooled chiller plants in Hong Kong, different design cooling medium temperatures may be used for certain chiller plants. For instance, 27°C may be used as the design condenser water inlet temperature for direct seawater cooled chillers. In such circumstances, the following correction factor (c_F) needs to be applied to yield the correct chiller power demand:

$$w_{CHR, Corrected} = c_F \, w_{CHR} \tag{4.9}$$

where w_{CHR} is the chiller power demand predicted by the model (Equation 4.8), and the correction factor c_F is to be evaluated as follows:

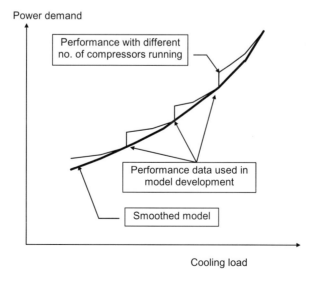

Fig. 4.4   Performance curve for a reciprocating chiller.

$$c_F = 1/w_{CHR}\left(1, T_{CIN,Design}\right) \tag{4.10}$$

Here, $w_{CHR}(1, T_{CIN, Design})$ is the model prediction at full-load ($q_{CHR} = 1$) and at the specific design inlet temperature of the cooling medium $T_{CIN, Design}$.

The power demand of a chiller under a particular combination of load and cooling medium supply temperature can then be evaluated from:

$$W_{CHR}\left(q_{CHR}, T_{CIN}\right) = c_F w_{CHR}\left(q_{CHR}, T_{CIN}\right) W_{CHR}\left(1, T_{CIN,Design}\right) \tag{4.11}$$

As the models were established on the basis of a collection of performance data for a series of chillers of different makes and unit sizes, their application is not restricted only to the specific chillers on which the model was based. However, deviations of the model predictions from the performance data of a particular chiller may be greater than a model developed specifically for that chiller. If a higher degree of accuracy is required, a specific model should be established for each chiller. Nevertheless, the generic models provide a convenient basis for building energy studies and are able to provide predictions with acceptable accuracy.

The model coefficients summarised in Table 4.2 are shown to seven significant figures. Such precision for coefficients of curve-fit models is often necessary for ensuring accurate model predictions. For instance, in a case of using a water pump model (a fourth order polynomial) to predict the water flow rate from measured pressure rise across a pump, it was found necessary to use coefficients with at least six significant figures (Yik & Burnett 1995). For the chiller models presented, it was found that the predictions will not be significantly affected even if only four significant figures are maintained for all the coefficients. However, the errors will become unacceptable if the coefficients are rounded to less than six decimal places.

Cooling towers

Cooling towers are widely used in air-conditioning systems for rejection of condenser heat from chillers. Heat rejection is accomplished by spraying the condenser water against a stream of air that is drawn by a fan to flow through the cooling tower such that a minute part of the water is evaporated. The heat of vaporisation is sustained by a drop in temperature of the water (which is the major mechanism for cooling down the condenser water) and the water that has been evaporated into vapour will be disposed of together with the air stream.

The rate of heat rejection achieved in a cooling tower is proportional to the range (defined as the difference in temperature between the inlet and exit water), which is dependent upon the approach (the difference between the exit water temperature and the wet-bulb temperature of the entering air), the mass flow rates of the air and water, and the design of the cooling tower. For a given cooling tower that operates with constant air and water mass flow rates, the leaving water temperature, which is the system variable that needs to be evaluated and used as an input to a chiller model (see Equation 4.8) in a plant performance simulation, is a function of the range and ambient air wet-bulb temperature, as depicted in Fig. 4.5.

Although an analytical model can be derived for modelling cooling towers, its use involves more complicated solution procedures. Where performance data for a cooling tower

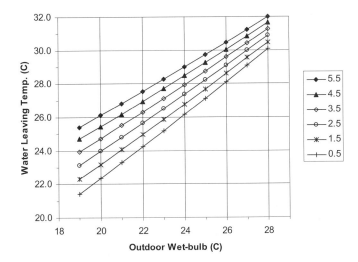

Fig. 4.5 Temperature of water leaving cooling tower (°C) at various ambient air wet-bulb temperatures (°C) and ranges (°C). (Material adapted by permission from ASHRAE handbook of Systems and Equipment (2001). © American Society of Heating, Refrigerating and Air-Conditioning Engineers, Inc., www. ashrae.org.).

similar to those shown in Fig. 4.5 are available, a curve-fit model that is more convenient to use can be derived for the cooling tower for prediction of the leaving water temperature. Equation 4.12, which is once again a bi-quadratic equation, has been shown to be an adequate mathematical form for this purpose:

$$T_{CEX} = f(R, T_{WB})$$
$$= c_0 + c_1 R + c_2 R^2 + c_3 T_{WB} + c_4 T_{WB}^2 + c_5 R T_{WB}$$
$$+ c_6 R^2 T_{WB} + c_7 R T_{WB}^2 + c_8 R^2 T_{WB}^2 \tag{4.12}$$

where:

$R = \text{range (°C)} = T_{CIN} - T_{CEX}$
$T_{CEX} = \text{exit water temperature (°C)}$
$T_{CIN} = \text{inlet water temperature (°C)}$
$T_{WB} = \text{ambient air wet-bulb temperature (°C)}$

It follows that the heat rejection rate (*HR*) is given by:

$$HR = m_w c_w R \tag{4.13}$$

where:

$m_w = \text{water mass flow rate (kg s}^{-1})$
$c_w = \text{specific heat capacity of water (kJ kg}^{-1}\text{ K}^{-1})$

As can be seen from Equation 4.12, the cooling tower model is an implicit equation (with R being a function of T_{CEX}), which needs to be solved together with the chiller model simultaneously in a simulation step.

Fans and pumps

Turbo-machinery like fans and pumps can be found in nearly all HVAC systems as drivers for fluid flow. Because the dynamics of fluid flow through a fan or pump is highly complicated, performances of fans and pumps often need to be determined through experimental measurements. Most manufacturers provide fan or pump performance for reference by design engineers in graphical form, typically including curves showing variations of fan/pump total pressure and power with flow rate at a specific running speed, as shown in Fig. 4.6.

Given performance curves similar to those shown in Fig. 4.6 for a specific fan or pump, performance data can be sampled from the curves and a curve-fit model can be derived for the fan or pump for the given (rated) speed (N_0), which will be a two-equation model, one representing the relation between the total pressure (TP) and the flow rate (Q) and the other the power (W) and the flow rate. A polynomial of order three or four will generally suffice for the total pressure and power models for a fan or pump. Equations 4.14a and 4.14b are third order polynomial models that may be used for modelling a fan or a pump at the rated speed:

$$TP = A_0 + A_1 Q + A_2 Q^2 + A_3 Q^3 \tag{4.14a}$$

$$W = B_0 + B_1 Q + B_2 Q^2 + B_3 Q^3 \tag{4.14b}$$

In order to widen the use of a fan/pump model for modelling fans/pumps of similar design but of different unit sizes, models in the normalised form, as shown in Equation 4.15, may be used in a simulation program:

$$q = \frac{Q}{Q_0} \tag{4.15a}$$

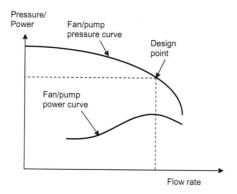

Fig. 4.6   Typical fan/pump performance curves.

$$tp = \frac{TP}{TP_0} = a_0 + a_1 q + a_2 q^2 + a_3 q^3 \tag{4.15b}$$

$$w = \frac{W}{W_0} = b_0 + b_1 q + b_2 q^2 + b_3 q^3 \tag{4.15c}$$

where:

Q = flow rate (m^3 s^{-1})
Q_0 = the design flow rate (m^3 s^{-1})
q = normalised flow rate
TP = fan/pump total pressure (Pa)
TP_0 = the design fan/pump total pressure (Pa)
tp = normalised fan/pump total pressure
W = fan/pump power (kW)
W_0 = the design fan/pump power (kW)
w = normalised fan/pump power
a_0 to a_3 = coefficients in the fan/pump total pressure model
b_0 to b_3 = coefficients in the fan/pump power model

The input to this fan/pump model for a specific fan or pump includes a_0 to a_3, b_0 to b_3, Q_0, TP_0 and W_0. Equations 4.15a to 4.15c can be converted back to Equations 4.16a and 4.16b for determination of the fan total pressure and fan power under a specific operation condition at the rated fan speed:

$$TP = a_0 TP_0 + a_1 \left(\frac{TP_0}{Q_0} \right) Q + a_2 \left(\frac{TP_0}{Q_0^2} \right) Q^2 + a_3 \left(\frac{TP_0}{Q_0^3} \right) Q^3 \tag{4.16a}$$

$$W = b_0 W_0 + b_1 \left(\frac{W_0}{Q_0} \right) Q + b_2 \left(\frac{W_0}{Q_0^2} \right) Q^2 + b_3 \left(\frac{W_0}{Q_0^3} \right) Q^3 \tag{4.16b}$$

Comparison of Equations 4.14 and 4.16 allows conversions between the coefficients in the two expressions for a fan/pump model to be seen. Unlike chillers, performances of pumps and fans are more variable, rendering it impossible to derive more widely applicable generic normalised models.

Where a variable speed fan or pump is used in a system, e.g. a variable air volume (VAV) system, or in the secondary loop of a two-loop chilled water pumping system, fan or pump performance under reduced operating speeds will need to be predicted in a simulation. If no detailed data about the performance of the fan or pump at various operating speeds are available, fan or pump performance may be approximately predicted based on the rated speed performance and the fan/pump laws.

The fan/pump laws that apply to two dynamically similar operating conditions of a fan or pump running at two different speeds, the rated speed N_0 and a reduced speed N^*, are as shown in Equation 4.17:

$$\frac{Q^*}{Q} = \frac{N^*}{N_0} \tag{4.17a}$$

$$\frac{TP^*}{TP} = \left(\frac{N^*}{N_0}\right)^2 \tag{4.17b}$$

$$\frac{W^*}{W} = \left(\frac{N^*}{N_0}\right)^3 \tag{4.17c}$$

Combining the fan/pump laws and the fan performance model for the rated speed (Equation 4.16) yields:

$$TP^* = a_0 TP_0 \left(\frac{N^*}{N_0}\right)^2 + a_1 \left(\frac{TP_0}{Q_0}\right)\left(\frac{N^*}{N_0}\right)Q^* + a_2 \left(\frac{TP_0}{Q_0^{\ 2}}\right)Q^{*2} + a_3 \left(\frac{TP_0}{Q_0^{\ 3}}\right)\left(\frac{N_0}{N^*}\right)Q^{*3} \tag{4.18a}$$

$$W^* = b_0 W_0 \left(\frac{N^*}{N_0}\right)^3 + b_1 \left(\frac{W_0}{Q_0}\right)\left(\frac{N^*}{N_0}\right)^2 Q^* + b_2 \left(\frac{W_0}{Q_0^{\ 2}}\right)\left(\frac{N^*}{N_0}\right)Q^{*2} + b_3 \left(\frac{W_0}{Q_0^{\ 3}}\right)Q^{*3} \tag{4.18b}$$

In a simulation, the fan/pump speed at a particular operating condition (N^*) can be determined according to the relevant control feedback signal, e.g. the differential pressure across the index branch in a chilled water network or the duct static pressure in a ducting network. After determining the fan/pump speed, Equations 4.18a and 4.18b may be used to predict the fan/pump total pressure and power according to the flow rate at that speed.

Although most new variable air volume (VAV) systems nowadays are equipped with variable speed fans, there are still many existing VAV systems that use constant speed fans with inlet guide vanes. Where the financial viability of converting VAV systems with inlet guide vane controls into variable fan speed controls needs to be justified, the annual fan energy use of the system with both types of system controls will need to be predicted. The fan performance at various flow rates and inlet guide vane angles can be modelled by the following bi-quadratic equations:

$$tp = a_0 + a_1 q + a_2 q^2 + a_3 \phi + a_4 \phi^2 + a_5 q\phi + a_6 q^2\phi + a_7 q\phi^2 + a_8 q^2\phi^2 \tag{4.19a}$$

$$w = b_0 + b_1 q + b_2 q^2 + b_3 \phi + b_4 \phi^2 + b_5 q\phi + b_6 q^2\phi + b_7 q\phi^2 + b_8 q^2\phi^2 \tag{4.19b}$$

where ϕ denotes the inlet guide vane position. This model would need to be derived by curve-fitting the fan performance data obtained from the manufacturer. When the rated fan total pressure, power and flow rate, and the operating flow rate and inlet guide vane position are known, the operating fan total pressure and power can be determined from:

$$TP = (a_0 + a_3\phi + a_4\phi^2)TP_0 + (a_1 + a_5\phi + a_7\phi^2)\frac{TP_0}{Q_0}Q$$

$$+ (a_2 + a_6\phi + a_8\phi^2)\frac{TP_0}{Q_0^{\ 2}}Q^2 \tag{4.20a}$$

$$W = \left(b_0 + b_3\phi + b_4\phi^2\right)W_0 + \left(b_1 + b_5\phi + b_7\phi^2\right)\frac{W_0}{Q_0}Q$$

$$+ \left(b_2 + b_6\phi + b_8\phi^2\right)\frac{W_0}{Q_0^2}Q^2 \qquad\qquad (4.20b)$$

VAV terminal boxes

In VAV systems, regulation of supply flow rate for temperature control in individual rooms is performed by VAV boxes (also called VAV terminals). A VAV box is basically an air damper with variable flow resistance. The flow resistance (K) is a function of the flow rate (Q) through the VAV box and the inclination angle of the flow regulation damper inside the box (θ, hereafter referred to simply as the damper angle, see Fig. 4.7). The total pressure drop across a VAV box is related to the flow rate through it by:

$$\Delta P = k\frac{\rho_a}{2}\left(\frac{Q}{A}\right)^2 \qquad\qquad (4.21)$$

where:

ΔP = total pressure loss across the VAV box (Pa)
k = pressure loss coefficient (—)
ρ_a = air density (kg m^{-3})
A = inlet cross-sectional area (m^2)
Q = volume flow rate (m^3 s^{-1})

Equation 4.22 has been shown to be a suitable model (Khoo *et al.* 1998) to relate the pressure loss coefficient k of a VAV box to the damper angle in the VAV box:

$$\ln k = a + b\theta^c \qquad\qquad (4.22)$$

where:

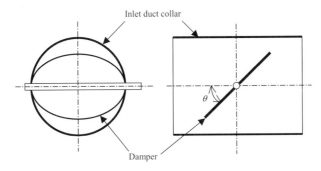

Fig. 4.7   Damper angle of flow regulation damper in a VAV box.

θ = damper angle (rad)

a, b and c = the model coefficients

Note that in this model, $\theta=0$ when the damper is fully open.

Therefore, modelling the performance of a VAV box requires a set of numerical values for the coefficients A, a, b and c. The coefficients a, b and c are dependent on the design of each specific VAV box and would need to be determined from curve fitting the pressure losses at various damper angles and flow rates obtained from either experimental measurements or, if available, data from the manufacturer.

Laboratory measurements were conducted for two commercial VAV box models of flow handling capacities most commonly used in office buildings in Hong Kong, and the values of the coefficients evaluated from the measurements are as summarised in Table 4.3. Figure 4.8 shows the characteristics of the two VAV box models, including the pressure loss coefficient at various damper angles and the flow rate at various damper angles under a range of pressure difference across the boxes.

Combining Equations 4.21 and 4.22 yields:

$$\Delta P = \exp(a+b\theta^c)\frac{\rho_a}{2A^2}Q^2 \tag{4.23}$$

Therefore, the flow resistance K of a VAV box can be determined by:

$$K = \exp(a+b\theta^c)\frac{\rho_a}{2A^2} \tag{4.24}$$

In using the above VAV box model, values of the coefficients (A, a, b and c) are the required input. In addition to the model coefficients, the follow input data should also be included:

- The maximum damper angle (fully closed position) θ_M or the damper angle corresponding to the minimum degree of opening θ_{Min} (rad)
- The effective maximum damper angle θ_{EM} (rad)

The damper angle corresponding to the minimum degree of opening, when set, will limit the VAV box from closing beyond this level, which is a means to ensure there will at least be a certain flow rate of air being supplied into an air-conditioned space to ensure a minimum degree of ventilation provision and to avoid dumping of the cold supply air directly onto occupants who are below the diffusers, even though the required supply flow rate to offset the room sensible load falls below this flow rate. The effective maximum damper angle θ_{EM} needs to be

Table 4.3   Coefficients for pressure loss coefficients models for two VAV boxes.

Model	Design flow rate (m^3 s^{-1})	a	b	c
1	0.45	−0.06	2.52	3.35
2	0.8	−0.0163	1.46	4.81

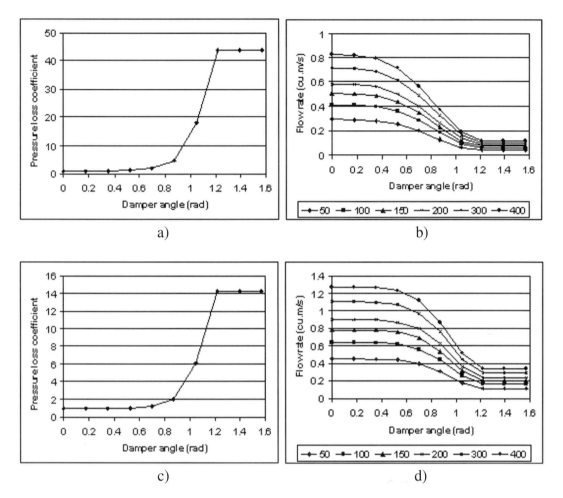

Fig. 4.8 VAV box model characteristics (legend denotes pressure, Pa). (a) and (b) Model 1 (c) and (d) Model 2

included because, as found from the laboratory measurement (Fig. 4.8), the pressure loss co-efficient k would level off when the damper has been positioned (closed) to and beyond a certain angle (about 70° or 1.2 rad). The corresponding flow rate at damper angles greater than this maximum effective angle could be regarded as 'leakage' from the VAV box.

Control valves

Similar to a VAV box, a control valve for regulating chilled or hot water flow through air-handling equipment may be regarded as a flow restriction with a variable flow resistance. The flow resistance varies with the degree of opening of the control valve, which is quantified by the displacement of the valve stem from the closed position, called the valve lift (Z).

For a control valve that is maintained at a fixed valve lift, the valve will behave as if it is a flow restriction with a fixed flow resistance, similar to an orifice. By regarding water as an incompressible fluid, the pressure drop across a flow restriction (ΔP_v) when water is flowing through it at the volume flow rate (V) can be described by:

$$\Delta P_v = k\frac{1}{2}\rho\left(\frac{V}{A}\right)^2 \tag{4.25}$$

where:

A is the net flow cross-sectional area
ρ is and the density of water
k is a proportionality constant

Hence, the flow rate through a control valve that is maintained at a fixed valve lift can be related to the pressure drop across the valve by:

$$V = C\sqrt{\Delta P_v} \tag{4.26}$$

In Equation 4.26, C may be regarded as a constant for a fixed valve lift, but obviously its value will vary with the valve lift. It follows that description of the operating characteristics of a control valve requires a relationship between C and Z, for the whole range of Z from the fully closed to the fully open position.

In practice, the characteristics of a control valve are described by its flow coefficient (K_v, also called capacity index), which is the value of C at the fully open position, and the inherent characteristic of the valve. The latter is a relation between the normalised flow rate (v) and the normalised valve lift (z) under a constant differential pressure across a control valve, as depicted by Equation 4.27 below:

$$v = f(z) \tag{4.27a}$$

where:

$$v = V/V_0 \quad (0 \leq v \leq 1) \tag{4.27b}$$

$$z = Z/Z_0 \quad (0 \leq z \leq 1) \tag{4.27c}$$

Here, V is the flow rate at an intermediate valve lift (Z) while incurring a pressure drop ΔP_v across the valve; V_0 is the flow rate at the fully open position ($Z = Z_0$) that will incur the same pressure drop ΔP_v (Equation 4.28); and z is the ratio of the valve lift (Z) to the fully-open valve lift (Z_0):

$$V_0 = K_v\sqrt{\Delta P_v} \tag{4.28}$$

For a series of control valves having the same inherent characteristic, only the common inherent characteristic and the flow coefficient (K_v) of each valve size are required for describing the performance of all the valves in the series. Figure 4.9 shows three common inherent characteristics of control valves. For chilled and hot water flow regulation in HVAC systems, the use control valves with equal percentage inherent characteristics will result in much better control performance than linear valves; while the quick opening valves are not suitable for this purpose, they are primarily for isolation of equipment from a piping circuit.

When the chilled or hot water flow rate through an air-handling unit that is regulated by a control valve needs to be predicted, a mathematical model for the control valve must be available. Equation 4.29 below can be used to model an equal percentage control valve, but the coefficients a and b would need to be evaluated from the inherent characteristic curve obtained from the manufacturer, such as that shown in Fig. 4.10. The non-linear curve-fit method described in Section 4.1 can be used for determining a and b in Equation 4.29 from sampled operating points from the inherent characteristics graph:

$$v = ae^{bz} \tag{4.29}$$

Note that Equation 4.29, with fixed values for a and b, will not be applicable when the valve lift approaches zero. This is because, at $z = 0$ (the fully closed position), $v = 0$ and this will require a to be zero. However, this will in turn require b to be infinitely large. Nonetheless, this restriction will not give rise to significant problems, as most valves will be subject to a minimum controllable flow rate below which the valve will be completely shut off, or, for certain valves, will just allow a minute flow rate to pass (called the let-by).

Equations 4.27b, 4.28 and 4.29 can be combined to allow the water flow rate through a control valve to be related to the two influencing system variables z and ΔP_v, as follows:

$$V = K_v ae^{bz} \sqrt{\Delta P_v} \tag{4.30}$$

Equation 4.30 may be used as a component model within a water pipe network model for detailed simulation of water flows in the network.

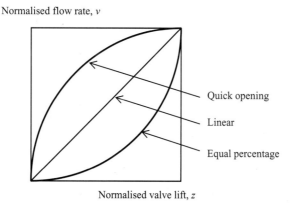

Normalised flow rate, v

Quick opening

Linear

Equal percentage

Normalised valve lift, z

Fig. 4.9   Three common inherent characteristics of control valves.

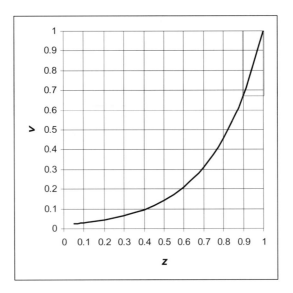

Fig. 4.10   Inherent characteristic of an equal percentage control valve.

4.3 A detailed steady-state cooling and dehumidifying coil model

Plate-finned chilled water cooling and dehumidifying coils (hereafter referred to simply as cooling coils) are essential components of central air-conditioning systems. Prediction of temperature and humidity inside air-conditioned spaces requires evaluation of the sensible and latent cooling capacities of cooling coils under varied operating conditions but most manufacturers publish rated performance only. Therefore, in-depth studies on air-conditioning system performance under part-load conditions require the use of a cooling coil model developed on the basis of fundamentals of heat and mass transfer.

Cooling coil performance prediction involves the use of empirical correlations to determine heat and mass transfer coefficients. Correlations that apply to flat-plate-fins have been well established (Rich 1973; McQuiston 1978) but most cooling coils in use nowadays are with corrugated or wavy fins (Fig. 4.11) for their better heat transfer performance. Since most corrugated and wavy fins are proprietary designs, few heat transfer data are available in the open literature. Beecher & Fagan (1987) derived correlations for both flat fins and corrugated fins of a saw-tooth pattern (Fig. 4.11a).

Later, Webb (1990) derived, on the basis of Beecher & Fagan's experimental data, a set of more concise correlations. Unfortunately, Beecher & Fagan's data is restricted to dry fin surfaces and to a limited range of coil configurations. More recently, Mirth & Ramadhyani (1993a, 1994) developed heat transfer correlations from results of experiments with five commercial coils with wavy fins. They reported that their correlations can predict air-side Nusselt numbers accurately for dry fin surfaces but with much greater uncertainties for wetted fin surfaces. Clearly, more extensive experimental studies on the performance of cooling and dehumidifying coils with corrugated and wavy fins are required before accurate correlations for this type of fin design can be derived.

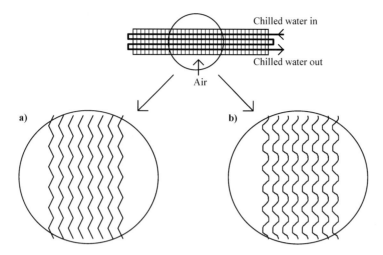

Fig. 4.11 A plate-finned cooling coil with a) corrugated and b) wavy fins.

A steady-state simulation model for cooling coils with corrugated plate fins has been developed (Yik *et al.* 1997) and used as a component model in a coupled building and air-conditioning system model (Yik *et al.* 1995). The model can predict sensible and total cooling capacities of cooling coils and the thermodynamic states of the air and chilled water. This section outlines the methods that were used to develop the cooling coil model. Box 4.2 summarises the meanings of symbols used in the coil model.

Geometric and flow parameters

In the model, a cooling coil is treated as a counter-flow heat exchanger (ARI 1991; ASHRAE 2000) (Fig. 4.12a), which has been shown to be a valid modelling approach if the number of tube passes exceeds three (Stevens *et al.* 1976). The heat transfer area of a cooling coil (A) is defined as the total exposed surface area of the fins and the fin collars (Fig. 4.12b) (hereafter referred to as the coil surface). The surface factor (F_s) used in evaluating A ($A = F_s N_r A_a$) is approximated by the ratio of the total exposed surface area of the fin and the fin collar of a typical cell in the cooling coil (ΔA) (Fig. 4.12d) to the frontal area of the cell (ΔA_a):

$$F_s = \Delta A / \Delta A_a = \left[2\left(x_a x_b - \pi D_o^2 / 4\right) + (s - y)\pi D_o\right] / x_a s \tag{4.31}$$

The air-side heat transfer coefficients are evaluated on the basis of the average air velocity (u_a) that would exist at the minimum 'free-flow' area (A_c, $A_c = \sigma A_a$) at a cross-sectional plane through the centre lines of a row of tubes in a cooling coil. The value of σ is also determined on the basis of a typical cell as follows:

$$\sigma = \Delta A_c / \Delta A_a = (x_a - D_o)(s - y) / x_a s \tag{4.32}$$

Chilled water flows across a cooling coil through a number of parallel water circuits formed by the array of tubes connected at their ends by U-bends. Many cooling coil manufacturers

Box 4.2 Symbols used in the cooling and dehumidifying coil model

A	heat transfer area of a coil (m^2)
A_a	frontal or face area of a coil (m^2)
A_c	core or minimum free flow area for air (m^2)
A_{dc}	heat transfer area of a dry coil or the dry portion of a partly wet coil (m^2)
A_{to}	total external surface area of the tubes (m^2)
A_{wc}	heat transfer area of a wet coil or the wet portion of a partly wet coil (m^2)
C	condensing factor (°C^{-1})
C_c	coil characteristics (°C kg J^{-1})
C_p	specific heat capacity (J kg^{-1} K^{-1})
D_i	internal diameter of coil tube (m)
D_o	outer diameter of the fin collar (m)
D_{va}	mass diffusivity of water vapour in air (m^2 s^{-1})
F_s	surface factor
Gz	Graetz Number
h'	specific enthalpy of saturated air at temperature T′ (J kg^{-1})
h_a	specific enthalpy of air (J kg^{-1})
h_d	total heat (or mass) transfer coefficient for wet coil surface (m s^{-1})
h_{fg}	latent heat of evaporation/condensation of water/water vapour (J kg^{-1})
h_i	heat transfer coefficient for internal tube surface (W m^{-2} °C^{-1})
h_o	sensible heat transfer coefficient for coil surface (W m^{-2} °C^{-1})
$J(s), J_m(s)$	correction terms for sensible and total heat transfer at wetted coil surface
j_s, j_m	Chilton–Colburn j factors for sensible heat and total heat transfer respectively
k_f	thermal conductivity of fin (W m^{-1} °C^{-1})
k_w	thermal conductivity of water (W m^{-1} °C^{-1})
m	mass flowrate (kg s^{-1})
N_{cir}	circuit number, ratio of number of water circuits to number tubes per row
N_p	number of fin corrugation patterns per longitudinal tube spacing
N_r	number of tube rows
N_{tpr}	number of tubes per row
Nu	Nusselt number (Nu$_a$ based on arithmetic mean temperature difference, Nu$_l$, based on log mean temperature difference)
P_d	fin pattern depth (m)
Pr	Prantl number
q	heat transfer rate (W)
r	outer radius of fin collars (m)
Re$_{D0}$	Reynolds number based on fin-collar outer diameter
Re$_{xb}$	Reynolds number based on the longitudinal tube spacing
Re$_s$	Reynolds number based on fin spacing
s	fin spacing (or fin pitch) (m)
SHR	sensible heat ratio
T	temperature (°C)
T'	dewpoint temperature (°C)

Box 4.2 *Continued*

U	overall heat transfer coefficient (W m^{-2} °C^{-1})
u_a	air flow velocity (m s^{-1})
u_w	water flow velocity (m s^{-1})
w	moisture content (kg/kg)
x_a	transverse tube spacing (in the direction perpendicular to air-flow direction) (m)
x_b	longitudinal tube spacing (in the air-flow direction) (m)
x_f	half width of one fin pattern (m)
y	fin thickness (m)
$\Delta\theta$	log mean temperature difference (°C)
ΔH	log mean enthalpy difference (kJ/kg)
θ	fin pattern angle (rad)
v	kinematic viscosity of air (m^2 s^{-1})
λ	under relaxation factor used in numerical solution schemes
ε	error or deviation (ε_r – relative deviation, ε_a – absolute deviation)
ρ_a	air density (kg m^{-3})
μ_a	air viscosity (kg m^{-1} s^{-1})
η_f	fin efficiency of the dry fin
η_s	surface effectiveness of the dry surface
η_{wf}	fin efficiency of the wet fin
η_{ws}	surface effectiveness of the wet surface
σ	contraction ratio

Subscripts

1	plane 1, air on-coil plane (water leaving-coil plane)
2	plane 2, air leaving-coil plane (water on-coil plane)
a	air or actual
b	dry/wet boundary plane
c	whole coil
dc	dry coil
f	flat-fin
wc	wet coil
wcs	wet coil sensible heat transfer
wct	wet coil total heat transfer
l	latent heat transfer
m	mass transfer or model prediction
p	corrugated (patterned) fin
r	reference
s	sensible heat transfer or coil surface
sat	saturated condition
t	total heat transfer
tol	tolerance limit
w	water, wet condition

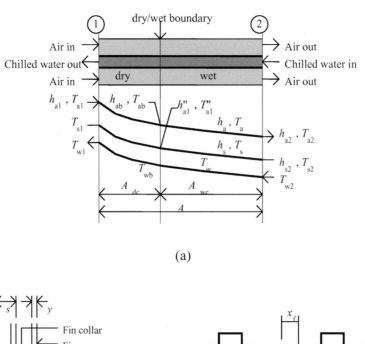

(a)

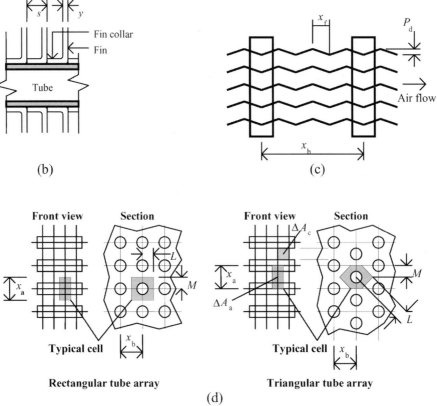

Fig. 4.12   (a) Counter-flow heat exchanger model. (b) Sectional view of a tube and fins. (c) Fin corrugation pattern. (d) Rectangular and triangular tube arrangement.

use the terms 'double', 'full', 'half' or 'quarter' circuits to describe coil circuit arrangements which, when translated into the numerical values of 2, 1, 0.5, or 0.25 respectively, will correspond to the ratio of the number of parallel water circuits to the number of tubes per row. This ratio (N_{cir}) is used in determining the water velocity (u_w) inside individual tubes as follows:

$$u_w = m_w \Big/ \left(\rho_w N_{cir} N_{tpr} \pi D_i^2 \big/ 4 \right) \tag{4.33}$$

The tube-side heat transfer coefficient is determined on the basis of this velocity.

Sensible and total heat transfer in a coil

For a completely dry coil (without condensation), the conventional log mean temperature difference method (Equation 4.34) is employed to model the coil:

$$q_{dc} = A_{dc} U_{dc} \Delta\theta_{dc} \tag{4.34}$$

For a completely wet coil (condensation starts right at the plane where the air enters the coil), the heat transfer between the air and the chilled water is modelled using Equations 4.35a and 4.35b. These equations refer respectively to the total and sensible heat transfer between the air and the coil surface, rather than between the air and the chilled water, but the steady rate of total heat transfer will be the same. The log mean temperature difference ($\Delta\theta_{wc}$) and log mean enthalpy difference (ΔH_{wc}) are evaluated from the temperatures and specific enthalpies of the air, and those of the saturated air-film above the layer of condensate over the coil surface, at the air entering and leaving planes:

$$q_{wct} = A_{wc} h_d \eta_{ws} \Delta H_{wc} \tag{4.35a}$$

$$q_{wcs} = A_{wc} h_o \eta_{ws} \Delta\theta_{wc} \tag{4.35b}$$

For a partly wet coil (i.e. condensation starts at an intermediate plane inside the coil, Fig. 4.12a), Equations 4.34 and 4.35 are used to determine the rates of heat transfer at the dry and the wet parts respectively. Their sum will be the total rate of heat exchange between the air and the chilled water. However, evaluation of the log mean difference terms $\Delta\theta_{dc}$, $\Delta\theta_{wc}$ and ΔH_{wc} (Equations 4.36 and 4.37) will involve the conditions at the interface between the dry and the wet parts of the coil (referred to here as the dry/wet boundary, Fig. 4.12a):

$$\Delta\theta_{dc} = \frac{(T_{a1} - T_{w1}) - (T_{ab} - T_{wb})}{\ln\{(T_{a1} - T_{w1})/(T_{ab} - T_{wb})\}} \tag{4.36}$$

$$\Delta\theta_{wc} = \frac{(T_{ab} - T_{sb}) - (T_{a2} - T_{s2})}{\ln\{(T_{ab} - T_{sb})/(T_{a2} - T_{s2})\}} \tag{4.37a}$$

$$\Delta H_{wc} = \frac{(h_{ab} - h_{sb}) - (h_{a2} - h_{s2})}{\ln\{(h_{ab} - h_{sb})/(h_{a2} - h_{s2})\}} \tag{4.37b}$$

The heat transfer rates at the dry and wet parts of a partly wet coil are related to the resulting changes in the states of the fluids by Equation 4.38:

$$q_{dc} = m_a Cp_a (T_{a1} - T_{ab}) = m_a (h_{a1} - h_{ab}) = m_w Cp_w (T_{w1} - T_{wb}) \qquad (4.38a)$$

$$q_{wct} = m_a (h_{ab} - h_{a2}) = m_w Cp_w (T_{wb} - T_{w2}) \qquad (4.38b)$$

$$q_{ct} = q_{dc} + q_{wct} = m_a (h_{a1} - h_{a2}) = m_w Cp_w (T_{w1} - T_{w2}) \qquad (4.38c)$$

Equations used to evaluate the heat transfer coefficients involved in Equations 4.34 to 4.38, including McQuiston's correlations for flat plate fins, are summarised below.

Sensible and total heat transfer coefficients

The overall heat transfer coefficient for a dry coil (U_{dc}), or the dry part of a partly wet coil, can be calculated by:

$$U_{dc} = 1 / [1/(\eta_s h_o) + A/(h_i A_i)] \qquad (4.39)$$

The thermal resistances of the tube wall and the fin collar can be ignored because they would be much smaller than the air- and water-side film resistances included in the equation. The fouling resistances at the fin, the fin collar and the internal tube wall surfaces are also discounted but can be included in the model easily for predicting performance of coils in use. Thermal resistance due to imperfect contact between the tube and the fin collar is not explicitly included because its effect is already included in the fin-side heat transfer coefficient (h_o) used in the model (McQuiston & Parker 1988). Furthermore, the assumption has been made that the internal surface of the tubes is smooth (i.e. there are no internal fins or turbulators in the tubes).

The water-side heat transfer coefficient (h_i) is evaluated using the Dittus–Boelter correlation where $n = 0.4$ for cooling coils ($n = 0.3$ in the case of heating coils) (Welty 1978):

$$h_i = 0.023 (k_w / D_i) \operatorname{Re}_{D_i}^{0.8} \operatorname{Pr}_w^n \qquad (4.40)$$

The air-side sensible and total heat transfer coefficients (h_o & h_d) for flat-fin surfaces are determined from the Chilton–Colburn j-factors (j_s and j_m):

$$h_o = j_s \rho_a u_a Cp_a \operatorname{Pr}^{-2/3} \qquad (4.41a)$$

$$h_d = j_m \rho_a u_a (v/D_{va})^{-2/3} \qquad (4.41b)$$

For flat plate fins, the j-factors are evaluated using McQuiston's correlations (McQuiston 1978; McQuiston & Parker 1988).

For a dry coil surface:

$$j_s = F_{j,N_r} (0.0014 + 0.2618\, JP) \qquad (4.42)$$

and for a wet coil surface:

$$j_s = F_{j,N_r}(0.0014 + 0.2618\,JP\,J(s)) \tag{4.43a}$$

$$j_m = F_{j,N_r}(0.0014 + 0.2618\,JP\,J_m(s)) \tag{4.43b}$$

where:

$$J(s) = (0.9 + 4.3 \times 10^{-5}\,\text{Re}_s^{1.25})[(s-y)/s] \tag{4.44a}$$

$$J_m(s) = (0.8 + 4.0 \times 10^{-5}\,\text{Re}_s^{1.25})[s/(s-y)]^4 \tag{4.44b}$$

$$JP = \text{Re}_{D_o}^{-0.4}(A/A_{to})^{-0.15} \tag{4.45}$$

$$F_{j,N_r} = \frac{1 - 1280 N_r\,\text{Re}_{x_b}^{-1.2}}{1 - 5120\,\text{Re}_{x_b}^{-1.2}} \tag{4.46}$$

$J(s)$ and $J_m(s)$ (Equation 4.44) account respectively for the effects of the condensate on the sensible and total heat transfer at the wet coil surface. Equation 4.44 applies to dropwise condensation only but filmwise condensation rarely occurs in coils in practice (McQuiston 1978).

Surface effectiveness

The effectiveness terms (η_s and η_{ws}) included in Equations 4.35 and 4.39 are to account for the effect of the temperature variation over the fin and fin collar surfaces on the rate of sensible and total heat transfer. For a dry coil, the surface effectiveness (η_s) is calculated from the fin efficiency (η_f) using (McQuiston & Parker 1988):

$$\eta_s = 1 - (A_f/A)(1 - \eta_f) \tag{4.47a}$$

$$\eta_f = \tanh(mr\phi)/(mr\phi) \tag{4.47b}$$

$$m = \sqrt{2h_o/(k_f y)} \tag{4.47c}$$

The fin efficiency for a wetted fin (η_{wf}) is evaluated by (McQuiston 1975; McQuiston & Parker 1988):

$$\eta_{wf} = \tanh(m_{wet} r\phi)/(m_{wet} r\phi) \tag{4.48a}$$

$$m_{wet} = \sqrt{(2h_o/k_f y)[1 + (Ch_{fg}/Cp_a)]} \tag{4.48b}$$

where:

$\phi = [(R_e/r) - 1][1 + 0.35\ln(R_e/r)])$
$R_e/r = 1.28\psi\sqrt{\beta - 0.2}$ for rectangular tube array
$R_e/r = 1.27\psi\sqrt{\beta - 0.3}$ for triangular tube array
$\psi = M/r$
$\beta = L/M$

$M \& L$ are as defined in Fig. 4.12d.

The parameter C (called the condensing factor: Chen 1991; Mirth & Ramadhyani 1993b) in Equation 4.48b was defined by McQuiston (McQuiston 1975; McQuiston & Parker 1988) to be a coefficient that relates the moisture content and temperature differences between the air and the saturated air film at the coil surface:

$$C = (w_s - w_a)/(T_s - T_a) \tag{4.49}$$

The value of C was assumed to be constant in deriving Equation 4.48 but it will actually vary from plane to plane in a coil. The recommended method to determine m_{wet} (McQuiston 1975) is to use the average of the two values of C, C_1 and C_2, which are to be evaluated on the basis of the on-coil and leaving-coil plane conditions respectively. However, it was found that using simply the value of C_2 for C would make very little difference in the predicted coil performance but would significantly reduce the amount of computational effort. Hence, in the cooling coil model, C is calculated based on the exit plane conditions only.

Similar to η_s, η_{ws} is evaluated using Equation 4.47 except that η_f is replaced by η_{wf}. The value of η_{ws} so obtained is regarded as a representative value that would be applicable to the entire wet heat transfer surface.

Use of Webb's correlations to account for effects of fin corrugation

The following are Webb's correlations (Webb 1990).
   For flat fins:

$$Nu_a = 0.4Gz^{0.73}\left(\frac{s-y}{D_0}\right)^{-0.23} N_r^{0.23} \quad (Gz \leq 25) \tag{4.50a}$$

$$Nu_a = 0.53Gz^{0.62}\left(\frac{s-y}{D_0}\right)^{-0.23} N_r^{0.31} \quad (Gz > 25) \tag{4.50b}$$

For corrugated fins:

$$Nu_a = 0.5Gz^{0.86}\left(\frac{x_a}{D_0}\right)^{0.11}\left(\frac{s-y}{D_0}\right)^{-0.09}\left(\frac{P_d}{x_b}\right)^{0.12}\left(\frac{2x_f}{x_b}\right)^{-0.34} \quad (Gz \leq 25) \tag{4.51a}$$

$$Nu_a = 0.83Gz^{0.76}\left(\frac{x_a}{D_0}\right)^{0.13}\left(\frac{s-y}{D_0}\right)^{-0.16}\left(\frac{P_d}{x_b}\right)^{0.25}\left(\frac{2x_f}{x_b}\right)^{-0.43} \quad (Gz > 25) \tag{4.51b}$$

where:

$$Gz = \frac{16 \, Re_s \, Pr_a \, \sigma x_a (s-y)(x_a x_b - \pi D_o^2 / 4)}{N_r \left[2(x_a x_b - \pi D_o^2 / 4) \sec \theta + (s-y) \pi D_o \right]^2}$$

(4.52)

$$x_f = x_b / 2 N_p$$

(4.53)

$$\sec \theta = \sqrt{x_f^2 + P_d^2} \Big/ x_f$$

(4.54)

The air velocity and some other coil parameters in Webb's correlations were defined slightly differently from those adopted here but appropriate conversions have already been made such that a consistent set of variables and parameters is used here.

The Nusselt number in Webb's correlations are based on arithmetic mean temperature differences (Nu_a). This Nusselt number was used by Beecher & Fagan (1987) in developing heat transfer correlations from their experimental results so as to minimise the effects of experimental errors in the temperature measurement. Webb's correlations were derived on the same basis. Therefore, the Nusselt number (Nu_a) evaluated using Webb's correlations has to be converted to the conventional Nusselt number based on the log mean temperature difference (Nu_l). According to Beecher & Fagan, the conversion is:

$$Nu_l = 0.25 \, Gz \ln \left\{ \frac{1 + 2 Nu_a / Gz}{1 - 2 Nu_a / Gz} \right\}$$

(4.55)

Beecher & Fagan's experiment covered only a limited range of coil configurations. In particular, all the corrugated surfaces tested belong to coils with three tube rows ($N_r = 3$) and a fixed transverse to longitudinal tube spacing ratio ($x_a / x_b = 1.15$). Also, a uniform plate surface temperature was maintained during the experiment, which implies that the heat transfer rates measured are valid for fins with 100% fin efficiency only. Therefore, heat transfer coefficients cannot be determined solely on the basis of Webb's (nor Beecher & Fagan's) correlations.

For two coils of identical configuration, except that the fins of one are corrugated and the other flat, the heat transfer coefficients of the two coils can both be evaluated using Webb's correlations, provided the configuration is within the range covered by Beecher & Fagan's experiment. Then, the effect of the fin corrugation pattern can be quantified by the ratio of the two heat transfer coefficients:

$$\frac{h_{op}}{h_{of}} = \frac{Nu_{lp}}{Nu_{lf}} = \frac{\ln \{[1 + 2 Nu_{ap} / Gz] / [1 - 2 Nu_{ap} / Gz]\}}{\ln \{[1 + 2 Nu_{af} / Gz] / [1 - 2 / Gz]\}}$$

(4.56)

However, if the assumption is made that the effects on the Nusselt number, due to variations in the number of tube rows, the ratio of transverse to longitudinal tube spacing and fin efficiency, are the same for flat and corrugated fins, this ratio can then be used as a factor to 'correct' the air-side heat transfer coefficient estimated using other heat transfer correlations for flat plate fins (e.g. McQuiston's correlations) that are free from the restrictions to which Webb's correlations are subjected. The coil model described here was developed on this basis.

Because Beecher & Fagan's experiment was on dry coil surfaces only, whether the 'correction factor' so determined could be applied to the wet surfaces in cooling and dehumidify-ing coils was unknown. The method used to verify whether this approach would lead to im-proved coil performance predictions was to develop a coil model and compare the model predictions against a set of known coil performance data. Three versions of the coil model have been developed; one without applying this correction (i.e. a flat fin model with heat transfer coefficients determined solely on the basis of McQuiston's correlations); one with the correction applied to both the sensible and total heat transfer coefficients; and the last with the correction applied to the sensible heat transfer coefficient only.

From the comparison between the predicted results with and without accounting for the fin corrugation effects (Yik *et al.* 1997), it was found that the sensible heat transfer performance enhancement due to the use of corrugated fins is greater than the possible deviations, which shows the significance of this effect. However, no significant effect on the total heat transfer was detected. Comparison of the predictions from these model versions with manufacturer's data showed that Webb's correlations derived from Beecher & Fagan's experimental data on dry surfaces can be used to adjust the sensible heat transfer coefficient estimated from Mc-Quiston's correlations for flat-finned coils, in modelling performance of cooling coils with corrugated plate fins. However, no adjustment should be applied to the total heat transfer coefficient.

Solution procedures for coil surface conditions and heat transfer

Due to the use of the log mean difference method, solving the equations involved becomes an implicit problem. The values of $\Delta\theta_{dc}$, $\Delta\theta_{wc}$ and ΔH_{wc} (Equations 4.36 and 4.37) have to be de-termined from the leaving coil air and water conditions, but these are the unknowns yet to be solved. For a partly wet coil, the conditions at the dry/wet boundary in the coil must be known as well. Also, different air-side sensible and total heat transfer coefficients have to be used in calculating the sensible and total heat transfer rates for the dry and wet coil surfaces (see above). However, whether a coil will be completely dry, completely wet or partly wet, and, for a partly wet coil, the dry and wet fractions of the heat transfer area, are not known at the out-set. Furthermore, evaluating the wet surface effectiveness η_{ws} for a completely or partly wet coil requires the surface condition at the air exit plane to be determined first, but this is also dependent on the unknown leaving coil air conditions.

In the model, the successive back substitution method is adopted to determine the cooling and dehumidification performance of a chilled water coil (Fig. 4.13). At the start of the simu-lation process, estimated values for the unknown leaving coil air and water conditions (T_{a2}, w_{a2} and T_{w1}) were assigned. On the basis of the inputs and the estimated values, the heat trans-fer coefficients can be evaluated.

In each iteration loop, the specific enthalpy of the air at the dry/wet boundary of the coil (h_{ab}, Fig. 4.12a) would be calculated (using Equation 4.57a) and used to check whether a coil would be completely dry or wet, or partly wet. This equation was derived on the basis that at the dry/wet boundary, the coil surface temperature would be equal to the dew point tempera-ture of the on-coil air:

$$h_{ab} = \left(T_{a1}'' - T_{w2} + Y_r h_{a1} + C_c h_{a1}''\right)/\left(C_c + Y_r\right) \tag{4.57a}$$

where:

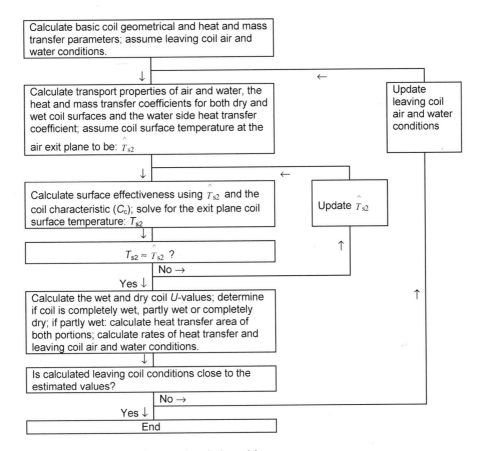

Fig. 4.13   Flowchart of calculation procedures in the model.

$$Y_r = \frac{m_a}{m_w Cp_w} = \frac{T_{w1} - T_{w2}}{h_{a1} - h_{a2}} = \frac{T_{wb} - T_{w2}}{h_{ab} - h_{a2}} \qquad (4.57b)$$

$$C_c = (T_s - T_w)/(h_a - h_s) \qquad (4.57c)$$

The parameter Y_r could be evaluated from the known mass flow rates of air and chilled water. Assuming that the thermal resistance of the tube and the fin collar is negligible, the contact resistance is accounted for by the air-side total heat transfer coefficient and, neglecting both the air and water side fouling resistances (the same assumptions were made in evaluating U_{dc}), the rate of heat transfer at an elemental section of the wet part of a coil can be described by:

$$\delta q_{wct} = \eta_{ws} h_d (h_a - h_s) \delta A_{ws} = h_i (A_i / A)(T_s - T_w) \delta A_{ws}$$

Comparing the above with Equation 4.57c shows that C_c may also be expressed with reference to the ratio of the heat transfer coefficients as follows:

$$C_c = \frac{T_s - T_w}{h_a - h_s} = \frac{\eta_{ws} h_d}{h_i (A_i / A)} \tag{4.58}$$

However, C_c cannot be evaluated until η_{ws} has been evaluated, which requires the wet coil surface temperature at the air exit plane to be determined first. Equation 4.58 therefore is a non-linear implicit equation of the unknown variable T_{s2}. In the model, T_{s2} is solved from Equation 4.58, in conjunction with Equations 4.47 to 4.49.

The following successive backward substitution scheme is adopted in the coil model to solve for T_{s2}:

(1)   An assumed value for T_{s2} (denoted by $\hat{T}_{s2}$) is first assigned. In the model, the temperature of the saturated air film above a wet coil surface is assumed to be equal to the coil surface temperature.

(2)   On the basis of the assumed coil surface temperature, the moisture content and the specific enthalpy of the saturated air film at the air exit plane can be determined using the following psychrometric relationships:

$$w_{s2} = w_{s2}\left(\hat{T}_{s2}\right) \tag{4.59}$$

$$h_{s2} = C_{p_d} \hat{T}_{s2} + w_{s2}\left(C_{p_s} \hat{T}_{s2} + h_{fg0}\right) \tag{4.60}$$

The relation between the moisture content and temperature of saturated air (Equation 4.59) is approximated by the following third order polynomial (Wang 1987):

$$w_{sat}(T) = a_0 + a_1 T + a_2 T^2 + a_3 T^3 \, (\text{kg/kg-dry air}; \, T \text{ in } ^\circ\text{C}) \tag{4.61}$$

where:

$a_0 = 3.7658 \times 10^{-3}$
$a_1 = 3.0517 \times 10^{-4}$
$a_2 = 4.648 \times 10^{-6}$
$a_3 = 3.787 \times 10^{-7}$

(3)   Based on the assumed leaving coil air conditions and the coil surface moisture content calculated, the value of C (Equation 4.49, based on the air exit plane), and hence η_{ws} (Equation 4.48) and then C_c (using the second expression in Equation 4.58) are evaluated.

(4)   Equation 4.58 can be re-arranged to become:

$$T_{s2} = T_{w2} + C_c (h_{a2} - h_{s2}) \tag{4.62}$$

Substituting Equation 4.59 (in the form of Equation 4.61) into 4.60, using T_{s2} as the independent variable and substituting the resultant expression for h_{s2} into Equation 4.62 yields a fourth order polynomial (as shown in Equation 4.63) from which T_{s2} can be solved by using the Newton–Raphson method:

$$b_0 + b_1 T_{s2} + b_2 T_{s2}^2 + b_3 T_{s2}^3 + b_4 T_{s2}^4 = 0 \tag{4.63}$$

(5)   If the absolute deviation ($|\varepsilon|$, Equation 4.64) between the assumed value ($\hat{T}_{s2}$) and the calculated value of T_{s2} (solved from Equation 4.63) is greater than a predetermined tolerance limit (ε_{tol}; Equation 4.65), an improved estimate for $\hat{T}_{s2}$ will be calculated (Equation 4.66). Procedures (2)–(4) will then be repeated starting from this improved estimate. This iterative calculation procedure will continue until $|\varepsilon|$ becomes smaller than the tolerance limit:

$$|\varepsilon| = \text{ABS}(T_{s2} - \hat{T}_{s2}) \tag{4.64}$$

$$\varepsilon_{tol} = 0.5(\varepsilon_r T_{s2} + \varepsilon_a) \tag{4.65}$$

$$\hat{T}_{s2}^{n+1} = \lambda T_{s2} + (1-\lambda)\hat{T}_{s2}^n \tag{4.66}$$

The weighting factors (λ and $1 - \lambda$) in Equation 4.66 for determining the value of $\hat{T}_{s2}$ for use in the $(n+1)^{th}$ round of iteration are both equal to 0.5. This under relaxation method was found necessary to avoid instability in the numerical solution scheme.

Having evaluated T_{s2}, C_c (while solving for T_{s2}) and Y_r, h_{ab} can be evaluated from Equation 4.57a and compared with the specific enthalpies of the on-coil air (h_{a1}) and the leaving-coil air (h_{a2}) (value assumed); the coil would be (ASHRAE 2000):

Partly wet if: $h_{a1} > h_{ab} > h_{a2}$
Completely wet if: $h_{ab} \geq h_{a1}$
Completely dry if: $h_{a2} \geq h_{ab}$

If the coil is found to be partly wet, the rate of sensible heat exchange between the air and the chilled water in the dry coil part can be determined using Equation 4.38a as follows:

$$q_{dc} = m_a C_{p_a}(T_{a1} - T_{ab}) \tag{4.67}$$

where T_{ab} can be determined from h_{ab}, using Equation 4.38a:

$$T_{ab} = T_{a1} - (h_{a1} - h_{ab})/C_{p_a} \tag{4.68}$$

Through Equation 4.34, A_{dc} and then A_{wc} can be determined:

$$A_{dc} = q_{dc}/(U_{dc}\Delta\theta_{dc}) \tag{4.69}$$

$$A_{wc} = A - A_{dc} \tag{4.70}$$

At this stage, all the variables required in calculating $\Delta\theta_{dc}$, $\Delta\theta_{wc}$ and ΔH_{wc} will have been evaluated or assigned with estimated values except T_{sb}, h_{sb} and T_{wb}. However, T_{sb} and h_{sb} can be determined from the known on-coil air condition because, at the dry/wet boundary,

$T_{sb} = T_{a1}^n$ and $h_{sb} = h_{a1}^n$ (Fig. 4.12a). Assuming that C_c determined based on the air exit plane will be applicable to the entire wet coil surface, Equation 4.58 can be applied to the dry/wet boundary, from which T_{wb} can be determined as follows:

$$T_{wb} = T_{sb} - C_c (h_{ab} - h_{sb}) \qquad (4.71)$$

Thus, the sensible and total heat transfer at the wet coil part can be determined using Equations 4.35 and 4.37. Having determined q_{dc}, q_{wcs} and q_{wct}, and thus q_c, the leaving coil air and water conditions (T_{a2}, w_{a2} and T_{w1}) can be calculated from Equation 4.38 and relevant psychrometric relationships. These values can then be compared with the assumed values. If the difference between the calculated and the estimated values for any one of the variables exceeds the respective pre-assigned tolerance limit (the convergence criterion for that variable), the weighted sum of the two values for each unknown variable can be taken as the improved estimate and used in the next round of calculation (similar to Equation 4.66), i.e. under-relaxation is used. The calculation procedures were repeated until the differences between the calculated and the estimated values became smaller than the convergence criteria.

The convergence criteria adopted include the maximum relative (ε_r) and absolute (ε_a) deviations, as follows:

For temperatures, $\varepsilon_r = 0.01$; $\varepsilon_a = 0.05\,^\circ C$
For moisture contents, $\varepsilon_r = 0.01$; $\varepsilon_a = 0.0001$ kg/kg

Similar calculation procedures, but involving just the equations related to a dry or wet surface, are adopted for modelling a completely dry or completely wet coil.

4.4 Modelling distribution networks

Water is an ideal medium for transport of heat energy because it has high specific heat capacity and low viscosity, is highly incompressible, and is generally abundantly available at low cost. Using water as a heat transport medium allows bigger (in capacity) and more efficient equipment to be used to centrally produce or reject the heat associated with provision of heating or cooling for different parts of a building, or to a group of buildings (e.g. in district heating or cooling systems). Its use lessens the restriction on layout arrangement of equipment and thus provides flexibility in design. This explains why most medium to large size buildings are equipped with centralised HVAC plants, with chilled or hot water distributed by pumps, via a piping system, to air-handling units at various locations in the building.

In an air-conditioning system, air is cooled and dehumidified by cooling coils through which the heat extracted from the air is transferred to the chilled water, which will carry the heat to the chillers when it is circulated back to the chiller plant through the piping system. The cooling output of the air-handling equipment is controlled by regulating the rate of chilled water flowing through its cooling coil by using a control valve. Therefore, in air-conditioning system performance studies, it is often necessary to predict water flow rates in a piping network that connects the central plant and the air-handling equipment.

In this section, the Hardy–Cross method for predicting steady-state water flow rate in a piping network is introduced. Being basically also a network comprising a set of flow resistances, the airflow rates in a ducting system can also be modelled using the same method.

Mathematical representation of elements in a network

A piping system comprises a collection of inter-connected elements that form a network of flow paths. Elements in a pipe network may include straight pipes and pipe fittings, such as bends, 'T' and 'Y' connections for splitting or merging flows to or from pipe branches, various types of valves for flow isolation or regulation, and components of equipment connected to it (e.g. chiller evaporators, cooling coils, etc.). A piping system often includes one or more pump(s) to drive the fluid to flow round the network.

As already discussed in Section 4.2, for a given pump with known characteristics, the pumping pressure can be related to the flow rate it delivers by a curve-fit model as follows:

$$\Delta p_{pump} = a_0 + a_1 Q + a_2 Q^2 + \ldots + a_m Q^m \tag{4.72}$$

Apart from pumps, every other element in a pipe network may be regarded as a flow resistance. The flow rate through a flow resistance (Q) can be related to the pressure drop across it (Δp) by the following expression:

$$\Delta p = KQ^n \tag{4.73}$$

where K, the flow resistance, and n, the flow exponent, are parameters that quantify the pressure drop characteristics of the element. The flow exponent n equals 1 when the flow is laminar, 2 when the flow is fully turbulent, and between 1 and 2 for transition flow conditions. Most elements have a constant K value but, valves, control valves in particular, have K values that depend on the degree of opening.

Flow resistances in series and parallel

The flow and pressure drop characteristics for elements connected in series or in parallel can be condensed into a single equivalent model as follows:

(1)   For N elements in series (Fig. 4.14a) flow rates through the elements are identical and the total pressure drop equals the sum of pressure drops through the elements (Equation 4.74):

$$\Delta p = \sum_{i=1}^{N} \Delta p_i \tag{4.74}$$

If all the elements have the same flow exponent n, the entire circuit may be represented by Equation 4.75 with an equivalent K given by:

$$K = \left(\sum_{i=1}^{N} K_i \right) \tag{4.75}$$

(2)   For N elements in parallel (Fig. 4.14b), all with the same flow exponent n, the total flow rate equals the sum of flow rates through the elements and the pressure drops through all the elements are identical. Therefore:

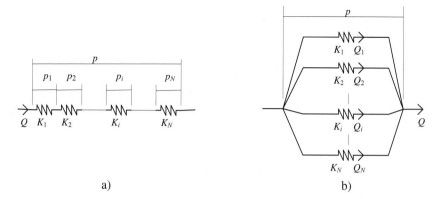

Fig. 4.14   Flow resistances: (a) in series; (b) in parallel.

$$Q = \sum_{i=1}^{N} Q_i \tag{4.76}$$

$$\Delta p = \Delta p_i \quad \text{for } i = 1, 2, 3, \ldots, N \tag{4.77}$$

The equivalent K value for a combined model that can represent the overall characteristics of N parallel elements can be determined from:

$$K^{-1/n} = \left(\sum_{i=1}^{N} K_i^{-1/n} \right) \tag{4.78}$$

Flow resistances in a piping system of relatively simple configuration can be condensed into an equivalent single element model by repeatedly applying the above series and parallel models to appropriate sections of the system. The total system flow rate can be determined when the combined model is coupled to a pump model. The flow rates through individual elements can then be determined by re-expanding the model. However, any changes to the elements in the pipe network will necessitate the derivation of a new model. Certain piping network design would also preclude the use of this simplistic method, e.g. a chilled water supply system with reverse return, as shown in Fig. 4.15.

For modelling piping systems with complex interconnections, more systematic and generally applicable methods are required. Such methods will typically involve solving a set of simultaneous equations, which are mathematical statements of pressure and flow balances for the flow-loop or nodes in the network.

The Hardy–Cross method

The Hardy–Cross method, developed in the 1930s, is extensively applied to analyse fluid flows in pipe networks. This is an iterative method, which is flexible and simple to implement and may even be done by hand calculations. In using the Hardy–Cross method, values of the flow characteristics (K's and n's) of all elements in the pipe network must be known and flow rates through the elements must each be assigned with an initial value (Q_{oi}). It is essential to ensure that the initial flow rates assigned satisfy continuity. This requires the sum of flow rates

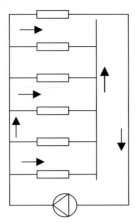

Fig. 4.15   A reverse return chilled water distribution system.

toward a connection point (a node) to be equal to the sum of flow rates leaving the same point.

For the ith element, the flow rate through which is initially assumed to be Q_{oi} whereas ΔQ is the change in flow rate required to bring the estimate to the true value Q_i, then:

$$Q_i = Q_{oi} + \Delta Q \tag{4.79}$$

Using Equation 4.73, the pressure drop is related to the flow rate as:

$$\Delta p_i = K_i (Q_{oi} + \Delta Q)^n \tag{4.80}$$

Expanding the right hand side of Equation 4.80 using a Taylor series expansion, we get:

$$\Delta p_i = K_i \{Q_{oi}^n + n Q_{oi}^{n-1} \Delta Q + [n(n-1)/2!] Q_{oi}^{n-2} \Delta Q^2 + \ldots\}$$

If ΔQ is small compared with Q_i, those terms within the braces on the right hand side of the above equation that include ΔQ terms of an order of 2 or higher may be dropped. Thus:

$$\Delta p_i = K_i (Q_{oi}^n + n Q_{oi}^{n-1} \Delta Q) \tag{4.81}$$

For a loop formed by several elements, such as the one shown in Fig. 4.16, flow in the clockwise direction is taken as positive, and anti-clockwise negative, while the 'pressure drop' incurred by a positive flow will be positive and that by a negative flow negative. Based on this convention, the algebraic sum of pressure drops through all the elements round the loop (from point a to b, c, d and back to a in Fig. 4.16) equals zero.

Hence, by applying Equation 4.81 to each element and summing up the pressure drops, we can write:

$$\sum_i \Delta p_i = 0 = \sum_i K_i (Q_{oi}^n + n Q_{oi}^{n-1} \Delta Q_i)$$

To cope with the situation that flows through elements in a loop may either be positive or negative flows, the above equation is rewritten as:

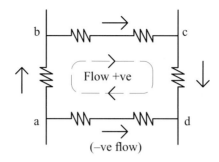

Fig. 4.16   Flow in a loop.

$$\sum_i K_i Q_{oi} |Q_{oi}|^{n-1} + \sum_i K_i n \Delta Q_i |Q_{oi}|^{n-1} = 0 \tag{4.82}$$

Accordingly, the required correction ΔQ to the estimated flow rates can be determined from:

$$\Delta Q = -\frac{\sum_i K_i Q_{oi} |Q_{oi}|^{n-1}}{\sum_i K_i n |Q_{oi}|^{n-1}} \tag{4.83}$$

The value of ΔQ obtained can then be applied to each element in the loop (using Equation 4.79) to correct the initial estimates and thus yield better estimates of the flow rates. Note that to maintain continuity of flow, whenever the flow rate of any element in a given loop is adjusted, the same adjustment must be applied to all other elements in the same loop (i.e. by the same ΔQ for all elements in the loop). Adjacent loops with elements in common with the loop will be affected and thus the effect of this change will propagate to other loops. By repeating this loop flow rate adjustment process, from one loop to another in sequence, flow rates through elements in the entire network can be updated.

Why the above convention needs to be used will become evident by considering a piping network as shown in Box 4.3.

Box 4.3 Flow and pressure drop round a loop

The pressure difference between nodes A and B in the figure on the next page are related to the pressure drops through the elements as follows:

$$\Delta P_{AB} = P_A - P_B = \Delta P_1 + \Delta P_2 + \Delta P_3 = \Delta P_4 \tag{i}$$

Assuming the flow exponents (n) of all elements are the same and equal to 2:

$$\Delta P_i = K_i Q^2 \tag{ii}$$

$$K_1 Q_1^2 + K_2 Q_2^2 + K_3 Q_3^2 = K_4 Q_4^2 \tag{iii}$$

Equation iii holds only if the flow rates are the correct flow rates.

Box 4.3 *Continued*

   If the estimated flow rates through the elements are $\hat{Q}_i$, for $i =$ 1, 2, 3 & 4, which may not be the correct flow rates, substituting these estimated flow rates into Equation iii will not lead to equality of the left and right hand sides of the equation. How-ever, the need to ensure mass conservation in the loop requires that, in making adjustments to the estimated flow rates to bring their values to the correct values, the same increment must be made to each flow rate in the same direction whilst flow rates in

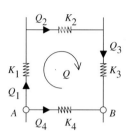

the opposite direction must each be reduced by the same amount, and vice versa. Such a correction would not upset the flow rate through any branch in the network outside the loop under concern.

   Assume that $\hat{Q}_i$, for $i = 1$, 2 and 3, each needs to be increased by ΔQ but $\hat{Q}_4$ needs to be reduced by ΔQ in order to satisfy Equation iii, then:

$$Q_i = \hat{Q}_i + \Delta Q \quad \text{for } i = 1, 2 \ \& \ 3 \tag{iv}$$

$$Q_4 = \hat{Q}_4 - \Delta Q \tag{v}$$

The squares of the flow rates are:

$$Q_i^2 = \left(\hat{Q}_i + \Delta Q\right)^2 = \hat{Q}_i^2 + 2\hat{Q}_i \Delta Q + \Delta Q^2 \quad \text{for } i = 1, 2 \text{ and } 3 \tag{vi}$$

$$Q_4^2 = \left(\hat{Q}_4 - \Delta Q\right)^2 = \hat{Q}_4^2 - 2\hat{Q}_4 \Delta Q + \Delta Q^2 \tag{vii}$$

Neglecting the ΔQ^2 terms in Equations vi and vii and substituting them into iii:

$$\sum_{i=1}^{3} K_i \hat{Q}_i^2 + 2\Delta Q \sum_{i=1}^{3} K_i \hat{Q}_i = K_4 \hat{Q}_4^2 - 2\Delta Q K_4 \hat{Q}_4$$

It follows that:

$$\sum_{i=1}^{3} K_i \hat{Q}_i^2 + 2\Delta Q \sum_{i=1}^{3} K_i \hat{Q}_i - K_4 \hat{Q}_4^2 + 2\Delta Q K_4 \hat{Q}_4 = 0 \tag{viii}$$

   The above shows that it is important to observe the direction of flow through each ele-ment in the loop such that the correct pressure drops can be determined. A specific con-vention for describing the flow rates can be adopted to make the calculation procedures more systematic and to help minimise mistakes. In this convention, the estimated flow rate for an element is quantified by its absolute magnitude in conjunction with a variable *Sign*, as follows:

$$\tilde{Q} = Sign \cdot \hat{Q} \tag{ix}$$

where:

Box 4.3 *Continued*

$\hat{Q}_i =$ the magnitude of the estimated flow rate, which will always be a positive value irrespective of the direction of the flow

and

Sign = +1 if the flow rate through the element is along the positive loop flow direction (applies to element 1, 2, and 3)

Sign = −1 if the flow rate through the element is opposite to the positive loop flow direction (applies to element 4)

It follows that:

$$K_i \hat{Q}_i^2 = K_i \tilde{Q}_i |\tilde{Q}_i| \quad \text{for } i = 1, 2 \text{ and } 3 \tag{x}$$

and:

$$-K_4 \hat{Q}_4^2 = K_4 \tilde{Q}_4 |\tilde{Q}_4| \tag{xi}$$

which can now be generalised into:

$$K_i \hat{Q}_i^2 = K_i \tilde{Q}_i |\tilde{Q}_i| \quad \text{for } i = 1, 2, 3 \text{ and } 4 \tag{xii}$$

Substituting Equations x to xi into viii, we get:

$$\sum_{i=1}^{4} K_i \tilde{Q}_i |\tilde{Q}_i| + 2\Delta Q \sum_{i=1}^{4} K_i |\tilde{Q}_i| = 0 \tag{xiii}$$

Hence:

$$\Delta Q = -\frac{\sum_{i=1}^{4} K_i Q_i |Q_i|}{2 \sum_{i=1}^{4} K_i |\tilde{Q}_i|} \tag{xiv}$$

It can now be seen that Equation xiv is just a special version of Equation 4.83. The latter is more general, which can cover elements with flow exponents that are different from 2 and there can be any number of elements in the loop.

The procedures involved in using the Hardy – Cross method for solving flow rates of elements forming a network with multiple loops can be summarised as follows:

(1)   Estimate initial flow rates for all elements that satisfy the continuity requirement.
(2)   Calculate ΔQ using Equation 4.83 for each loop in the network.

(3)   Update estimates of flow rate for all elements using Equation 4.79. Note that for an
       element that belongs to two adjacent loops, flow corrections for both loops apply to the
       element.
(4)   Check if ΔQs for all loops are smaller than the predetermined deviation limit.
(5)   Repeat steps (2) to (4) if deviation limit is exceeded in any loop.
(6)   When converged solutions have been obtained, pressure drops through individual ele-
       ments can be determined using Equation 4.80.

Loop with a pump element

For a loop that includes a pump (Fig. 4.17), the algebraic sum of pressure drops across all elements except the pump will be equal to the pressure rise across the pump. Thus, Equation 4.82
is modified as:

$$\sum_i K_i Q_{oi} |Q_{oi}|^{n-1} + \Delta Q_i \sum_i K_i n |Q_{oi}|^{n-1} - \Delta p_{pump} = 0 \tag{4.84}$$

for $i = 1$ to the total number of elements in the loop but excluding the pump.

   If the pump characteristics can be modelled by a cubic equation as shown in Equation 4.85
and an initial estimate Q_{opump} for the pump flow rate has been made, which needs to be corrected to yield the true pump flow rate as shown in Equation 4.86, substituting Equation 4.86
into 4.85 and neglecting terms containing ΔQ^2 or ΔQ^3 yields Equation 4.87:

$$\Delta p_{pump} = a_0 + a_1 Q + a_2 Q^2 + a_3 Q^3 \tag{4.85}$$

$$Q_{pump} = Q_{opump} + \Delta Q \tag{4.86}$$

$$\Delta p_{pump} = a_0 + a_1 Q_{opump} + a_2 Q_{opump}^2 + a_3 Q_{opump}^3$$
$$+ \left(a_1 + 2a_2 Q_{opump} + 3a_3 Q_{opump}^2\right) \Delta Q \tag{4.87}$$

Substituting Equation 4.87 into 4.84 and solving for ΔQ yields:

$$\Delta Q = \frac{\left[Sign(Q_{opump})\right]\left(a_0 + a_1 |Q_{opump}| + a_2 Q_{opump}^2 + a_3 |Q_{opump}|^3\right) - \sum K_i Q_{oi} |Q_{oi}|^{n-1}}{\sum K_i n |Q_{oi}|^{n-1} - \left(a_1 + 2a_2 |Q_{opump}| + 3a_3 Q_{opump}^2\right)} \tag{4.88}$$

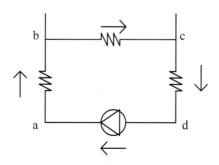

Fig. 4.17   Flow in a loop with a pump.

Note that the term $Sign(Q_{opump})$ and the absolute value sign for some terms of Q_{opump} have been included in Equation 4.88 to take account of direction of pump flow, where:

$Sign(Q_{opump}) = +1$ if Q_{opump} is along the positive loop-flow direction (clockwise in this example).

$Sign(Q_{opump}) = -1$ if Q_{opump} is opposite to the loop-flow direction (anti-clockwise in this example).

Network with a constant pressure branch

For the case where the pressure difference across two nodes in the loop is always maintained at a constant level (Fig. 4.18), the method to update the flow rate estimate for elements in the loop is similar to that for the case with a pump element except that a constant term rather than a pump pressure-flow equation is used to represent the constant pressure element (which may or may not physically exist).

The pressure balance equation for such a loop becomes:

$$\sum_i K_i Q_{oi} |Q_{oi}|^{n-1} + \Delta Q_i \sum_i K_i n |Q_{oi}|^{n-1} - \Delta p_f = 0 \tag{4.89}$$

and the equation for updating the flow rate estimates is:

$$\Delta Q = \frac{\Delta p_f - \Sigma k_i Q_{oi} |Q_{oi}|^{n-1}}{\Sigma K_i n |Q_{oi}|^{n-1}} \tag{4.90}$$

Here, the sign of Δp_f should be positive if pressure increases in the positive loop-flow direction, i.e. clockwise for the condition shown in Fig. 4.18, and negative otherwise.

Note that if the constant pressure difference is the result of the working of a control system, e.g. in a variable speed pumping system that has the pump speed regulated such that the differential pressure across the index branch is maintained at a constant level, it is not advisable to apply the above method to the loop that contains the constant differential pressure branch. Doing so would require flow rate corrections to be made to all affected loops to ensure no violation of continuity. Rather, the flow rate through the index branch can be determined in the normal way but the differential pressure control across the index branch can be determined

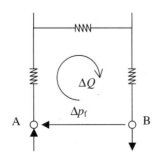

Fig. 4.18   Flow in a loop with a constant pressure difference across two nodes.

after the flow rate correction, and then used to update the pump speed before progressing to the next round of the iterative calculation.

4.5 Modelling air-conditioning systems

Before concluding this chapter, the modelling methods adopted in an air-conditioning plant simulation program (Yik & Burnett 1998; Yik & Chan 1998) are briefly described to illustrate how component models can be integrated to model the performance of an air-conditioning system. The mathematical expressions used in the program to model the performance of equipment like chillers, pumps, cooling towers, heat exchangers, air-handling units, etc. are as summarised in Table 4.4.

The program includes a library of models for typical air-conditioning equipment, which is basically a database of the model coefficients, but the user may modify the coefficients to more accurately represent the performance of particular equipment. The built-in models were established by regression methods on the basis of manufacturers' performance data. In order to extend the applicability of a model to cover similar equipment of different capacities, equipment performance data were normalised into fractional values of the rated conditions, before regression methods were applied to evaluate the coefficients.

The procedures adopted in the program to calculate the electricity consumption of an air-conditioning system are as follows.

Air-side equipment power consumption

For constant air volume (CAV) systems or fan-coil systems, the air supply flow rate and the fan power demand will be fixed at their respective design values irrespective of the space cooling load. Fan energy use of such systems therefore will be calculated simply by integrating the rated power over the total hours of operation of each fan.

For a variable air volume (VAV) system, the air supply rate from the air-handling unit (AHU) is proportional to the instantaneous sensible cooling load of the air-conditioned zones. Calculation of the total supply flow rate from the AHU will take into account whether or not the space sensible cooling load in an individual zone has dropped below the level where the VAV boxes will maintain the supply flow rate at the pre-set minimum level. Under such conditions, the minimum supply flow rate will be taken as the supply flow rate for the zone concerned. The supply flow rate demand of all zones will be summed to yield the total flow rate to be delivered by the AHU.

Besides the supply flow rate, the fan power demand of a VAV AHU is dependent on the fan inlet guide vane position or running speed of the fan (Table 4.4), as appropriate to the flow rate regulation method adopted. Rather than using a detailed duct network model, a fan modulation curve, which represents the locus of the fan operating point (the fan total pressure at a particular supply flow rate) as it varies under the control of the duct static pressure control system (Fig. 4.19), is used to model approximately the combined characteristics of the fan, the supply ductwork and the control system. On the basis of this model and the calculated supply flow rate, the required fan total pressure can be determined. The inlet guide vane position or fan speed can then be solved from the fan pressure characteristic curve in the fan model ($P(V_s, \phi)$).

Table 4.4 Component models in an air-conditioning plant simulation program.

Model	Math. Mod.*	Dependent variable (z)	Independent variables (x, y or θ)	Remarks
Chiller (model 1)	Eq. T1	z: fractional power consumption	x: fractional cooling load	Effects of condenser air/water inlet temp. variations ignored or included in the correlation.
Chiller (model 2)	Eq. T2	z: fractional power consumption	x: fractional cooling load; y: condenser air/water inlet temp.	
Constant speed pump	Eq. T1	z: fractional power consumption	x: fractional flowrate	
Variable speed pump (model 1)	Eq. T1	z: fractional pump total pressure	x: fractional water flowrate	Pump speed to be solved using pump affinity laws.
	Eq. T1	z: fractional power consumption	x: fractional water flowrate	
Variable speed pump (model 2)	Eq. T2	z: fractional pump total pressure	x: fractional water flowrate; y: fractional pump rotational speed	
	Eq. T2	z: fractional power consumption	x: fractional water flowrate; y: fractional pump rotational speed	
Pump modulation curve	Eq. T1	z: fractional system total pressure	x: fractional system flowrate	
Seawater temperature	Eq. T1	z: seawater temperature	x: day number in the year (1 to 365)	
Cooling tower	Eq. T2	z: leaving tower water temperature	x: ambient air wet-bulb temperature; y: cooling tower range	
Heat exchanger	Eq. T3	z: heat transfer rate	θ: log mean temperature difference	A and U evaluated based on design conditions and are assumed to be constants under part load conditions.
Constant speed fan	Eq. T1	z: fractional power consumption	x: fractional air flow rate	
Constant speed variable flow fan with inlet guide vane control	Eq. T2	z: fractional fan total pressure	x: fractional air flow rate; y: fractional inlet guide vane position	
	Eq. T2	z: fractional lower consumption	x: fractional air flow rate; y: fractional inlet guide vane position	
Variable speed fan (model 1)	Eq. T2	z: fractional fan total pressure	x: fractional air flow rate; y: fractional fan rotational speed	Fan speed to be solved using fan similarity laws.
		z: fractional lower consumption	x: fractional air flow rate; y: fractional fan rotational speed	
Variable speed fan (model 2)	Eq. T1	z: fractional fan total pressure	x: fractional air flow rate	
	Eq. T1	z: fractional lower consumption	x: fractional air flow rate	
Cooling coil	Eq. T1	z: fractional chilled water flow rate	x: fractional coil total load	
Fan modulation curve	Eq. T1	z: fractional system total pressure	x: fractional system total flow rate	

*Model equations: $z = a_0 + a_1 x + a_2 x^2 + \ldots + a_n x^n$; n to be specified in input file (T1)

$z = c_0 + c_1 x + c_2 x^2 + c_3 y + c_4 y^2 + c_5 xy + c_6 x^2 y + c_7 xy^2 + c_8 x^2 y^2$ (T2)

$z = A U \theta$, where A = heat transfer area; U = overall heat transfer coefficient and θ = log mean temperature difference. (T3)

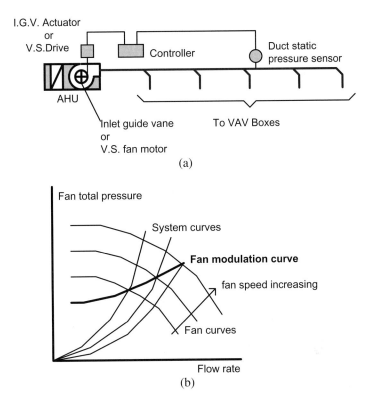

I.G.V. Actuator
or
V.S.Drive

Controller

Duct static
pressure sensor

AHU

Inlet guide vane
or
V.S. fan motor

To VAV Boxes

(a)

Fan total pressure

System curves

Fan modulation curve

fan speed increasing

Fan curves

Flow rate

(b)

Fig. 4.19   Duct static pressure control and fan modulation curve of a VAV system.

This will allow the fan power to be calculated using the fan power characteristic curve in the fan model ($W_{fan}(V_s, \phi)$). Figure 4.20 shows the calculation procedures involved.

The plant simulation program includes the use of fan similarity laws as an optional variable speed fan model for predicting part-load fan power of VAV AHUs. If this simplified method is chosen, the fan model required will simply be the equations relating the fan total pressure and the power to the flow rate under the rated speed. The fan speed at a reduced supply flow rate will be determined on the basis of the fan modulation curve and the fan laws.

The required chilled water flow rate through the cooling coil in an AHU or a fan coil unit is calculated using a simplified regression model, which expresses the chilled water flow rate as a function of the total coil load from all the zones served by the AHU.

The total cooling load due to treatment of the fresh air supply for ventilation, and the corresponding requirement on fan power and chilled water supply to the coil, are calculated on the basis of the given fresh air supply requirement and the concurrent outdoor weather conditions.

Central chiller plant power consumption

The space total cooling load for individual zones in the building, the fresh air load and the chilled water demand from individual air-handling equipment will be summed up to yield the total cooling load on the chiller plant and the total flow rate to be delivered by the pumps. The

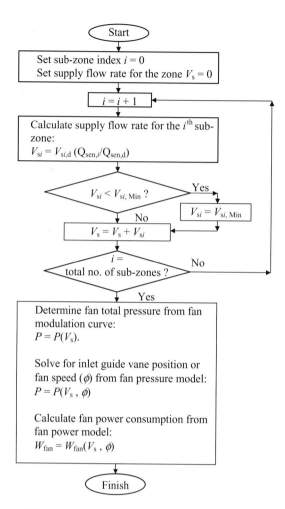

Fig. 4.20   VAV system modelling procedures.

chiller and pump power will be determined from their respective models. Similar to VAV AHUs, the power demand of variable speed, variable flow pumps will be predicted through the use of a pump modulation curve, in conjunction with the pump model. Also, the use of pump similarity laws for predicting the energy consumption of variable speed pumps is included as a pump model option.

Where the chiller model with variable condenser water inlet temperature is selected, the cooling tower model, the heat exchanger model and the seawater temperature model will be used to determine the inlet water temperature at the chiller condensers (Fig. 4.21) before chiller power consumption is predicted (Fig. 4.22).

An option has been included in the program to model the effect of incorporating an optimised sequencing control strategy for the chiller plant. When this option is selected, the

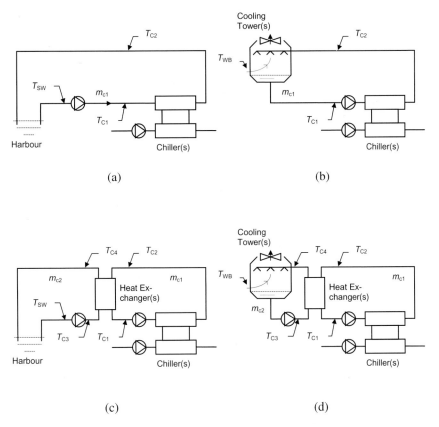

Fig. 4.21 Schematic diagrams of chiller plants.(a) Direct seawater-cooled. (b) Water-cooled using a cooling tower.(c) Indirect seawater-cooled. (d) Indirect water-cooled using a cooling tower.

program will search for the optimal combination of equipment that will lead to the lowest energy consumption to be put into operation to cope with a particular cooling load.

The total power demand for various types of equipment and for all equipment in the building is then summed up and the calculation will progress for another hour until calculation for the entire simulation period is finished.

The variables in Figs. 4.21 and 4.22 are:

T_{C1} to T_{C4}:	Condenser water temperatures (see Fig. 4.21)
m_{c1} and m_{c2}:	Condenser water/seawater mass flow rates (see Fig. 4.21)
T_{SW}:	Seawater temperature
T_{Amb} and T_{WB}:	Ambient air dry-bulb and wet-bulb temperatures
q_E, q_C and W_C:	Cooling output, heat rejection and power demand of chiller
R_G:	Cooling tower range
A_{HX}, U_{HX} and θ:	Heat transfer area, overall heat transfer coefficient and LMTD of heat exchanger

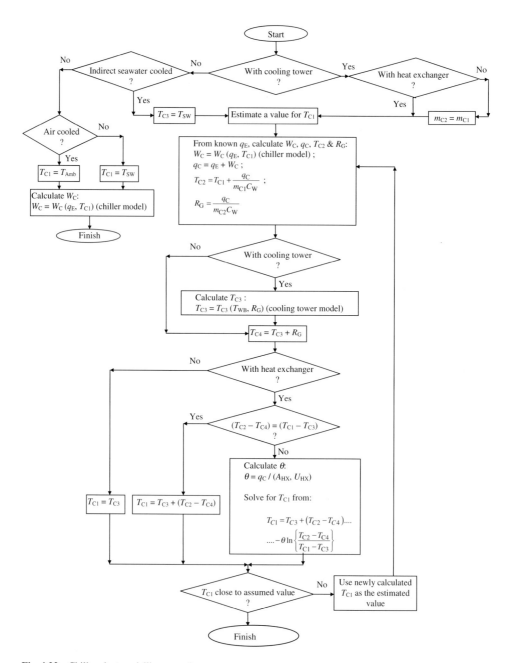

Fig. 4.22   Chiller plant modelling procedures.

References

ARI (1991) *Standard 410 – Standard for Forced-circulation air-cooling and air-heating coils.* Air-conditioning and Refrigeration Institute, Arlington, VA.

ARI (1992) *ARI Standard 550–92 – Centrifugal and Rotary Screw Water-Chilling Packages.* Air-conditioning and Refrigeration Institute, Arlington, VA.

ASHRAE (2000) *ASHRAE HVAC Systems and Equipment Handbook.* American Society of Heating, Refrigerating and Air-conditioning Engineers, Inc., Atlanta, GA.

Austin, S.B. (1991) Optimum chiller loading. *ASHRAE Journal* 33(7) pp.40–3.

Beecher, D.T. & Fagan, T.J. (1987) Effects of fin pattern on the air-side heat transfer coefficient in plate finned-tube heat exchangers. *ASHRAE Transactions* 93(2) pp.1961–84.

Chen, L.T. (1991) Two-dimensional fin efficiency with combined heat and mass transfer between water-wetted fin surface and moving moist air stream. *International Journal Heat Fluid Flow* 12(1) pp.71–6.

Khoo, I., Levermore, G.J. & Letherman, K.M. (1998) Variable-air-volume terminal units I: Steady-state models. *Building Services Engineering Research & Technology* 19(3) pp.155–62.

McQuiston, F.C. (1975) Fin efficiency with combined heat and mass transfer. *ASHRAE Transactions* 81(1) pp.350–5.

McQuiston, F.C. (1978) Correlation of heat, mass and momentum transport coefficients for plate-fin-tube heat transfer surfaces. *ASHRAE Transactions* 84(1) pp.294–309.

McQuiston, F.C. & Parker, J.D. (1988) *Heating, Ventilating, and Air-conditioning Analysis and Design.* John Wiley, New York.

Mirth, D.R. & Ramadhyani, S. (1993a) Comparison of methods of modelling the air-side heat and mass transfer in chilled-water cooling coils. *ASHRAE Transactions* 99(2) pp.285–9.

Mirth, D.R. & Ramadhyani, S. (1993b) Prediction of cooling-coil performance under condensing conditions. *International Journal Heat Fluid Flow* 14(4) pp.391–400.

Mirth, D.R. & Ramadhyani, S. (1994) Correlations for predicting the air-side Nusselt numbers and friction factors in chilled-water cooling coils. *Experimental Heat Transfer* 7(2) pp.143–62.

Rich, D.G. (1973) The effect of fin spacing on the heat transfer and friction performance of multi-row, smooth plate and tube heat exchangers. *ASHRAE Transactions* 79(2) pp.37–45.

Stevens, R.A., Fernandez, J. & Woolf, J. (1976) Mean temperature difference in one-, two- and three-pass crossflow heat exchangers. *ASME Transactions* 79(3) pp.287–97.

Wang, S.K. (1987) *Air-conditioning.* Hong Kong Polytechnic, Hong Kong.

Webb, R.L. (1990) Air-side heat transfer correlations for flat and wavy plate fin-and-tube geometries. *ASHRAE Transactions* 96(2) pp.445–9.

Welty, J.R. (1978) *Engineering Heat Transfer.* John Wiley, New York.

Yik, F.W.H. & Burnett, J. (1995) An experience of energy auditing on a central air-conditioning plant in Hong Kong. *Journal of Energy Engineering* 92(2) 6–30.

Yik, F.W.H. & Burnett, J. (1998) BECON: A program for predicting building energy consumption for air-conditioning buildings. *Transactions HKIE* 5(3) pp.89–94.

Yik, F.W.H. & Chan, T.K. (1998) Air-conditioning system modelling in BECON. *Proceedings, CIBSE National Conference, Bournemouth* pp.40–9.

Yik, F.W.H. & Lam, V.K.C. (1998) Chiller models for plant design studies. *Building Services Engineering Research & Technology* 19(4) pp.233–41.

Yik, F.W.H., Underwood, C.P. & Chow, W.K. (1995) Simultaneous modelling of heat and moisture transfer and air-conditioning system in buildings. *Proceedings, Building Simulation '95 Conference. Madison.*

Yik, F.W.H., Underwood, C.P. and Chow W.K. (1997) Chilled-water cooling and dehumidifying coils with corrugated plate fins: Modelling method. *Building Services Engineering Research & Technology* 18(1) pp.47–58.

Chapter 5
Modelling Control Systems

The methods described in the previous chapter are applicable to the design and analysis of building systems and equipment over long time horizons for which the plant can be assumed to respond (at least in relation to the accompanying building envelope) in a steady state manner. For many analyses, a steady state description of plant and system components accompanied with a fully dynamic description of the building envelope leading to a so-called 'quasi-steady' method has utility. Examples include plant selection and sizing, system option appraisals, control strategies and the analysis of part load plant behaviour over seasons. For the analysis of control system design and stability, however, a fully dynamic description of the plant is essential in order to capture the time-varying behaviour of state variables and, in some cases, model parameters. In this case, the plant requires to be modelled with time explicitly defined as an independent variable.

5.1 Distributed system modelling

Distributed systems in buildings break down into flow elements and heat exchange elements. Since, in most cases, incompressible flow is either prevalent or can be approximated to, then, for control system modelling, steady flow behaviour in systems can be assumed, perhaps with allowance for transport delays where necessary. With the same argument applying to flow-generating equipment such as fans and pumps, then the methods described in Chapter 4 are applicable to the modelling of flow elements as far as control system analyses are concerned. Heat exchange equipment on the other hand demands a more detailed level of treatment in order to capture the time-varying nature of state variables and, in many cases, the influence of non-stationary parameters.

Systems involving heat and mass transfer

In this section, a method for modelling the response of a distributed heat exchanger involving air with heat and mass transfer on one side and water on the other is described. This exemplar deals with one of the more complex cases in building systems such as is found in air cooling processes and cooling towers. The system can be idealised as a basis for modelling, as shown in Fig. 5.1.

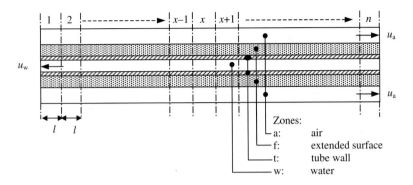

Fig. 5.1 Idealised heat exchange component.

Heat and mass balances can be written for each of the four zones as follows. It is assumed that the thermal resistances of the tube wall and extended surface materials can be neglected. Therefore, if the extended surface temperature gradient is accounted for by an effectiveness term, heat transfer from air will be with reference to the extended surface root temperature (T_f) and heat transfer from water will be with reference to the tube wall surface temperature (T_t). Thus for the air zone:

$$\rho_a A_{af} \Delta l \frac{\partial h_a}{\partial t} = -\rho_a u_a A_{af} \Delta l \frac{\partial h_a}{\partial l} - M_a D_O \Delta l (h_a - h_{as}(T_f)) \tag{5.1}$$

where:

ρ_a = density of air (kg m^{-3})
A_{af} = effective free flow area on the air side (m^2)
Δl = element width (Figure 5.1) (m)
h_a = enthalpy of air (kJ kg^{-1})
u_a = air velocity with respect to the equivalent free flow area (m s^{-1})
M_a = effective heat and mass transfer coefficient (kg m^{-2} s^{-1})
D_o = air side 'wetted' perimeter of the heat transfer surface (m)
$h_{as}(T_f)$ = enthalpy of air saturated at the extended surface root temperature, T_f (kJ kg^{-1})

With substitutions for air enthalpies ($h_a = c_{pa} T_a + w_a h_{fg}$; $h_{as}(T_f) = c_{pa} T_f + w_{as}(T_f) h_{fg}$), Equation 5.1 becomes:

$$c_{pa} \frac{\partial T_a}{\partial t} + h_{fg} \frac{\partial w_a}{\partial t} = -u_a c_{pa} \frac{\partial T_a}{\partial l} - u_a h_{fg} \frac{\partial w_a}{\partial x} - \dots$$
$$\dots \frac{M_a D_o}{\rho_a A_{af}} \left[c_{pa} (T_a - T_f) + h_{fg} (w_a - w_{as}(T_f)) \right] \tag{5.2}$$

where:

c_{pa} = specific heat capacity of air (kJ kg^{-1} K^{-1})
T_a = air dry bulb temperature (K)

h_{fg} = latent heat of vapourisation of water (kJ kg^{-1})
w_a = air moisture content (kg kg^{-1})
$w_{as}(T_f)$ = moisture content of air saturated at the extended surface temperature, T_f (kg kg^{-1})

Note that the effective mass transfer coefficient can be obtained either directly from experimental data or from the following:

$$M_a = \eta_T \frac{h'_o}{c_{pa} Le^{\frac{2}{3}}}$$
(5.3)

where η_T is the wetted extended surface (fin) effectiveness, h'_o is the overall surface convection coefficient (kW m^{-2} K^{-1}) and Le is the Lewis number. Procedures for obtaining η_T and Le can be found in the literature (e.g. McQuiston *et al.* 2000).

A mass balance attributable to moisture transport can be expressed as follows:

$$\rho_a A_{af} \Delta l \frac{\partial w_a}{\partial t} = -\rho_a u_a A_{af} \Delta l \frac{\partial w_a}{\partial l} - M_a D_o \Delta l (w_a - w_{as}(T_f))$$

i.e.:

$$\frac{\partial w_a}{\partial t} = -u_a \frac{\partial w_a}{\partial l} - \frac{M_a D_o}{\rho_a A_{af}} (w_a - w_{as}(T_f))$$
(5.4)

Multiplying Equation 5.4 through by h_{fg} and subtracting the result from Equation 5.2 results in the sensible heat balance for the heat exchanger experiencing both heat and mass transfer:

$$\frac{\partial T_a}{\partial t} = -u_a \frac{\partial T_a}{\partial l} - \frac{M_a D_o}{\rho_a c_{pa} A_{af}} (T_a - T_f)$$
(5.5)

An energy balance for the water side will be:

$$\rho_w A_{wf} c_{pw} \Delta l \frac{\partial T_w}{\partial t} = \rho_w u_w A_{wf} c_{pw} \Delta l \frac{\partial T_w}{\partial l} - h_i D_i \Delta l (T_w - T_t)$$

i.e.:

$$\frac{\partial T_w}{\partial t} = u_w \frac{\partial T_w}{\partial l} - \frac{h_i D_i}{\rho_w c_{pw} A_{wf}} (T_w - T_t)$$
(5.6)

where:

ρ_w = density of water (kg m^{-3})
A_w = the equivalent free flow area on the water side (m^2)

c_{pw} = specific heat capacity of water (kJ kg^{-1} K^{-1})
u_w = water velocity with respect to the equivalent free flow area (m s^{-1})
h_i = tube internal surface convection coefficient on the water side (kW m^{-2} K^{-1})
D_i = water side 'wetted' perimeter of the heat transfer surface (m)
T_w = water temperature (K)
T_t = tube wall temperature (K)

To close the heat exchanger model, energy balances can now be defined for the tube wall material and the extended surface core. For low tube wall extended surface thicknesses:

$$\rho_t c_{pt} D_i d_t \Delta l \frac{\partial T_t}{\partial t} = -h_i D_i \Delta l (T_t - T_w) - \frac{D_i \Delta l}{R_{tf}} (T_t - T_f)$$

i.e.:

$$\frac{\partial T_t}{\partial t} = -\frac{h_i}{\rho_t c_{pt} d_t} (T_t - T_w) - \frac{(T_t - T_f)}{\rho_t c_{pt} d_t R_{tf}} \tag{5.7}$$

where:

ρ_t = density of tube wall material (kg m^{-3})
c_{pt} = specific heat capacity of the tube wall material (kJ kg^{-1} K^{-1})
d_t = tube wall thickness (m)
R_{tf} = thermal resistance between the tube wall and extended surface core (m^2 K kW^{-1})
T_f = temperature of the extended surface (fin) core (K)

and, for the extended surface material:

$$\rho_f c_{pf} D_o d_f \Delta l \frac{\partial T_f}{\partial t} = -M_a D_o \Delta l (h_{as}(T_f) - h_a) - \frac{D_i \Delta l}{R_{tf}} (T_f - T_t)$$

Substituting for the air enthalpies as in Equation 5.2, this becomes:

$$\frac{\partial T_f}{\partial t} = -\frac{M_a}{\rho_f c_{pf} d_f} \left[c_{pa}(T_f - T_a) + h_{fg}(w_{as}(T_f) - w_a) \right] - \frac{D_i}{\rho_f c_{pf} D_o d_f R_{tf}} (T_f - T_t) \tag{5.8}$$

where:

ρ_f = density of extended surface material (kg m^{-3})
c_{pf} = specific heat capacity of the extended surface material (kJ kg^{-1} K^{-1})
d_f = extended surface wall thickness (m)
R_{tf} = thermal resistance between the tube wall and extended surface core (m^2 K kW^{-1})

The governing equations for this case are Equations 5.4–5.8. A partial discretisation of certain of the governing equations (Equations 5.4–5.6) can be applied to deal with the spatial terms, reducing the governing equations to a set of ordinary differential equations, which are then amenable to solution using most numerical integration methods. Using one-sided differencing, the following apply for a general variable, ϕ, dependent on space, l:

$$\frac{d\phi}{dl} \cong \frac{\phi_x - \phi_{x-1}}{\Delta l} \qquad \text{(backward in space)} \tag{5.9}$$

$$\frac{d\phi}{dl} \cong \frac{\phi_{x+1} - \phi_x}{\Delta l} \qquad \text{(forward in space)} \tag{5.10}$$

For counter flow heat exchange letting the convention $x(1)$, $x(2)$ ... $x(n)$ apply from inlet to outlet on the air side, a backward-in-space approximation can be used for air side spatial derivatives whereas a forward-in-space approximation is used on the water side.

Making the above substitutions, the governing equations can be summarised as follows:

$$\frac{dw_{a,x}}{dt} = -\frac{u_a}{\Delta l}\left(w_{a,x} - w_{a,x-1}\right) - \frac{M_a D_o}{\rho_a A_{af}}\left(w_{a,x} - w_{as,x}\left(T_{f,x}\right)\right) \tag{5.11}$$

$$\frac{dT_{a,x}}{dt} = -\frac{u_a}{\Delta l}\left(T_{a,x} - T_{a,x-1}\right) - \frac{M_a D_o}{\rho_a c_{pa} A_{af}}\left(T_{a,x} - T_{f,x}\right) \tag{5.12}$$

$$\frac{dT_{w,x}}{dt} = \frac{u_w}{\Delta l}\left(T_{w,x+1} - T_{w,x}\right) - \frac{h_i D_i}{\rho_w c_{pw} A_{wf}}\left(T_{w,x} - T_{t,x}\right) \tag{5.13}$$

$$\frac{dT_{t,x}}{dt} = -\frac{h_i}{\rho_t c_{pt} d_t}\left(T_{t,x} - T_{w,x}\right) - \frac{\left(T_{t,x} - T_{f,x}\right)}{\rho_t c_{pt} d_t R_{tf}} \tag{5.14}$$

$$\frac{dT_{f,x}}{dt} = -\frac{M_a}{\rho_f c_{pf} d_f}\left[c_{pa}\left(T_{f,x} - T_{a,x}\right) + h_{fg}\left(w_{as,x}\left(T_{f,x}\right) - w_{a,x}\right)\right] - \cdots$$

$$\cdots \frac{D_i}{\rho_f c_{pf} D_o d_f R_{tf}}\left(T_{f,x} - T_{t,x}\right) \tag{5.15}$$

At the inlet zones to the heat exchanger ($x = 1$, air side; $x = n$, water side) inlet water and air temperatures and moisture content are defined by the boundary conditions. These boundary variables are defined as T_{wi}, T_{ai} and w_{ai} respectively. Output state variables from the model will be $w_{a,n}$, $T_{a,n}$ and $T_{w,1}$ – the outlet air moisture content, dry bulb temperature and water temperature respectively. For linear modelling applications the parameters of the governing equations are constant and each equation can be expressed in the convenient matrix differential equation or state-space form:

$$\dot{\mathbf{T}} = \mathbf{A}\mathbf{T} + \mathbf{B}\mathbf{u}$$

in which $\dot{\mathbf{T}}$ is a vector of n derivatives, $\mathbf{A}$ is an $n \times n$ matrix of coefficients, $\mathbf{T}$ is a vector of n state variables, $\mathbf{B}$ a matrix of input coefficients and $\mathbf{u}$ is a vector of inputs. For example, Equation 5.11 can be summarised as follows:

$$
\begin{bmatrix} w_{a,1} \\ w_{a,2} \\ w_{a,3} \\ \cdot \\ \cdot \\ \cdot \\ w_{a,n-1} \\ w_{a,n} \end{bmatrix} =
\begin{bmatrix}
A_{11} & 0 & 0 & \cdot & & & \cdot & & 0 \\
A_{21} & A_{22} & 0 & \cdot & & & \cdot & & 0 \\
0 & A_{32} & A_{33} & 0 & \cdot & & \cdot & & 0 \\
\cdot & & & & & & & & \\
\cdot & & & & & & & & \\
\cdot & & & & & & & & \\
\cdot & \cdot & \cdot & \cdot & \cdot & A_{n-1,n-2} & A_{n-1,n-1} & 0 \\
\cdot & \cdot & \cdot & \cdot & \cdot & 0 & A_{n,n-1} & A_{n,n}
\end{bmatrix} + \ldots
$$

$$
\ldots
\begin{bmatrix}
B_{11} & B_{12} & 0 & \cdot & & \cdot & & 0 \\
0 & 0 & B_{23} & 0 & \cdot & \cdot & & 0 \\
0 & 0 & 0 & B_{34} & 0 & \cdot & & 0 \\
\cdot & \cdot & \cdot & \cdot & \cdot & \cdot & & \cdot \\
\cdot & & & & & & & \\
\cdot & & & & & & & \\
\cdot & \cdot & \cdot & \cdot & \cdot & 0 & B_{n-1,n} & 0 \\
\cdot & \cdot & \cdot & \cdot & \cdot & & 0 & B_{n,n+1}
\end{bmatrix}
\times
\begin{bmatrix}
w_{ai} \\
w_{as,1}(T_{f,1}) \\
w_{as,2}(T_{f,1}) \\
w_{as,3}(T_{f,1}) \\
\cdot \\
\cdot \\
w_{as,n-1}(T_{f,n-1}) \\
w_{as,n}(T_{f,n})
\end{bmatrix}
\qquad (5.16)
$$

based upon which coefficients for the **A** and **B** matrices are given below:

$$
A_{n,n} = -\left(\frac{u_a}{\Delta l} + \frac{M_a D_o}{\rho_a A_{af}} \right)
$$

$$
A_{2,1} = A_{3,2} \ldots A_{n,n-1} = \frac{u_a}{\Delta l}
$$

$$
B_{1,1} = \frac{u_a}{\Delta l}
$$

$$
B_{1,2} = B_{2,3} \ldots B_{n,n+1} = \frac{M_a D_o}{\rho_a A_{af}}
$$

Similar structures are possible for Equations 5.12–5.15.

Systems involving sensible heat transfer

The case involving combined heat and mass transfer is applicable to certain air cooling processes and aspects of it also apply to cooling towers. However, in most applications mass transfer is absent, or only partially applicable (for example in air cooling processes, condensing heat transfer will only take place once the air has been lowered to its dew point, which may be some distance into the body of the heat exchanger). The case involving sensible heat transfer only is simpler than that for combined heat and mass transfer in that $\frac{\partial w_a}{\partial t}$ and $\frac{\partial w_a}{\partial l}$ are now zero (and $w_{as} = w_a$ = constant). Governing Equation 5.4 is therefore now inapplicable and the combined heat and mass transfer coefficient, M_a, is replaced in Equations 5.5 and 5.8 with the effective air side surface convection coefficient, h_o. Specifically, Equation 5.5 becomes:

$$\frac{\partial T_a}{\partial t} = -u_a \frac{\partial T_a}{\partial l} - \frac{h_o D_o}{\rho_a c_{pa} A_{af}} (T_a - T_f) \tag{5.17}$$

where $h_o = \eta_S h'_o$ in which h'_o is the overall surface convection coefficient and η_S is the extended surface (fin) effectiveness for sensible heat transfer. Equation 5.8 becomes:

$$\frac{\partial T_f}{\partial t} = -\frac{h_o}{\rho_f c_{pf} d_f} (T_f - T_a) - \frac{D_i}{\rho_f c_{pf} D_o d_f R_{tf}} (T_f - T_t) \tag{5.18}$$

Experimental results for h_o have been reported extensively in the literature. For example Gomaa *et al.* (2002) give results for chilled water cooling coils based on flat, corrugated and 'turbulated' plate fin details. They found that h_o mainly depended on u_a and only very weakly depended on fluid properties. Thus for constant air volume plants, modelling with a constant value of h_o is likely to be acceptable in many cases.

Reduced-order methods

The model described by Equations 5.11–5.15 can be relaxed in two ways for all but the most precise applications. First, typical rates of change on the air side of heat exchange processes are sufficiently high to be of little practical interest due to a combination of low free flow volume on the air side and the inherently low thermal capacity of air. This makes it possible to put $\dfrac{dw_{a,x}}{dt} = \dfrac{dT_{a,x}}{dt} = 0$ and Equations 5.11 and 5.12 reduce to the following:

$$w_{a,x} = \frac{u_a \rho_a A_{af}}{u_a \rho_a A_{af} + M_a D_o \Delta l} w_{a,x-1} + \frac{M_a D_o \Delta l}{u_a \rho_a A_{af} + M_a D_o \Delta l} w_{as,x}(T_{f,x}) \tag{5.19}$$

$$T_{a,x} = \frac{u_a \rho_a c_{pa} A_{af}}{u_a \rho_a c_{pa} A_{af} + M_a D_o \Delta l} T_{a,x-1} + \frac{M_a D_o \Delta l}{u_a \rho_a c_{pa} A_{af} + M_a D_o \Delta l} T_{f,x} \tag{5.20}$$

Second, by neglecting both the thermal resistance of the tube material and, likewise, the resistance between tube wall and extended surface, the fin core temperature takes on the same value as the tube wall temperature (i.e. $T_{t,x} = T_{f,x}$). Thus Equation 5.14 disappears and Equations 5.13 and 5.15 become:

$$\frac{dT_{w,x}}{dt} = \frac{u_w}{\Delta l}(T_{w,x-1} - T_{w,x}) - \frac{h_i D_i}{\rho_w c_{pw} A_{wf}}(T_{w,x} - T_{f,x}) \tag{5.21}$$

and:

$$\frac{dT_{f,x}}{dt} = -\frac{M_a}{\rho_f c_{pf} d_f}\left[c_{pa}\left(T_{f,x}-T_{a,x}\right)+h_{fg}\left(w_{as,x}\left(T_{f,x}\right)-w_{a,x}\right)\right]-\ldots$$

$$\ldots \frac{h_i D_i}{\rho_f c_{pf} D_o d_f}\left(T_{f,x}-T_{w,x}\right) \tag{5.22}$$

If the application involves sensible heat exchange only, then Equation 5.19 is inapplicable, Equation 5.21 for the water zone applies and Equations 5.20 and 5.22 for the air and material zones are adjusted to the following:

$$T_{a,x} = \frac{u_a \rho_a c_{pa} A_{af}}{u_a \rho_a c_{pa} A_{af}+h_o D_o \Delta l}T_{a,x-1}+\frac{h_o D_o \Delta l}{u_a \rho_a c_{pa} A_{af}+h_o D_o \Delta l}T_{f,x} \tag{5.23}$$

$$\frac{dT_{f,x}}{dt} = -\frac{h_o}{\rho_f c_{pf} d_f}\left(T_{f,x}-T_{a,x}\right)-\frac{h_i D_i}{\rho_f c_{pf} D_o d_f}\left(T_{f,x}-T_{w,x}\right) \tag{5.24}$$

Finally, the issue of the choice of n requires to be addressed. Clearly n requires to be large enough to capture all of the dynamics of interest with an acceptable level of accuracy and precision but not so large as to introduce unnecessary complexity in the model and, correspondingly, high computational cost. A general procedure for the choice of n is not possible but some insights are offered here for the typical case of air to water heat exchange processes of relevance to building systems.

Consider the simplified case described above for sensible heat exchange only. The describing differential equations for water and material zones (5.21 and 5.24) can be summarised as follows (adopting the dot notation for the time rate of change, d/dt):

$$\dot{T}_{w,x} = \frac{T_{f,x}}{\tau_{wa}}+\frac{T_{w,x+1}}{\tau_{wb}}-\frac{T_{w,x}}{\tau_{wc}} \tag{5.25}$$

$$\dot{T}_{f,x} = \frac{T_{w,x}}{\tau_{fa}}+\frac{T_{a,x}}{\tau_{fb}}-\frac{T_{f,x}}{\tau_{fc}} \tag{5.26}$$

from which two sets of time constants, (τ_{wa}, τ_{wb}, τ_{wc} and τ_{fa}, τ_{fb}, τ_{fc}), are given by:

$$\tau_{wa} = C_w/(h_i D_i);\ \tau_{wb} = \Delta l/u_w;\ \tau_{wc} = C_w \Delta l/(u_w C_w+h_i D_i \Delta l)$$

and:

$$\tau_{fa} = C_f D_o/h_i D_i;\ \tau_{fb} = C_f/h_o;\ \tau_{fc} = C_f D_o/(h_i D_i+h_o D_o)$$

in which $C_w = \rho_w c_{pw} A_{wf}$ and $C_f = \rho_f c_{pf} d_f$.

Perhaps the simplest way to summarise the dynamic characteristics of a linear system or set of systems is through a reference to its dominant time constant. Normally (though by no means exclusively), this will be the largest time constant evident in the model since it dominates the stability of a system when, for instance, under closed loop control. Thus for the above case it is typically possible to put:

$$\tau_{dom} = \max(\tau_{wa}, \tau_{wb}, \tau_{wc}, \tau_{fa}, \tau_{fb}, \tau_{fc}) \tag{5.27}$$

For linear heat exchange processes common in building equipment applications involving air and water, time constants can range from a few seconds to a few minutes. Thus Equations 5.25 and 5.26 together with the auxiliary algebraic equation for air temperature (Equation 5.23) can be solved for a range of parameter values that reduce to a time constant set whose values span a few seconds to a few minutes and for various values of n. Results of these test simulations for n taking on values of 1, 2, 3, 5 and 10 are given in Figs 5.2 and 5.3. Figure 5.2 gives results of unit step response tests for an example case in which $\tau_{dom} = 30\,s$ and Fig. 5.3 gives results of the percentage root-mean-square error in response for $n = 1, 2, 3, 5$ with respect to $n = 10$ expressed as a function of the dominant time constant. Evident from these typical results are the following:

- A model order corresponding to $n = 1$ would appear to be inadequate for all but the crudest simulations.
- Simulations involving disturbances in water boundary conditions will lead to less accurate results at low n than will be the case for disturbances in air boundary conditions.
- There appears to be little improvement in accuracy to be gained by increasing n beyond 10.
- Modelling with $n = 3$ (or higher) will in most cases lead to acceptable results with RMS errors of less than 5% for all cases involving low-to-moderate time constants.

Reduced order descriptions of this kind can alternatively be derived by splitting the heat exchange system of interest into one or more series zones and writing energy balances for them. This approach is referred to as lumped-parameter modelling and it is assumed that each fluid zone behaves as a 'continuously stirred tank'. The resulting equations are found to be identical to those defined in the above.

Dealing with non-linearities

Like h_o, a constant value of h_i may be assumed for many practical modelling cases. Indeed, h_i is usually large compared with h_o and heat transfer is thus dominated by the latter. However, in many applications, control over heat transfer is effected by varying the mass flow rate of water and, hence u_w and, in common with h_o, h_i is mainly dependent on u_w. In these circumstances, two issues become evident:

- h_i requires to be modelled.
- Non-linearity is introduced to the model partly due to the manner in which h_i is modelled and partly due to the products implicit in the term $\dfrac{u_w}{\Delta l}(T_{w,x+1} - T_{w,x})$ (e.g. Equation 5.21).

Whilst results for the Nusselt number for fully developed turbulent flow in tubes are well reported in the literature, for example the improved correlation of Petukov (reported in

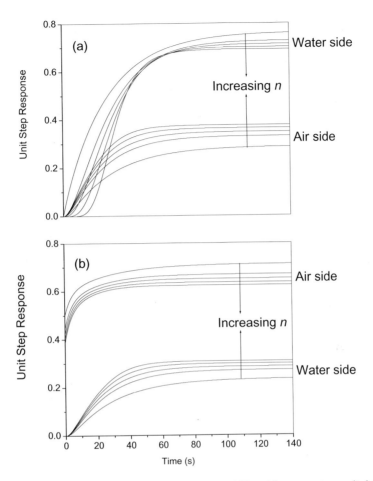

Fig. 5.2   Unit step responses in water and air outlet temperatures. (a) Water inlet temperature excited. (b) Air inlet temperature excited.

Holman 1997), results at low Reynolds number are also required. Based on the latter for high Reynolds number with Hausen's correlation for low Reynolds number (Holman 1997), the following overall fit-to-data is suggested for modelling h_i in situations when u_w is expected to vary:

$$\mathrm{Nu}_{d_{eq}} = A + B\left(\mathrm{RePr}^{\frac{1}{3}}\right) + C\left(\mathrm{RePr}^{\frac{1}{3}}\right)^2 \tag{5.28}$$

where $\mathrm{Nu}_{d_{eq}}$ is the Nusselt number with respect to the equivalent tube diameter, d_{eq}, Re is the Reynolds number and Pr is the Prandtl number. The constants A, B and C are given by (respectively) 2.5537, 3.1800×10^{-3} and 4.7255×10^{-8}. Equation 5.28 is applicable to (negligible $< \mathrm{Re} \leq 10000$).

For most control system simulations of relevance to building plant, interest tends to be confined to a limited operating range and a simplified form of Equation 5.28 can be assumed together with an assumption of constant fluid properties. The same applies in situations where h_o requires to be modelled (e.g. variable air volume applications) – see for example

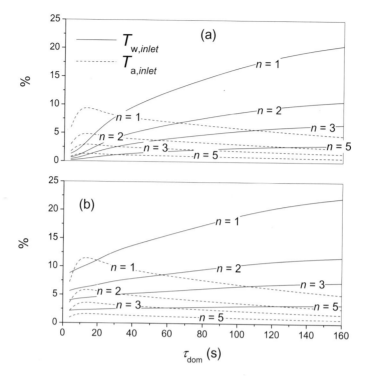

Fig. 5.3   RMS response errors (%). (a) Air inlet temperature excited. (b) Water inlet temperature excited.

Gomaa *et al.* (2002). Thus reduced expressions for both h_i and h_o are possible:

$$h_i = N_i u_w^{n_i} \tag{5.29}$$

$$h_o = N_o u_a^{n_o} \tag{5.30}$$

With these substitutions for h_i and h_o in the reduced equations for sensible heat exchange, several sources of non-linearity become evident:

$$\dot{T}_{w,x} = \frac{N_i D_i u_w^{n_i}}{C_w} T_{f,x} + \frac{u_w}{\Delta l} T_{w,x+1} - \left(\frac{u_w}{\Delta l} + \frac{N_i D_i u_w^{n_i}}{C_w} \right) T_{w,x} \tag{5.31}$$

$$\dot{T}_{f,x} = \frac{N_i D_i u_w^{n_i}}{C_f D_o} T_{w,x} + \frac{N_o u_a^{n_o}}{C_f} T_{a,x} - \left(\frac{N_o u_a^{n_o}}{C_f} + \frac{N_i D_i u_w^{n_i}}{C_f D_o} \right) T_{f,x} \tag{5.32}$$

$$T_{a,x} = \frac{N_o D_o \Delta l u_a^{n_o}}{\left(u_a C_a + N_o D_o \Delta l u_a^{n_o} \right)} T_{f,x} + \frac{u_a C_a}{\left(u_a C_a + N_o D_o \Delta l u_a^{n_o} \right)} T_{a,x-1} \tag{5.33}$$

in which $C_a = \rho_a c_{pa} A_{af}$.

A non-linear form of model can be 'linearised' around some nominal operating point through the use of a Taylor series expansion. Truncating the series after the first-order term a function of m variables, $f(\phi_1,\phi_2,\ldots\phi_m)$, can be expanded as follows:

$$f(\phi_1,\phi_2,\ldots\phi_m) = f(\bar{\phi}_1,\bar{\phi}_2,\ldots\bar{\phi}_m) + (\phi_1 - \bar{\phi}_1) \times \frac{\partial}{\partial\phi_1}f(\bar{\phi}_1,\bar{\phi}_2,\ldots\bar{\phi}_m) + \ldots$$

$$\ldots(\phi_2 - \bar{\phi}_2) \times \frac{\partial}{\partial\phi_2}f(\bar{\phi}_1,\bar{\phi}_2,\ldots\bar{\phi}_m) + \ldots(\phi_m - \bar{\phi}_n) \times \frac{\partial}{\partial\phi_m}f(\bar{\phi}_1,\bar{\phi}_2,\ldots\bar{\phi}_m)$$

in which the over-bar notation implies evaluation based on variables taking on steady-state values at the initial (known) operating conditions. Recasting all dependent variables as deviations from these initial steady state operating conditions (i.e. the general deviation variable is: $\phi' = \phi - \bar{\phi}$) enables the steady-state terms to be dropped:

$$f(\phi_1,\phi_2,\ldots\phi_m) = \phi_1' \frac{\partial}{\partial\bar{\phi}_1}f(\bar{\phi}_1,\bar{\phi}_2,\ldots\bar{\phi}_m) + \ldots$$

$$\ldots\phi_2' \frac{\partial}{\partial\bar{\phi}_2}f(\bar{\phi}_1,\bar{\phi}_2,\ldots\bar{\phi}_m) + \ldots + \phi_m' \frac{\partial}{\partial\bar{\phi}_m}f(\bar{\phi}_1,\bar{\phi}_2,\ldots\bar{\phi}_m) \quad (5.34)$$

Expressing the derivatives of each term in Equations 5.31 through 5.33, and combining these as implied by Equation 5.34 leads to the locally linearised model for sensible heat exchange involving air and water. Expressing the result in state-space form:

$$
\begin{bmatrix} \dot{T}'_{w,1} \\ \dot{T}'_{w,2} \\ \cdot \\ \cdot \\ \cdot \\ \dot{T}'_{w,n} \end{bmatrix} =
\begin{bmatrix}
a_{w(1,1)} & a_{w(1,2)} & 0 & \cdot & \cdot & 0 \\
0 & a_{w(2,2)} & a_{w(2,3)} & 0 & \cdot & 0 \\
\cdot & \cdot & \cdot & \cdot & \cdot & \cdot \\
\cdot & \cdot & \cdot & \cdot & \cdot & \cdot \\
\cdot & \cdot & \cdot & \cdot & \cdot & \cdot \\
0 & \cdot & \cdot & \cdot & 0 & a_{w(n,n)}
\end{bmatrix}
\times
\begin{bmatrix} T'_{w,1} \\ T'_{w,2} \\ \cdot \\ \cdot \\ \cdot \\ T'_{w,n} \end{bmatrix} + \ldots
$$

$$
\ldots
\begin{bmatrix}
b_{w(1,1)} & b_{w(1,2)} & 0 & \cdot & \cdot & 0 \\
b_{w(2,1)} & 0 & b_{w(2,3)} & 0 & \cdot & 0 \\
\cdot & \cdot & \cdot & \cdot & \cdot & \cdot \\
\cdot & \cdot & \cdot & \cdot & \cdot & \cdot \\
\cdot & \cdot & \cdot & \cdot & \cdot & \cdot \\
b_{w(n,1)} & 0 & \cdot & \cdot & b_{w(n,n+1)} & b_{w(n,n+2)}
\end{bmatrix}
\times
\begin{bmatrix} u'_w \\ T'_{f,1} \\ \cdot \\ \cdot \\ T'_{f,n} \\ T'_{wi} \end{bmatrix} \quad (5.35)
$$

$$
\begin{bmatrix} \dot{T}'_{f,1} \\ \dot{T}'_{f,2} \\ \cdot \\ \cdot \\ \cdot \\ \dot{T}'_{f,n} \end{bmatrix} =
\begin{bmatrix}
a_{f(1,1)} & 0 & \cdot & \cdot & \cdot & 0 \\
0 & a_{f(2,2)} & 0 & \cdot & \cdot & 0 \\
\cdot & \cdot & \cdot & \cdot & \cdot & \cdot \\
\cdot & \cdot & \cdot & \cdot & \cdot & \cdot \\
\cdot & \cdot & \cdot & \cdot & \cdot & \cdot \\
0 & \cdot & \cdot & \cdot & 0 & a_{f,(n,n)}
\end{bmatrix}
\times
\begin{bmatrix} T'_{f,1} \\ T'_{f,2} \\ \cdot \\ \cdot \\ \cdot \\ T'_{f,n} \end{bmatrix} + \ldots \quad (5.36)
$$

$$
\cdots
\begin{bmatrix}
b_{f(1,1)} & b_{f(1,2)} & b_{f(1,3)} & 0 & \cdot & 0 & b_{f(1,n+3)} & 0 & \cdot & 0 \\
b_{f(2,1)} & b_{f(2,2)} & 0 & b_{f(2,4)} & 0 & \cdot & \cdot & 0 & b_{f(1,n+4)} & 0 \\
\cdot & \cdot & \cdot & \cdot & \cdot & \cdot & \cdot & \cdot & \cdot & \cdot \\
\cdot & \cdot & \cdot & \cdot & \cdot & \cdot & \cdot & \cdot & \cdot & \cdot \\
b_{f(n,1)} & b_{f(n,2)} & 0 & \cdot & 0 & b_{f(n,n+2)} & \cdot & \cdot & \cdot & b_{f(n,2n+2)}
\end{bmatrix}
\times
\begin{bmatrix}
u'_w \\ u'_a \\ T'_{w,1} \\ \cdot \\ \cdot \\ \cdot \\ T'_{w,n} \\ T'_{a,1} \\ \cdot \\ \cdot \\ T'_{a,n}
\end{bmatrix}
$$

And closure is given by a set of algebraic equations for zone air temperatures:

$$
\begin{bmatrix} T'_{a,1} \\ T'_{a,1} \\ \cdot \\ \cdot \\ T'_{a,1} \end{bmatrix}
=
\begin{bmatrix}
a_{a(1,1)} & 0 & \cdot & \cdot & 0 \\
0 & a_{a(2,2)} & 0 & \cdot & 0 \\
\cdot & & & & \cdot \\
\cdot & & & & \cdot \\
\cdot & & & & a_{a(n,n)}
\end{bmatrix}
\times
\begin{bmatrix} T'_{f,1} \\ T'_{f,1} \\ \cdot \\ \cdot \\ T'_{f,1} \end{bmatrix}
+ \ldots
$$

$$
\cdots
\begin{bmatrix}
b_{a(1,1)} & b_{a(1,2)} & 0 & \cdot & \cdot & 0 \\
b_{a(2,1)} & 0 & b_{a(2,3)} & 0 & \cdot & 0 \\
\cdot & & & & & \cdot \\
\cdot & & & & & \cdot \\
b_{a(n,1)} & 0 & \cdot & \cdot & 0 & b_{a(n,n+1)}
\end{bmatrix}
\times
\begin{bmatrix} u'_a \\ T'_{ai} \\ T'_{a,1} \\ \cdot \\ T'_{a,n-1} \end{bmatrix}
\qquad (5.37)
$$

The coefficients are given by the following for $x = 1, 2, 3 \ldots n$.
  For the water zone:

$$
\begin{cases}
a_{w(x,x)} = \left(C_w \bar{u}_w + D_i \Delta l N_i \bar{u}_w^{n_i} \right) / \Delta l C_w \\
a_{w(x,x+1)} = \bar{u}_w / \Delta l
\end{cases}
$$

$$
\begin{cases}
b_{w(x,1)} = \left[\bar{T}_{f(x)} n_i D_i \Delta l N_i \bar{u}_w^{n_i-1} + \bar{T}_{w(x+1)} C_w - \bar{T}_{w(x)} \left(C_w + n_i D_i \Delta l N_i \bar{u}_w^{n_i-1} \right) \right] / \Delta l C_w \\
\left(\bar{T}_{w(n+1)} = T_{wi} \right) \\
b_{w(x,x+1)} = D_i N_i \bar{u}_w^{n_i} / C_w \\
b_{w(n,n+2)} = A_{w(x,x+1)}
\end{cases}
$$

For the heat exchange material zone:

$$
a_{f(x,x)} = \left(D_o N_o \bar{u}_a^{n_o} + C_f D_i N_i \bar{u}_w^{n_i} \right) / D_o C_f
$$

$$
\begin{cases}
b_{f(x,1)} = n_i D_i N_i \bar{u}_w^{n_i-1} \left(\bar{T}_{w(x)} - \bar{T}_{f(x)} \right) / D_o C_f \\
b_{f(x,2)} = n_o N_o \bar{u}_a^{n_o-1} \left(\bar{T}_{a(x)} - \bar{T}_{f(x)} \right) / C_f \\
b_{f(x,x+2)} = D_i N_i \bar{u}_w^{n_i} / D_o C_f \\
b_{f(x,n+x+2)} = N_o \bar{u}_a^{n_o} / C_f
\end{cases}
$$

And for the air zone:

$$a_{a(x,x)} = D_o \Delta l N_o \bar{u}_a^{n_o} / \left(C_a \bar{u}_a + D_o \Delta l N_o \bar{u}_a^{n_o} \right)$$

$$\begin{cases} b_{a(x,1)} = \dfrac{(1-n_o)D_o C_a \Delta l N_o \bar{u}_a^{n_o}}{\left(C_a \bar{u}_a + D_o \Delta l N_o \bar{u}_a^{n_o} \right)^2} \left(\bar{T}_{a(x-1)} - \bar{T}_{f(x)} \right) \quad \left(\bar{T}_{a(0)} = T_{ai} \right) \\ b_{a(x,x+1)} = C_a \bar{u}_a / \left(C_a \bar{u}_a + D_o \Delta l N_o \bar{u}_a^{n_o} \right) \end{cases}$$

The over-bar notation in all cases refers to variables evaluated at initial steady-state conditions.

Transfer functions

The application of deviation variables ensures that all state variables have initial conditions defined at $0, (t = 0)$; thus the simple substitution $\phi(s) \cdot s = \phi$ in which s is the Laplace variable enables a transfer function form of model to be generated as an alternative to the use of a state-space form. This form of model is a convenient alternative to state-space when model order is low.

As an illustration, transfer functions are expressed for the simplified air-to-water model described above for the case in which $n = 1$. The governing equations for this case will be as follows based on the coefficients as defined for the state-space case:

$$T'_{wo}(s) \cdot s = a_{w(1,2)}T'_{wi}(s) + b_{w(1,1)}u'_w(s) + b_{w(1,2)}T'_f(s) - a_{w(1,1)}T'_{wo}(s) \tag{5.38}$$

$$T'_f(s) \cdot s = b_{f(1,1)}u'_w(s) + b_{f(1,2)}u'_a(s) + b_{f(1,3)}T'_{wo}(s) + b_{f(1,6)}T'_{ao}(s) - a_{f(1,1)}T'_f(s) \tag{5.39}$$

$$T'_{ao}(s) = a_{a(1,1)}T'_f(s) + b_{a(1,1)}u'_a(s) + b_{a(1,2)}T'_{ai}(s) \tag{5.40}$$

With a little algebra, the transfer function set whose output is T'_{wo} can be expressed as:

$$\frac{T'_{wo}(s)}{u'_w(s)} = \frac{A_1 s + A_2}{s^2 + B_1 s + B_2} \tag{5.41}$$

$$\frac{T'_{wo}(s)}{u'_a(s)} = \frac{A_3}{s^2 + B_1 s + B_2} \tag{5.42}$$

$$\frac{T'_{wo}(s)}{T'_{wi}(s)} = \frac{A_4 s + A_5}{s^2 + B_1 s + B_2} \tag{5.43}$$

$$\frac{T'_{wo}(s)}{T'_{ai}(s)} = \frac{A_6}{s^2 + B_1 s + B_2} \tag{5.44}$$

In the above, the numerator coefficients of s are:

$$\begin{cases} A_1 = b_{w(1,1)} \\ A_2 = b_{w(1,2)}b_{f(1,1)} + b_{w(1,1)}\left(a_{f(1,1)} - a_{a(1,1)}b_{f(1,6)} \right) \end{cases}$$

$$A_3 = b_{w(1,2)}\left(b_{f(1,2)} + b_{f(1,6)}b_{a(1,1)}\right)$$

$$\begin{cases} A_4 = a_{w(1,2)} \\ A_5 = a_{w(1,2)}\left(a_{f(1,1)} - a_{a(1,1)}b_{f(1,6)}\right) \end{cases}$$

$$A_6 = b_{w(1,2)}b_{f(1,6)}b_{a(1,2)}$$

and the denominator coefficients of *s* are:

$$\begin{cases} B_1 = a_{w(1,1)} + a_{f(1,1)} - a_{a(1,1)}b_{f(1,6)} \\ B_2 = a_{w(1,1)}\left(a_{f(1,1)} - a_{a(1,1)}b_{f(1,6)}\right) - b_{w(1,2)}b_{f(1,3)} \end{cases}$$

Alternatively, if outlet air temperature is of interest, Equations 5.38–5.40 can be re-arranged to give the transfer function set for T'_{ao}:

$$\frac{T'_{ao}(s)}{u'_w(s)} = \frac{A_7 s + A_8}{s^2 + B_1 s + B_2} \tag{5.45}$$

$$\frac{T'_{ao}(s)}{u'_a(s)} = \frac{A_9 s^2 + A_{10} s + A_{11}}{s^2 + B_1 s + B_2} \tag{5.46}$$

$$\frac{T'_{ao}(s)}{T'_{wi}(s)} = \frac{A_{12}}{s^2 + B_1 s + B_2} \tag{5.47}$$

$$\frac{T'_{ao}(s)}{T'_{ai}(s)} = \frac{A_{13} s^2 + A_{14} s + A_{15}}{s^2 + B_1 s + B_2} \tag{5.48}$$

for which the numerator coefficients of *s* are:

$$\begin{cases} A_7 = a_{a(1,1)}b_{f(1,1)} \\ A_8 = a_{a(1,1)}\left(a_{w(1,1)}b_{f(1,1)} + b_{w(1,1)}b_{f(1,3)}\right) \end{cases}$$

$$\begin{cases} A_9 = b_{a(1,1)} \\ A_{10} = a_{a(1,1)}b_{f(1,2)} + b_{a(1,1)}\left(a_{w(1,1)} + a_{f(1,1)}\right) \\ A_{11} = a_{w(1,1)}a_{a(1,1)}b_{f(1,2)} + b_{a(1,1)}\left(a_{w(1,1)}a_{f(1,1)} - b_{w(1,2)}b_{f(1,3)}\right) \end{cases}$$

$$A_{12} = a_{w(1,2)}a_{a(1,1)}b_{f(1,3)}$$

$$\begin{cases} A_{13} = b_{a(1,2)} \\ A_{14} = b_{a(1,2)}\left(a_{w(1,1)} + a_{f(1,1)}\right) \\ A_{15} = b_{a(1,2)}\left(a_{w(1,1)}a_{f(1,1)} - b_{w(1,2)}b_{f(1,3)}\right) \end{cases}$$

Figures 5.4(a) & 5.4(b) show the transfer function sets for both T'_{wo} and T'_{ao} expressed in customary block diagram structures. It should be evident that for situations in which acceptable

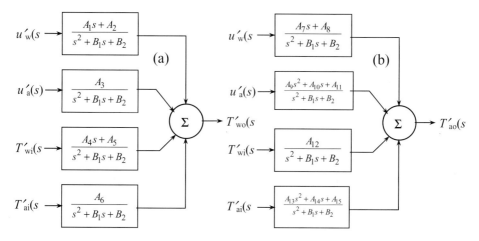

Fig. 5.4   Block diagram model of air-to-water heat exchange process. (a) Water outlet temperature. (b) Air outlet temperature.

model performance requires that n be greater than 1 (and, certainly, 2) the algebra involved in expressing transfer functions will become intractable. In such cases the state-space form of model described earlier will be the preferred choice.

As a comparison, unit step responses for all output variables in this linearised modelling approach are compared for the typical heat exchanger of 'moderate' thermal capacity considered earlier ($\tau_{dom} = 30\,\text{s}$). Here, the linearised state-space model with $n = 3$ represented by Equations 5.35–5.37 is compared with the linearised transfer function model (i.e. $n = 1$) depicted by Equations 5.41–5.48 in Figures 5.5 and 5.6. In common with the earlier findings (i.e. Fig. 5.2) the inadequate asymptotic performance inherent in the low order case ($n = 1$) is evident.

5.2 Modelling control elements

Control valves

Control valves are used in HVAC applications to regulate heat emission through flow rate. The relationship between these two variables can be linear, moderately linear or highly non-linear. Steam heat exchange processes, for example, exhibit linear emission characteristics at partial flow and certain chilled water applications such as chilled ceilings that have a narrow water-side temperature differential can exhibit only moderate non-linearity. Conventional hot water heating and chilled water cooling applications on the other hand exhibit severe non-linearity between heat transfer and water flow rate. An attempt is made to compensate for any non-linearity through the choice of control valve characteristic. The control valve characteristic expresses the relationship between the heating or cooling fluid passed by the valve and the valve stem position. This characteristic can be expressed in two ways – the inherent characteristic relates to the valve characteristic in the absence of connected system effects (i.e. system pressure variations) whereas the installed characteristic includes these effects.

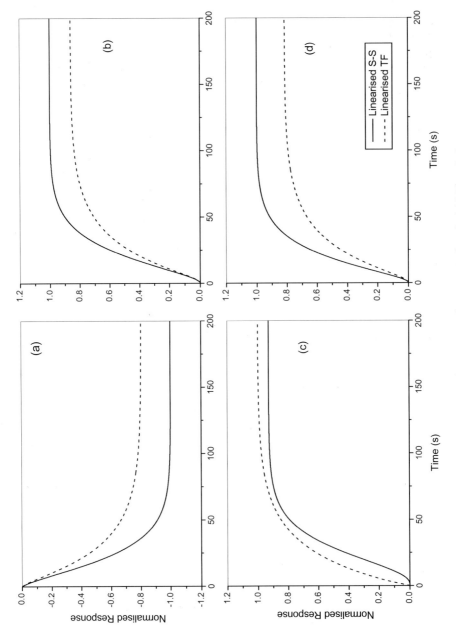

Fig. 5.5 Typical water outlet temperature responses. (a) u'_w excited. (b) u'_a excited. (c) T'_{wi} excited. (d) T'_{ai} excited.

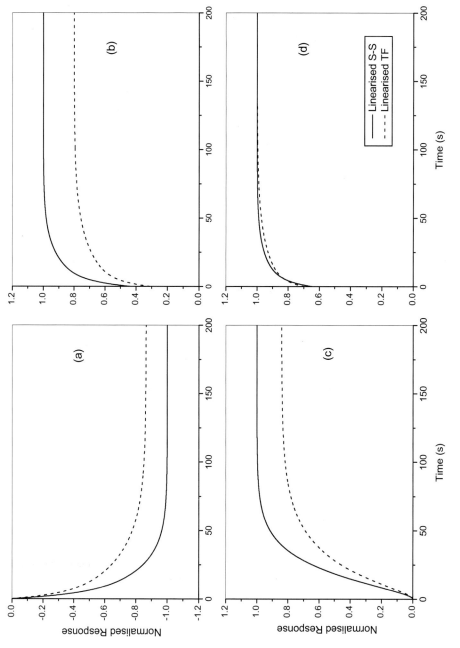

Fig. 5.6 Typical air outlet temperature responses. (a) u'_w excited. (b) u'_a excited. (c) T'_{wi} excited. (d) T'_{ai} excited.

Figure 5.7 illustrates the inherent characteristic of the two main valve characteristics – linear and equal percentage. The characteristic is maintained over a specific range of valve flow rate and thus the valve rangeability, R_v, is defined as:

$$R_v = V_{nom}/V_{min} \tag{5.49}$$

where V_{nom} is the nominal or rated volume flow rate passed by the valve when fully open and V_{min} is the volume flow rate at the minimum point on the maintained characteristic.

It is possible to show (e.g. Underwood 1999) that the inherent characteristic, G_v, expressed as the normalised flow rate of a control valve, can be obtained from the following for an equal percentage valve:

$$G_v = R_v^{(p_v-1)} \tag{5.50}$$

and, for a valve with a linear inherent characteristic:

$$G_v = \frac{1}{R_v} \times [1 + p_v \times (R_v - 1)] \tag{5.51}$$

where p_v is the valve stem position ($0 \leq p_v \leq 1$) corresponding to the range of valve operation for which the characteristic is maintained.

For that part of the operating valve position for which the characteristic is not maintained the flow/position relationship is uncertain, though this will generally reflect a very small range of valve positions between fully closed and some position, p_{v0} (Fig. 5.7). A pragmatic approach might be to assume that this region be represented by a linear relationship between flow and stem position from some cut-off flow rate, V_0, which represents any leakage across the valve at closure. Introducing L_v as the valve leakage fraction ($= V_0/V_{nom}$) and re-arranging the above with this linear assumption gives the following complete inherent characteristic for an equal percentage valve:

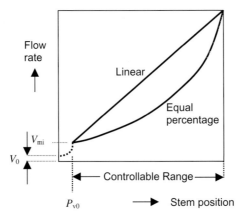

Fig. 5.7   Inherent valve characteristics.

$$G_v = R_v^{(p_v - 1)/(1 - p_{v0})} \qquad (p_{v0} \le p_v \le 1)$$

$$G_v = L_v + \frac{p_v}{p_{v0}} \times (1/R_v - L_v) \quad (0 \le p_v \le p_{v0}) \qquad (5.52)$$

and, for a linear valve:

$$G_v = (1 + p_v R_v - p_{v0} R_v - p_v)/[R_v \times (1 - p_{v0})] \qquad (p_{v0} \le p_v \le 1)$$

$$G_v = L_v + \frac{p_v}{p_{v0}} \times (1/R_v - L_v) \qquad (0 \le p_v \le p_{v0}) \qquad (5.53)$$

For a control valve in a system, pressure variations can be accounted for by defining an installed characteristic, G_v' (Underwood 1999):

$$G_v' = G_v \times [G_v^2 \times (1 - N_v) + N_v]^{-\frac{1}{2}} \qquad (5.54)$$

where N_v is the valve authority, defined as:

$$N_v = \Delta P_v / (\Delta P_v + \Delta P_s) \qquad (5.55)$$

in which ΔP_v and ΔP_s are the nominal valve and system pressure differentials respectively, at nominal design conditions. The system pressure differential is with respect to that part of the circuit over which the control valve is required to influence control – Fig. 5.8 illustrates this for applications involving three-port control valves.

A typical value for N_v is 0.5 (further details can be found in CIBSE 2000).

Finally, the flow rate passed by the valve will be:

$$V = V_{nom} \cdot G_v' \qquad (5.56)$$

Control dampers

For control dampers in ventilation plant, the choice of inherent characteristic will normally be linear and, in common with control valves, a damper authority N_d can be defined:

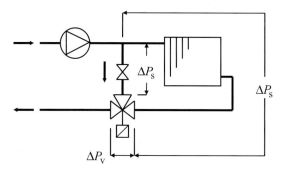

Fig. 5.8   Valve authority pressure differentials for three-port valves.

$$N_d = \Delta P_d / (\Delta P_d + \Delta P_s) \tag{5.57}$$

where ΔP_d and ΔP_s are the damper pressure difference at the nominal (rated) volume flow rate and the connecting system pressure difference in that part of the ducting network for which the damper is required to regulate flow.

Little work has been reported on the subject of damper characteristics for ventilation applications; Legg (1986) gives perhaps the fullest range of results as well as reviewing earlier results by other workers; and Hung & Lam (2003) have considered the case of single blade dampers in square ducts. It is customary to express the damper position in terms of degrees of blade rotation, α:

$$\alpha = p_d(\alpha_0 - 90) + 90 \tag{5.58}$$

where α_0 is the damper blade angle when fully open (typically $0° \leq \alpha_0 \leq 10°$) and p_d is the entering damper positioning signal ($0 \leq p_d \leq 1$). Legg (1986) then recommends the following for the damper inherent characteristic:

$$G_d = \exp[b(\alpha_0 - \alpha)/2] \tag{5.59}$$

where b is an empirical constant which depends on the damper blade profile, number of blades and blade action (opposed or parallel travel). Typically $0.075 < b < 0.11$ (Legg 1986).

The installed characteristic for a damper can then be determined using the same form of expression use earlier for control valves:

$$G'_d = G_d[G_d^2(1 - N_d) + N_d] \tag{5.60}$$

Applications for control dampers in ventilation systems more or less fall into three areas:

- Mixing dampers in central air handling plants for variable fresh and recirculated air (Fig. 5.9(a))
- Face and bypass dampers for regulating the capacity of cooling coils and heat recovery plant (Fig. 5.9(b))
- Throttling and variable air volume applications

For mixing dampers, the damper authority for the fresh and exhaust dampers based on Equation 5.57 will be:

$$N_{d(f,e)} = \Delta P_{d(f,e)} / (P_{s(f,e)} + \Delta P_{d(f,e)}) \tag{5.61}$$

in which the subscript (f,e) refers to either fresh or exhaust as appropriate and P_s is the system static pressure between inlet or discharge as appropriate. For the recirculating air damper:

$$N_{dr} = \Delta P_{dr} / (P_{se} + P_{sf} + \Delta P_{dr}) \tag{5.62}$$

Damper angles for fresh air and exhaust air dampers will climb with increasing control signal to effect free cooling whereas the recirculating damper will fall:

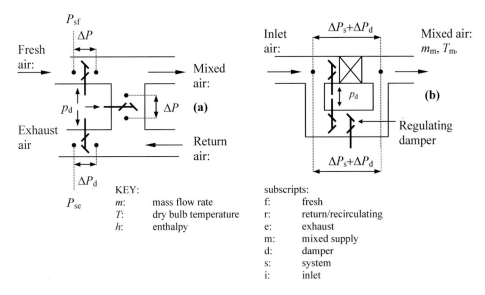

Fig. 5.9   Typical control damper configurations. (a): Mixing. (b): Face and bypass.

Fresh/exhaust air dampers: $\alpha_{(f,e)} = p_d \times (\alpha_{0(f,e)} - 90) + 90 \quad (p_{d0} \le p_d \le 1)$ (5.63)

Recirculating air damper: $\alpha_r = (1 - p_d) \times (\alpha_{0r} - 90) + 90 \quad (p_{d0} \le p_d \le 1)$ (5.64)

where p_{d0} is an (optional) minimum damper position, which can be set to ensure an acceptable minimum fresh air input by the plant. To effect a limit to free cooling (sometimes referred to as 'economy cycle control') an additional constraint can be set:

$$p_d = p_{d0} \quad \text{when} \quad h_f > h_r \tag{5.65}$$

G_d and G'_d can now be calculated for each control damper using Equations 5.59 and Equation 5.60.

Plant variables can now be expressed:

$$m_f = G'_{df} \cdot \rho \cdot V_{nom} \tag{5.66}$$

$$m_r = G'_{dr} \cdot \rho \cdot V_{nom} \tag{5.67}$$

$$m_m = m_f + m_r \tag{5.68}$$

$$m_e = m_m - m_r \tag{5.69}$$

$$T_m = (m_f T_f + m_r T_r)/m_m \tag{5.70}$$

$$h_m = (m_f h_f + m_r h_r)/m_m \tag{5.71}$$

For face and bypass dampers the regulating damper in the bypass path (Fig. 5.9(b)) ensures that the resistance through both paths is identical at equivalent flow rates. Thus the treatment of this case will be very similar to the above procedure for the fresh air and recirculating air dampers.

In both of these cases, the overall air flow rate and pressure difference across each damper set will be near constant provided that the correct choice of damper authority is made; (a treatment of damper authority can be found in Underwood (1999). In the third application mentioned above – that of throttling control such as is applicable with variable air volume (VAV) systems, the pressure/volume relationship varies with control damper position. Thus an alternative method is required in which the flow and pressure difference across the damper can be independently determined. Khoo *et al.* (1998) experimentally evaluated a variety of VAV terminals and recommended a model of the following form for the terminal pressure loss coefficient, ζ:

$$\zeta = \exp(A + B \cdot \alpha^a + C \cdot \alpha^b) \tag{5.72}$$

where A, B, C, a, and b are empirical constants. Thus a relationship between pressure and flow rate can be determined using the methods described in Chapter 4. However, when cross checking the performance of the above model with a method based on the results of Legg (1986), substantial differences where found to exist. It is therefore recommended that the constants in both sources be treated with caution and, wherever possible, original test data should be obtained specific to each control damper modelling case to be investigated.

Actuators

Two considerations permit a model to be expressed for the dynamic behaviour of valve or damper positioning devices (actuators). The first of these relates to the running speed of the actuator and the second relates to any slackness take-up in the linkage mechanism between the actuator and the device being positioned. This slackness results in a hysteresis effect between the output position of the actuator, p', (which forms the input position of the linkage mechanism) and the output position of the linkage mechanism, p, (which forms the input or actual position of the valve or damper) – Fig. 5.10.

The running speed, S, of an actuator can vary; for example, the use of controller-actuators where the running speed is slowed when the controlled variable is near set-point, or the use of dual mono-directional motor positioners in which it is possible to have differing running speeds for differing positioning directions. For general-purpose positioners, it is reasonable to assume that the actuator positioning rate can be matched to the incoming control signal rate subject to the restriction of the actuator maximum running speed S, as follows:

$$p'(t) = \Delta t \cdot S^+ + p(t-1) \quad \left(\frac{du}{dt} > S^+\right)$$

$$p'(t) = \Delta t \cdot S^- + p(t-1) \quad \left(\frac{du}{dt} < S^-\right)$$

$$p'(t) = u(t) \quad \left(S^- \leq \frac{du}{dt} \leq S^+\right) \tag{5.73}$$

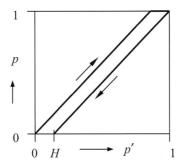

Fig. 5.10 Actuator hysteresis.

where $p'(t), p'(t-1)$ are current and previous time row values of actuator output position respectively, $u(t)$ is the current incoming control signal, Δt is the time interval between control signal updates and S^+, S^- are the actuator running speeds in each positioning direction (usually $S^+ = S^-$ for bi-directional actuators).

With a change in direction of actuator output position, p', any slack in the linkage mechanism, H, is required to be relieved before a movement in the linkage and, hence, damper or valve can be realised. Usually, H will be small (less than 1% of the nominal flow rate of the valve or damper) but can increase in time with component wear. It is evident from Fig. 5.10 that for an increasing signal:

$$p = (p' - H)/(1 - H) \tag{5.74a}$$

and for a declining signal,

$$p = p'/(1 - H) \tag{5.74b}$$

A turning point in the actuator position can be detected at $\dfrac{dp'}{dt} = 0$ (in sampled-data or stepwise numerical modelling applications this will be evident from a change in sign between successive values of $\Delta p'/\Delta t$). Thus an algorithm for dealing with hysteresis effects in the actuator linkage mechanism can be expressed as follows:

$$p(t) = p(t-1) \quad \text{when} \quad \left(\frac{dp'}{dt} = 0 \quad \text{and} \quad |p'(t) - p'(t-1)| \le H \right) \tag{5.75}$$

otherwise:

$$p(t) = (p'(t) - H)/(1 - H) \qquad \text{when} \quad ((p'(t) - p(t-1)) > H)$$

or:

$$p(t) = p'(t)/(1 - H) \qquad \text{when} \quad ((p'(t) - p(t-1)) < -H)$$

Variable speed drives

Though a wide variety of variable speed drives are available (mainly for applications involving fans and pumps – see for example Underwood 1999) current practice is beginning to con-

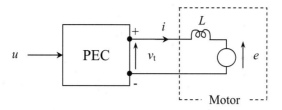

Fig. 5.11   PEC-driven motor drive.

verge on the use of power electronic converter (PEC) devices that act directly on either a.c. (induction or synchronous) or d.c. motors – a simplified layout is given in Fig. 5.11. Most of the PECs used in a.c. motor-driven building plant use pulse width modulation to construct a variable frequency output to drive the motor and it is reasonable to suppose for modelling purposes that processing delay will be minimal and that the ultimate output will have a very close linear association with applied input signal. That is:

$$v_t = K_{PEC} \times u \tag{5.76}$$

where v_t is the output voltage applied to the motor terminals, K_{PEC} is the PEC gain and u is the control signal applied to the PEC.

The applied motor voltage at steady state is required to balance with the motor-induced e.m.f. plus winding losses so that:

$$\frac{di}{dt} = \frac{v_t - (e + R_m i)}{L}$$

where i is the motor current, e the induced e.m.f., R_m the resistance in the motor winding circuit and L is the inductance of the winding circuit.

Since the motor e.m.f. varies approximately as the rotor speed, it is possible to express the equation for motor current as:

$$\frac{di}{dt} = \frac{v_t - (k_e \omega + R_m i)}{L} \tag{5.77}$$

where ω is the angular rotor speed (i.e. $\omega = 2\pi N$ where N is the rotor speed in revolutions per second) and k_e is a constant.

The electromagnetic torque delivered by the motor is given by the following if friction losses are neglected:

$$T_{em} = T_L + J\frac{d\omega}{dt}$$

in which T_L is the load torque and J is the motor inertia and since the electromagnetic torque can be assumed to vary approximately as the winding current, i:

$$k_T i = T_L + J\frac{d\omega}{dt} \tag{5.78}$$

where k_T is a constant.

Rationalising the above leads to a simple linear model for a direct variable speed motor:

$$\frac{di}{dt} = \left(LK_{PEC}u - k_e\omega - R_m i\right)/L \tag{5.79}$$

$$\frac{d\omega}{dt} = \left(k_T i - T_L\right)/J \tag{5.80}$$

For simulations around a nominal operating point, i, ω, u and T_L can be expressed as deviation variables (i', ω', u', T_L') and Laplace transforms can be taken enabling the above to be expressed as the block diagram model of Fig. 5.12 suitable for linear control system analysis.

Transport delays

Transport lags are frequently encountered in building system modelling due to the distributed nature of inter-connected plant and systems. A transport lag represents a pure time delay of distance divided by velocity time units arising from the time taken for a signal to travel between two points, i.e.:

$$t_{delay} = l/u$$

To model this as a transfer function simply apply a unit delay operator scaled by t_{delay}:

$$\frac{\phi_o'(s)}{\phi_i'(s)} = \exp\left(-t_{delay}s\right) \tag{5.81}$$

where ϕ_i', ϕ_o' represent deviations in input signal and output signal respectively. It is customary to represent Equation 5.81 with the first order term of the Taylor series (i.e. first-order Padé

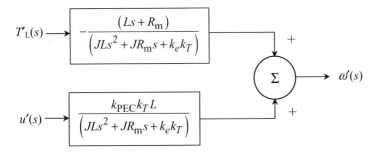

Fig. 5.12   Block diagram model of a variable speed drive mechanism.

approximation) making the delay operator more manageable for block diagram algebra treatment:

$$\exp\left(-t_{\text{delay}}s\right) \cong \frac{1 - t_{\text{delay}}s/2}{1 + t_{\text{delay}}s/2} \tag{5.82}$$

Sensor response

Two issues in sensor response as far as modelling for control applications is concerned relate to sensor linearity and sensor time response. In applications related to building heat and fluid flow, most measurement systems exhibit non-linearity between the applied input and corresponding output signals. Equations 5.83 and 5.84 for example, give the resistance – temperature relationships for typical thermistor and resistance temperature sensors respectively:

$$R = R_0 \exp[\beta(1/T - 1/T_0)] \tag{5.83}$$

$$R = R_0\left(1 + \alpha_1 T + \alpha_2 T^2 + \ldots \alpha_n T^n\right) \tag{5.84}$$

(where R is the electrical resistance across the sensor element, R_0 is a reference resistance, T is the sensed temperature, T_0 a reference temperature and $\beta, \alpha_1 \ldots \alpha_n$ are constants). Similar expressions are possible for capacitance type humidity sensors frequently used in building services applications, and for flow measurements many measurements systems make use of the pressure drop across a device such as an orifice or venturi. This then requires a square root function to translate velocity. Commercial sensors are generally supplied for application with all the appropriate signal conditioning needed to give a more or less linear relationship between the applied input and corresponding output signals, so from a modelling point of view it will frequently be acceptable to assume that the predicted state variables from the model require no further adjustment other than for time response considerations.

As to time response, many applications involve measuring variables in rapidly flowing streams such as in ducts or pipes, and this will usually result in a very responsive measurement. One commonly encountered situation in which sensor time response does become an issue, however, is when measurements are taken in slow moving or nominally still air, such as is the case with room air temperature measurements. This situation can be described using a first order model with delay time:

$$\frac{\phi_o'(s)}{\phi_i'(s)} = \frac{\exp\left(-t_{\text{delay}}s\right)}{(\tau s + 1)} \tag{5.85}$$

where ϕ_o', ϕ_i' are sensor output and input signals respectively, t_{delay} is the delay time of the sensor and τ the sensor time constant.

Practical values of t_{delay} and τ vary from a few seconds to a few minutes depending mainly on velocity of flow passing over the sensing head. Adams & Holmes (1977) recommended values for these parameters for 'traditional' thermistors and resistance temperature detectors (RTDs) tending to have slow thermal response (given here in the form of curve fits to Adams & Holmes' data as reported by Underwood 1999):

Thermistor: $t_{\text{delay}} = 19.37 - 8.93u_{\text{a}} + 2.426u_{\text{a}}^2 + 0.2825u_{\text{a}}^3 + 0.0117u_{\text{a}}^4$ (5.86)

Thermistor: $\tau = 151.5 - 70.8u_{\text{a}} + 18.04u_{\text{a}}^2 - 2.023u_{\text{a}}^3 + 0.0809u_{\text{a}}^4$ (5.87)

RTD: $t_{\text{delay}} = 21.69 - 8.518u_{\text{a}} + 1.942u_{\text{a}}^2 - 0.1974u_{\text{a}}^3 + 0.0073u_{\text{a}}^4$ (5.88)

RTD: $\tau = 198.9 - 95.4u_{\text{a}} + 23.16u_{\text{a}}^2 - 2.5u_{\text{a}}^3 + 0.097u_{\text{a}}^4$ (5.89)

where u_{a} is the velocity of air passing over the sensing head. Nowadays, the advent of semi-conductor-based sensors results in more responsive signal processing, and lower values of model parameters than those implied above are likely to be appropriate. Instrument manufacturers should always be consulted wherever possible for details of time response data.

For applications involving measurements in rapidly moving air or water streams t_{delay} will be very low and τ, as a maximum, will be a few seconds and frequently a few milliseconds.

5.3 Modelling control algorithms

Conventional fixed-parameter controllers

The classical proportional plus integral plus derivative (PID) controller can be represented by the following:

$$u = u_0 + K \times \left(\varepsilon + \frac{1}{t_1} \int_{t=0}^{t=\infty} \varepsilon dt + t_{\text{D}} \frac{d\varepsilon}{dt} \right)$$ (5.90)

where:

u = control signal (conventionally $0 \le u \le 1$)
u_0 = control signal when the plant is operating at a stationary condition ($\varepsilon = 0$)
K = controller gain
ε = control error ($\varepsilon = (r - \phi)$)
t_1 = integral action time
t_{D} = derivative action time
t = time

A transfer function form can be expressed with respect to deviations in error and control signal as:

$$\frac{u(s)}{\varepsilon(s)} = \frac{K}{s} \times \left(t_{\text{D}} s^2 + s + \frac{1}{t_1} \right)$$ (5.91)

In practice the PID control algorithm is implemented in discretised form and thus for many modelling applications it will be appropriate to use this form. At the jth sampling time instant:

$$u(j) = a \cdot \varepsilon(j) - b \cdot \varepsilon(j-1) + c \cdot \varepsilon(j-2) + u(j-1) \qquad (5.92)$$

where:

$$a = \frac{K}{T \cdot t_{\mathrm{I}}}\left(T^2 + T \cdot t_1 + t_1 \cdot t_{\mathrm{D}}\right)$$

$$b = \frac{K}{T}\left(T + 2t_{\mathrm{D}}\right)$$

$$c = \frac{K}{T} t_{\mathrm{D}}$$

in which T is the sampling time interval.

Besides modelling a more realistic implementation of the PID controller, there are two distinct advantages attributable to the discretised form of Equation 5.92. First, the subtraction in time implied by discretisation results in the elimination of the steady-state signal value u_0 and, thus, the need to set a value for it. Second, in the continuous forms of the PID algorithm (Equations 5.90 and 5.91) the integral term acts on the integral of the error with respect to time, and this can result in wind-up of the integral term during periods of prolonged error. The effect of wind-up is that the control signal saturates at either of its limits until a sufficient period of error with a sign opposite to that which caused the wind-up saturation in the first place is applied in order to eradicate the large amount of accumulation in the integral term. In the discretised form of the PID algorithm the 'integral' term acts on the current value of error only and, thus, wind-up is avoided.

It is therefore recommended that, when using the continuous form of the PID algorithm such as is given by Equations 5.90 and 5.91, a suitable saturation limit is applied to the integral term.

Adaptive control systems

In the conventional controller described above, the controller parameters K, t_{I} and t_{D} require to be set. Most controllers today possess, at least to a limited extent, an ability to self-tune or adapt. Essentially, this amounts to a capability within the control algorithm to adjust its own parameters to suit the environment within which it operates. Figure 5.13 illustrates the algorithm that is most frequently encountered for the control of some variable, ϕ, in response to a set point r. A model of the plant (implemented online in a practical interpretation of the controller) is used to predict plant response one sampling interval ahead and this prediction is used to determine adjustments to the parameters of a controller that minimise a cost function.

It is possible to write down an expression for the controller. A general form of linear controller with tunable parameter sequences, K_1, K_2, K_3, is frequently used:

$$K_1 u(t) = K_2 r(t) - K_3 \phi(t) \qquad (5.93)$$

where, for sequences of up to $n(k_1)$, $n(k_2)$, $n(k_3)$ non-negligible coefficients forming K_1, K_2 and K_3 respectively:

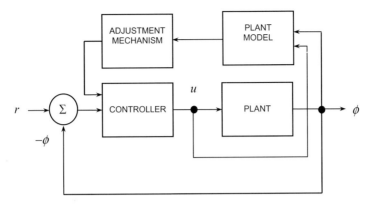

Fig. 5.13 Structure of an adaptive controller.

$$K_1 = 1 + k_{1,1}z^{-1} + \ldots k_{1,n(k_1)}z^{-n(k_1)}$$

$$K_2 = k_{2,0} + k_{2,1}z^{-1} + \ldots k_{2,n(k_2)}z^{-n(k_2)}$$

$$K_3 = k_{3,0} + k_{3,1}z^{-1} + \ldots k_{3,n(k_3)}z^{-n(k_3)}$$

and z is the unit time delay operator (by definition, $z = \exp(sT)$).

Online implementation requires a model of the plant to be generated (and updated) using measurements of plant input (u) and output (ϕ). Thus to simulate this controller, the related plant model conditions would be used to supply these data. Online system models can be generated using autoregressive (AR) methods, a common approach being the 'ARMAX' structure ('X' implies the presence of an exogenous input and the model responds to a moving average value of its input) because this allows one or more disturbance inputs, $e(t)$:

$$\phi(t) = u(t)\frac{B}{A} + e(t)\frac{C}{A} \tag{5.94}$$

where A, B, C are time series sequences of coefficients:

$$A = 1 + a_1 z^{-1} +, \ldots a_{n_A} z^{-a_{n_A}}$$

$$B = b_0 + a_1 z^{-1} +, \ldots b_{n_B} z^{-b_{n_B}}$$

$$C = c_0 + c_1 z^{-1} +, \ldots c_{n_C} z^{-c_{n_C}}$$

The sequences A and B require to be estimated whereas C will be known from disturbance measurements.

Adaptation or self-tuning of the controller now amounts to two operations.

(1) System identification; essentially estimating the sequences A, B.
(2) Controller adaptation; using the parameters of the system model and an adjustment mechanism, determine the coefficients of the controller, K_1, K_2, K_3.

These two operations are carried out on-line during each sampling interval.

Recursive estimation

For adaptive control applications the plant variables can be observed periodically in real time, and it is therefore possible (and desirable) to obtain a fresh estimate of the plant model parameters at each time step. Of essence is the need to carry out the computations associated with adaptive control comfortably within each sampling interval. To save computational effort, adaptive controllers make use of recursive least squares (RLS) estimation such that measurement observations made at time t are used to obtain plant parameter estimates for time $t + T$. Thus computational efficiency is achieved by requiring the storage of data from the previous time step only and further computational efficiency is achieved by avoiding the need for a complicated matrix inversion operation through the use of a 'matrix inversion lemma'. The method is summarised below; for a detailed treatment, readers are referred to Åström & Wittenmark (1989) or Ljung (1987).

Letting the sequences A and B be represented by:

$$\hat{\theta} = \left[a_1\ a_2\ \dots a_{n_A}\ b_1\ b_2\ \dots b_{n_B} \right]^{\mathrm{T}}$$

then an RLS estimate at time t can be expressed as:

$$\hat{\theta}(t) = \hat{\theta}(t+T) + \mathbf{K}(t)\left[\varphi(t) - \mathbf{X}^{\mathrm{T}}(t)\hat{\theta}(t-T) \right] \tag{5.95}$$

where the Kalman gain, $\mathbf{K}(t)$, is given by:

$$\mathbf{K}(t) = \frac{\mathbf{P}(t-1)\mathbf{X}(t)}{\lambda + \mathbf{X}^{\mathrm{T}}(t)\mathbf{P}(t-T)\mathbf{X}(t)}$$

and:

$$\mathbf{P}(t) = \left(\mathbf{I} - \mathbf{K}(t)\mathbf{X}^{\mathrm{T}}(t) \right) \times \mathbf{P}(t-T)\lambda^{-1}$$

The method assumes that parameters, once estimated, are constant. In many cases a time weighting of earlier data is desirable. A typical need for this is where parameters change slowly, and estimation then becomes unresponsive to a subsequent rapid change. The constant λ is introduced to deal with this, causing current data to have unit weighting but previous sampled data to be diminished exponentially. Thus λ is an exponential-forgetting factor. Its values lie, typically, in the range 0.95–0.99. One point of caution here is that exponential-forgetting only works when the system variables receive continuous excitation. Where the system or process spends long periods without excitation (for example systems which are designed mainly to accommodate periodic set point changes only) then discarding old data tends to lead to uncertainty in the estimated parameters – a condition known as estimator wind-up (Åström and Wittenmark 1989).

A recursive least squares algorithm can be summarised as follows.

STEP 1:

At time $= t$, form $\mathbf{X}(t)$ based on measurements of $\phi(t)$, $u_1(t-T)$, $\dots$, $u_m(t-mT)$ (i.e. for up to m time series inputs)

STEP 2:
Form $\mathbf{P}(t)$ in Equation 5.95
STEP 3:
Form $\mathbf{K}(t)$ in Equation 5.95
STEP 4:
Update $\hat{\boldsymbol{\theta}}(t)$ using Equation 5.95
STEP 5:
At time $= t + T$, go back to STEP 1

An increasingly popular alternative to AR modelling is to resort to the use of an artificial neural network (ANN) for prediction. In some cases, ANNs can provide better and more enduring results than those given by AR methods (e.g. Gouda *et al.* 2002 – for more general details of ANNs applicable to building energy plant see Underwood 1999).

Minimum variance control

A variety of methods are available for controller adaptation (for a review of these related to building plant, see Underwood 1999) but perhaps the most frequently applied of the first generation methods for adaptive control is the minimum variance controller (MVC), which seeks to minimise a cost function, J:

$$J = E\big(\phi^2(t + mT)\big) \tag{5.96}$$

where m is the input–output delay in sampling intervals and E is a polynomial in the delay operator z).

Let the predicted output from the system be $\phi(t)$ and the corresponding prediction error be $\phi_\varepsilon(t)$ such that: $\phi(t) = \phi(t) + \phi_\varepsilon(t)$. Then:

$$J = E\big(\hat{\phi}(t + mT) + \phi_\varepsilon(t + mT)\big)^2 = E\hat{\phi}^2(t + mT) + E\phi_\varepsilon^2(t + mT) + 2E\big(\hat{\phi}(t + k) + \phi_\varepsilon(t + k)\big)$$

For the linear controller of the form of Equation 5.93 the final term in the above disappears since only the non-linear terms will contribute to J. Also $\phi_\varepsilon(t + mT)$ is not influenced by the controller, merely by the model predictor; hence for minimum variance control it is necessary and sufficient to require that:

$$J = E\hat{\phi}^2(t + mT) = 0 \quad \text{or:} \quad \hat{\phi}(t + mT) = 0 \tag{5.97}$$

Hence for the general ARMAX case:

$$\phi(t + mT) = \frac{B}{A}u(t) + \frac{C}{A}e(t + mT) \tag{5.98}$$

It is possible to show that:

$$\hat{\phi}(t + mT, t) = BFu(t) + G\phi(t) = 0$$

in which $\hat{\phi}(t+mT, t)$ implies that the estimate of $\phi(t+mT)$ is based on data received up to, and including, t. From this the minimum variance controller can now be defined:

$$u(t) = -\frac{G}{BF}\phi(t)$$
(5.99)

where the polynomial coefficients F and G can be found from the polynomial identity (Wellstead & Zarrop 1991):

$$C = AF + z^{-m}G$$
(5.100)

and take the form:

$$F = 1 + f_1 z^{-1} + \ldots f_{m-1} z^{-(m-1)}$$
$$G = g_0 + g_1 z^{-1} + \ldots g_{n_G} z^{-n_G} \quad (n_G = \max(n_A - 1, n_C - m))$$

The result is a controller which gives the following output (Wellstead & Zarrop 1991):

$$\phi(t) = Fe(t)$$
(5.101)

and minimum output variance:

$$J_{\min} = \left(1 + f_1^2 + \ldots f_{m-1}^2\right) \times \sigma_e^2$$
(5.102)

(where σ_e^2 is the disturbance variance).

When implemented with respect to an error signal the MVC output becomes:

$$u(t) = \frac{G}{BF}(r(t) - \phi(t))$$
(5.103)

Model reference adaptive control

Model reference adaptive control (MRAC) is an interesting alternative to the use of an MVC scheme in that the controller seeks to force the plant to follow a fixed reference model. The reference model generally takes the form of one (or a set of) difference equations that calculate desired plant response from on-line measurements of current and previous plant input and output signals. The error between the model-predicted response and actual measurement is used as a basis for adjusting the controller parameters. This approach can be used when a reliable model of the plant can easily be expressed and to which the actual plant response can conceivably be matched.

Suppose the model generates an output ϕ_m then the error between the model output and the actual plant output will be:

$$\varepsilon_\phi = \phi - \phi_m$$
(5.104)

and the general MRAC system can therefore be realised as shown in Fig. 5.14.

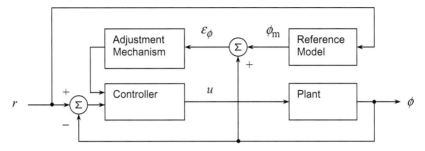

Fig. 5.14   Structure of a MRAC.

Of essence, the MRAC adjustment mechanism requires to minimise ε_ϕ through the setting of an adjustable controller parameter set, K. One simple approach is the gradient method (Landau 1979) in which the controller parameters are adjusted according to a sensitivity derivative $\dfrac{\partial \varepsilon_\phi}{\partial K}$ such that:

$$\frac{dK}{dt} = -\gamma \varepsilon_\phi \frac{\partial \varepsilon_\phi}{\partial K} \tag{5.105}$$

This assumes that the controller parameters can be changed according to the negative gradient of ε_ϕ^2, scaled by a constant, γ, which is called the adaptation gain. If it can be assumed that K will vary more slowly than other system variables then $\dfrac{\partial \varepsilon_\phi}{\partial K}$ can be found quite easily with the assumption that K is 'constant' (or at least for the duration of a sampling interval).

Fuzzy logic control

Considerable interest in fuzzy logic control (FLC) of HVAC plant throughout the 1990s has more or less eclipsed progress in many of the areas of conventional adaptive control because of the potential ease with which a FLC can be defined for a given application as well its robustness.

A FLC has three mechanisms (Fig. 5.15):

- A fuzzy input mechanism or 'fuzzifier'
- A fuzzy inference mechanism
- A fuzzy output mechanism or 'defuzzifier'

The fuzzifier accepts one or more inputs as 'crisp' numerical values and maps these onto one or more fuzzy input sets (FIS). The fuzzy inference mechanism then maps the resulting FIS onto a single aggregated fuzzy output set (FOS) using rule-based reasoning and the defuzzifier translates the aggregated FOS into a crisp output (i.e. control signal). Thus a FLC simply maps an input space onto an output space through the mechanism of a set of linguistic rules.

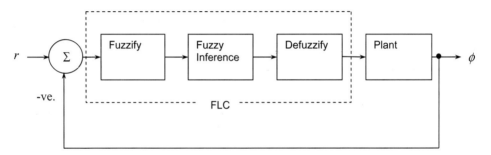

Fig. 5.15 General structure of a FLC.

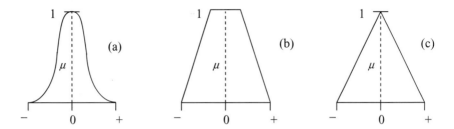

Fig. 5.16 Typical fuzzy membership functions. (a): Gaussian; (b): Trapezoidoidal; (c): Triangular.

Membership functions and fuzzy sets

The mapping of a crisp point onto a FIS is achieved through fuzzy membership functions. A fuzzy membership function normally takes a value anywhere in the interval 0–1 (though a binary membership or 0 or 1 is possible). Three of the more common types of membership function μ used in fuzzy systems are shown in Fig. 5.16 (the triangular function being most common at least as far as applications in HVAC control are concerned).

These membership functions have the advantage that they can intrinsically achieve a certain amount of input noise suppression. A set of fuzzy membership functions defined across a universe of discourse U forms a fuzzy set.

Fuzzy inference and defuzzification

The fuzzy inference mechanism is the heart of a FLC consisting of a rulebase of IF – THEN rules and a means for translating the (one or more) fuzzy input sets onto an aggregated fuzzy output set based on these rules. Each rule in the rulebase takes the following general form:

$$\text{Rule}_{x,y}: \text{ IF } (U_x \text{ is } A_x) \text{ AND} \ldots \quad \ldots \text{AND} \left(U_y \text{ is } A_y\right) \text{ THEN } (V \text{ is } B) \tag{5.106}$$

where A_x, A_y are fuzzy subsets of the universes of discourse U_x, U_y (i.e. $A_x \subset U_x$ and $A_y \subset U_y$) and B is a subset of the output space, V (i.e. $B \subset V$). The logical OR operator may also appear (and frequently does) in these expressions of rule construction.

There are several mechanisms for inferring fuzzy outputs from fuzzy inputs based on the defined rules – see for example Wang (1997). One of the most common and enduring one is

Mamdani inference (Mamdani & Assilian 1975). First, AND and OR combinations in the antecedents of the rules are resolved using normal Boolean logic with fuzzy membership degrees preserved, then the min operator is used to resolve AND, and the max operator used to resolve OR as follows:

$$A \text{ AND } B \rightarrow \min(A, B)$$
$$A \text{ OR } B \;\; \rightarrow \max(A, B) \tag{5.107}$$

Second, the result from the (resolved) antecedent is applied to the consequent in such a way as to ensure that the consequent enjoys the same degree of fuzzy membership as the antecedent. The following simple illustration shows how a single rule can thus be inferred based on the fuzzy system of Fig. 5.17(a) consisting of two fuzzy input sets, U_1 and U_2, and fuzzy output set, V. Suppose the following rule applies:

IF $(U_1$ is $A_{11})$ AND $(U_2$ is $A_{22})$ THEN $(V$ is $B_1)$

Suppose also that, on some occasion, crisp input points lie at $U_1(-0.5)$ and $U_2(0.6)$. From the FIS it is noted that the membership degree of A_{11} is 0.5 and for A_{22} it is 0.8. The resolved antecedent will therefore be:

$$\min(A_{11}, A_{22}) = \min(0.5, 0.8) \rightarrow 0.5$$

This degree of membership is 'fired' at the FOS resulting in a trapezoidal patch of the B_1 output fuzzy subset with a height of 0.5 representing the required fuzzy output – Fig. 5.17(b) illustrates the inference.

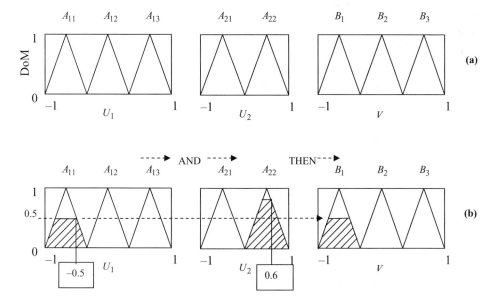

Fig. 5.17   Fuzzy inference illustration.

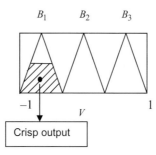

Fig. 5.18   Defuzzification to a crisp output.

Finally, the output is defuzzified by translating the patch of output subset to a crisp point. In practice the output will be an aggregate fuzzy set consisting of all patches of fuzzy outputs fired from each rule added together. The crisp point can be deduced using a number of methods – the average of the maximum patch of the aggregated set; the largest maximum; the smallest maximum; the centroid, to name four such methods. Probably the most common is the centroid, in which centre of the area of the aggregate patch is located by an integration method, though it does involve a calculation effort for each output produced. From the centre point, the crisp output is then read off the horizontal axis as shown in Fig. 5.18 for this illustration.

5.4 Solution schemes

Special difficulties arise in the solution of the ordinary differential equations that arise from the problems discussed in this chapter due to the tendency for modelling cases to be characterised by mixed dynamics. A brief review follows on the nature of the difficulties and their remedies but for more detailed information concerning solution schemes the reader should consult one of the specialist texts on the subject such as Press *et al.* (2002). In the above, most of the methods have generally reduced to sets of ordinary differential equations or transfer functions each of which, in general, can be expressed in the following form:

$$\frac{d\phi}{dt} = f(t, \phi)$$

The simplest way to advance a numerical solution over the interval, h, is as follows:

$$\phi_{n+1} = \phi_n + hf(t_n, \phi_n)$$

in which n refers to points along the solution continuum. In fact this is Euler's method and is equivalent to the forward-in-space discretisation adopted for the partial discretisation of the distributed heat exchange equations discussed earlier (i.e. Equation. 5.10). In practice, Euler's method is likely to prove insufficiently accurate as a general method of solution unless a very low value of interval h is used. Additionally, for many of the mixed dynamics problems encountered in control system simulation, this method will often prove to be unstable. Nonetheless, Euler's method is the basic building block for most of the numerical schemes available for advancing solutions to sets of differential equations.

As a first choice of solution scheme to most problems of this kind, the fourth order Runge–Kutta scheme can be selected:

$$\phi_{n+1} = \phi_n + k_1/6 + k_2/3 + k_3/3 + k_4/6 \tag{5.108}$$

where:

$$
\begin{aligned}
k_1 &= hf(t_n, \phi_n) \\
k_2 &= hf(t_n + h/2, \phi_n + k_1/2) \\
k_3 &= hf(t_n + h/2, \phi_n + k_2/2) \\
k_4 &= hf(t_n + h, \phi_n + k_3)
\end{aligned}
$$

The special problem that arises with sets of differential equations is the likelihood of variations in scales – some equations prompting quicker rates of change than others. This in fact is quite common in control system simulations of relevance to building energy problems. For example, the general heat exchange model developed in Section 5.1 has a ratio of time constants (τ_{dom}/τ_{min}) which varies from approximately 6 to approximately 25 for the various example cases used in the illustration (Fig. 5.2). Equation sets of this type are referred to as 'stiff' sets of equations and care needs to be taken over the choice of solution scheme and to ensure that h is dedicated to the quickest scale present in the set. In turn this leads to high computational overheads, which prompt the desire to consider the use of variable step solvers in which h is allowed to vary from a low value during periods of rapid rates of change in one or more equations to a high value during periods of low activity.

Approaches to the solution of stiff sets of equations are generally based on higher order semi-implicit methods with automatic stepsize adjustment. Rosenbrock methods that are based on generalisations of the Runge–Kutta method are very appealing because they are relatively simple and give good accuracy for all but the higher order problems. A Rosenbrock method seeks a solution to:

$$\varphi(t_0 + h) = \varphi_0 + \sum_{i=1}^{s} c_i \mathbf{k}_i \tag{5.109}$$

The corrections, $\mathbf{k}_i$, are obtained by solving a set of s linear equations:

$$\left(1 - \gamma h \frac{\partial \mathbf{f}}{\partial \varphi}\right) \mathbf{k}_i = hf\left(\varphi_0 + \sum_{j=1}^{i-1} \alpha_{ij} \mathbf{k}_j\right) + h \frac{\partial \mathbf{f}}{\partial \varphi} \sum_{j=1}^{i-1} \gamma_{ij} \mathbf{k}_j \quad (i = 1, \dots, s) \tag{5.110}$$

$\mathbf{f}$ denotes a vector of derivative functions and $\dfrac{\partial \mathbf{f}}{\partial \varphi}$ is a Jacobian matrix of partial derivatives of derivative functions. $\gamma, c_j, \alpha_{ij}, \gamma_{ij}$ are constants. Note that when $\gamma = \gamma_{ij} = 0$ the above reduces to a standard Runge–Kutta scheme. An important feature of stiff integration schemes is the mechanism for stepsize adjustment. A frequently used implementation of Rosenbrock is due to Kaps & Rentrop (1979) in which two estimates from Equation 5.109 are computed – a 'real' one and a further, lower order, estimate using different values of c_i. Differences between the two sets of estimates can be used to estimate the problem truncation error that in turn can be used for stepsize adjustment. Kaps & Rentrop also concluded that the smallest viable value for s is 4 resulting in a generalised fourth order method. Other methods for treating stiff sets of equations are the predictor – corrector methods, most of which are based on the backward-differentiation schemes of Gear (1971).

References

Adams, S. & Holmes, M.J. (1977) *Determining Time Constants for Heating and Cooling Coils – BSRIA Technical Note TN6/77.* Building Services Research and Information Association, Bracknell.

Åström, K.J. & Wittenmark, B. (1989) *Adaptive Control.* Addison-Wesley, Reading, MA.

CIBSE (2000) *Guide H: Building Control Systems.* Chartered Institution of Building Services Engineers, London.

Gear, C.W. 1971 *Numerical Initial Value Problems in Ordinary Differential Equations.* Prentice Hall, Englewood Cliffs, NJ.

Gomaa, A.G., Underwood, C.P., Bond, T.J. & Penlington, R. (2002) Experimental investigation of sensible heat transfer in fin-and-tube cooling coils. *Building Services Engineering Research & Technology* 23(3) pp.197–204.

Gouda, M.M., Danaher, S. & Underwood, C.P. (2002) Application of an artificial neural network for modelling the thermal dynamics of a building's space and its heating system. *Mathematical and Computer Modelling of Dynamical Systems* 8(3) pp.333–44.

Holman, J.P. (1997) *Heat Transfer.* McGraw-Hill, New York.

Hung, C.Y.S. & Lam, H.N. (2003) Pressure loss characteristics of thin single-blade flat dampers for square airflow branch ducts. *International Journal of Heating, Ventilating, Air-conditioning and Refrigerating Research* 9(3) pp. 327–45.

Kaps, P. & Rentrop, P. (1979) Generalized Runge–Kutta methods of order four with stepsize control for stiff ordinary differential equations. *Numerische Mathematik* 33 pp.55–68.

Khoo, I., Levermore, G.J. & Letherman, K.M. (1998) Variable-air-volume terminal units I: Steady-state models. *Building Services Engineering Research & Technology* 19(3) pp.163–70.

Landau, Y.D. (1979) *Adaptive Control – The Model Reference Approach.* Marcel Dekker, New York.

Legg, R.C. (1986) Characteristics of single and multi-blade dampers for ducted air systems. *Building Services Engineering Research &Technology* 17(4) pp.129–45.

Ljung, L. (1987) *System Identification: Theory for the User.* Prentice-Hall, Englewood Cliffs, NJ.

Mamdani, E.H. & Assilian, S. (1975) An experiment in linguistic synthesis with a fuzzy logic controller. *International Journal Man – Machine Studies* 7(1) pp.1–13.

McQuiston, F.C., Parker, J.D. & Spitler, J.D. (2000) *Heating, Ventilating and Air Conditioning.* John Wiley, New York.

Press, W.H., Teukolsky, S.A., Vetterling, W.T. & Flannery, B.P. (2002) *Numerical Recipes in C++.* Cambridge University Press, Cambridge.

Underwood, C.P. (1999) *HVAC Control Systems.* E. & F.N. Spon, London.

Wang, L.-X. (1997) *A Course in Fuzzy Systems and Control.* Prentice-Hall, Englewood Cliffs, NJ.

Wellstead, P.E. & Zarrop, M.B. (1991) *Self-tuning Systems – Control and Signal Processing.* John Wiley, Chichester.

Chapter 6

Modelling in Practice I

The previous chapters have considered methods and procedures for modelling energy and energy-related problems in buildings. The application of these methods is usually made to one or more of the following areas of building technology analysis:

- Research and development
- Design and design analysis
- 'Troubleshooting' and post-occupancy problem solving

Of these investigative categories, design and design analysis represents the most inclusive problem space since the other categories relate to problem-specific situations. Figure 6.1 gives a possible view of the Design Pathway as far as energy and environmental considerations related to buildings are concerned (the conditions included in this pathway imply optimality but need not amount to any more than the satisfaction of 'acceptable' design criteria). Historically, modelling methods and the associated computer codes for implementing these methods have concentrated on describing the various building energy flow paths in the absence of plant and control – essentially restricted to analysis of problems in passive design. Modelling technologies for plant and control developed later and, consequently, the representation of plant and control in current building modelling programs is mixed. The least representative of these enabled nothing more than perfect set point tracking whilst the most comprehensive programs enabled a more or less dynamic (or certainly quasi-steady-state) description of plant and control with the result of much greater program utility.

However, as the Design Pathway suggests, there is good reason to view many building energy problems, at least initially, in a passive context within which the attributes of the building envelope can be dealt with independent of plant. Plant design and analysis issues can then be dealt with after this preliminary stage. Ghiaus (2003) for example, focuses on the internal temperature of a 'free running' building as a basis for decisions regarding the sizing and selection of plant.

Table 6.1 summarises the various approaches available for treating each step in the Design Pathway, cross-referenced to sections in earlier chapters that give details of the relevant methodology. The purpose of this chapter is to review recent progress in the various areas of modelling applicable to the Design Pathway – in effect a 'hierarchy of methods'. Then, in Chapter 7, interrelationships between the various methods will be explored together with

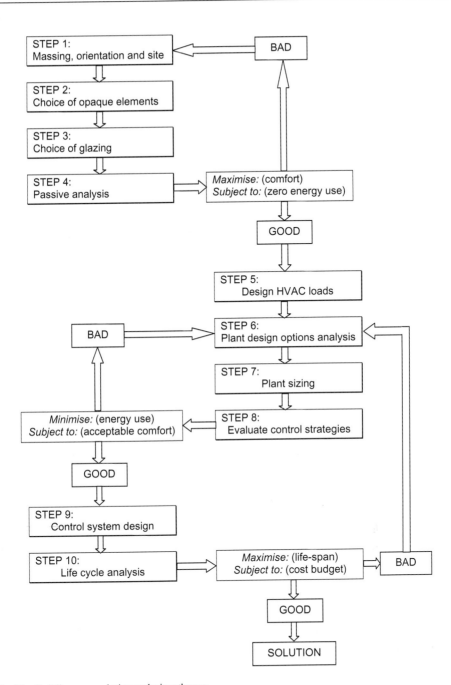

Fig. 6.1   Building energy design analysis pathways.

Table 6.1   Hierarchy of energy modelling methods.

Treatment	Modelling Options	Applicable Pathway Steps
Conduction heat transfer	Conventional time domain (1.3)	1, 2, 4, 5
	Response factor method (1.3)	
	Transfer function method (1.3)	
	Radiant time series (2.3)	
	Frequency response (1.3)	
	Numerical – finite difference (2.1)	
	Lumped capacitance (1.4)	
Window modelling	Simplified – fixed parameters (2.2)	3, 4, 5
	Optical properties (2.2)	
Short-wave radiation distribution	Simplified surface-targeted (2.2)	4, 5
	Mapped (2.2)	
Long-wave radiation distribution	Simplified surface fractions (2.2)	4, 5
	Rigorous (2.2)	
Surface convection	Constant coefficients (3.2)	4, 5
	Modelled coefficients (3.2)	
Moisture mass transfer	Zone moisture balance only (2.4)	2, 4, 5
	Fabric moisture sorption (3.1)	
Air mass balance	Constant air change rate	1, 3, 4, 5
	Zonal ventilation model (3.3)	
Air diffusion	Perfectly mixed	4
	Stratified multi-zonal (3.3)	
	Stratified – rigorous (3.2)	
Secondary plant and control	Constant set point tracking	5, 6, 7, 8, 9
	Steady-state model (4.2, 4.3, 4.5)	
	Dynamic-state model (5.1, 5.2, 5.3)	
Distribution	Not treated (i.e. close-coupled)	6, 7, 8, 9
	Steady-state models (4.4)	
	Dynamic-state models (5.2)	
Primary plant and control	Sum of loads	5, 6, 7, 8, 9
	Steady-state models (4.2, 4.3, 4.5)	
	Dynamic-state model (5.1, 5.2, 5.3)	

progress in the development of integrated programs for solving building energy-related problems – a 'convergence of methods'.

6.1 Developments in general

Many of the areas described above have continued to receive ongoing development and advancement. This section reviews some the recent progress.

Building elements

Davies (1999) gives a review of practice in building element conduction problems contrasting 'transfer' methods (i.e. response factor-type approaches – see Section 1.3), with numerical (finite difference) methods (see Section 2.1) and 'exact' analytical methods (i.e. conventional time domain methods – see Section 1.3). Wang & Chen (2001) use frequency domain regression methods to estimate Laplace transfer function-based transient wall conduction models that enjoy the same advantages of low computational cost and inherent stability as analytical or 'exact' methods. For dealing with thermal bridges in transient whole wall conduction heat transfer, Carpenter *et al.* (2003) propose a simple 'equivalent wall' modelling method based on creating a fictitious additional layer such that the transient response of the resulting multi-layer wall is similar to that of the real construction complete with thermal bridges.

Progress continues to be made on the development of reduced-order models of construction elements. Often these problems are cast as analogous electric circuits (see Section 1.4). Fraisse *et al.* (2002) for instance, report on the treatment of multi-layer walls represented by a three-resistance and four-capacitance circuit and the aggregation of several such multi-layer constructions. Ménézo *et al.* (2002) on the other hand use a linear systems aggregation technique in the context of state representations of the various components of an envelope model in order to effect order reductions in each component with corresponding reductions in computational cost.

The special case of three-dimensional conduction heat transfer through ground floor earth-contact slabs has been addressed mostly by developing the ground-coupled region as a full three-dimensional problem. Deru *et al.* (2003) for example, couple a three-dimensional finite element heat conduction model for the earth-contact zone of a building to an existing model of the building (the results of the work forming the basis of a standard test for building energy program evaluation – see Section 7.3). A less computationally expensive approach based on incorporating a form of the response factor method (see Section 1.3) to earth-contact heat transfer is reported by Davies *et al.* (2001). A full review of the subject can be found in Adjali *et al.* (1998).

The influence of surface convection coefficients in building heat transfer rates is acknowledged to vary from application to application, especially when mechanical ventilation plant is used during part of the day and switched off for the remainder. Computational cost prevents this coupling from being explored by allowing computational fluid dynamics to participate dynamically in the model. Beausoleil-Morrison (2001) proposes a method for treating mixed convection in rooms by drawing on convection correlations reported extensively in the literature, and goes on to develop an adaptive algorithm that allows convection equations to be switched between modelling time steps (Beausoleil-Morrison 2002). Convection due to flows within the cavities of construction elements has been studied by Manz (2003).

Recent progress on the modelling of glazing systems and windows has focused on extending data on the optical properties of coated glass (Karlsson & Roos 2000), modelling asymmetric radiant sensation and its impact on comfort (Zmeureanu *et al.* 2003), improving the modelling of convection and radiation in windows with blinds (Phillips *et al.* 2001) and the development of advanced glazing systems responsive to both daylight maximisation and seasonally responsive heat transfer (Clarke *et al.* 1998).

Heat and mass transfer

Applications involving combined heat and mass transfer have, until only recently, addressed the issue of coupling these methods into the substantive building model. Liesen & Pedersen (1999) reported on the integration of the rigorous heat and mass transfer equations for building elements into the transfer function-based building energy model used widely in North

America. Rode & Grau (2003) report on the integration of a vapour diffusion model into a finite-volume-based whole building model; and Clarke *et al.* (1999) have incorporated the heat and mass transfer equations into the finite-difference-based simulation ESP-r (applying it in this case to the investigation of mould growth in buildings).

Besides developments such as the above, problems involving moisture sorption have tended to rely on ad hoc methods. This is partly due to computational cost and partly because many building energy problems can be defined and solved with the absence of a detailed knowledge of latent effects. Manzan & Saro (2002) for instance, considered the coupling of the heat and mass transfer equations with a computational fluid dynamics program in order to study passive cooling effects due to water evaporation from a wetted roof surface. Mendes *et al.* (2003) use a simplified form of heat and mass transfer model to investigate the effects of moisture content in building elements on conduction heat transfer loads. Lucas *et al.* (2002) developed simulations based on experimental data to investigate the deterioration of surface coatings on building materials due to condensation.

Building envelopes

Recent practice in North America has focused on a new formulation of the conduction transfer function method – the radiant time series (RTS) method introduced by Spitler *et al.* (1997) for the purpose of calculating design cooling loads (see Section 2.3). This method recognises that, for cooling loads as distinct from load simulations, a daily periodic peak load set is needed and an explicit periodic rather than iterative approach can be applied. Figures 6.2a and 6.2b contrast the RTS method with the zone heat balance method (Section 2.3). In the latter, periodic inputs are dealt with by iterating on the flux histories until a steady periodic response is obtained, whereas the RTS method separates the convection and radiant components using a periodic response factor series to enable the periodic loads to be calculated directly (Spitler & Fisher 1999a). The RTS method therefore enjoys improvements in computational cost and stability for applications requiring design cooling loads. Periodic response factors for use with the RTS method are given by Spitler & Fisher (1999b). There is a clear solution parallel between the RTS method and the frequency response-based admittance method used in the UK for cooling load calculations (see Section 1.3) as can be seen in Fig. 6.2c. Fuller details of these transfer methods of calculating building energy loads can be found in Pedersen *et al.* (2003).

Some attention has been placed on an alternative class of 'grey box' model with reduced parameter sets (Braun & Chaturvedi 2002; Déqué *et al.* 2000). 'Black box' models that bypass the need for physical descriptions of systems (e.g. artificial neural networks – see for instance Gouda *et al.* 2002) require large amounts of training data, whereas deterministic 'white box' models tend to be highly parameterised. Grey box models generally consist of hybrid transfer-type model formats with parameters 'trained' from measurement data resulting in significantly reduced parameters but, correspondingly, a reduced range of applicability. In related work, Braun *et al.* (2001) apply such models to the analysis of system configuration as a function of various building thermal mass strategies.

Pinney & Parand (1991) report on the issue of 'preconditioning' in numerical simulations of building thermal response – the simulation time taken for all dependent variables to converge to results that are unaffected by their initial conditions and thereby proceed on to yield accurate results. Results generated during preconditioning are usually discarded. Preconditioning times can be significant in most applications but very protracted for buildings with high thermal capacity. Pinney & Parand considered preconditioning, node placement and initial condition-setting for numerical models (finite difference – see Section 2.1), and concluded amongst other things that 19 simulation days are required for asymptotic convergence for a variety of modelling applications. Later work by Crowley & Hashmi (2000) confirmed their findings as well as proposing an alternative direct solver for finite difference conduction problems which requires fewer starting values and leads to improved accuracy over the conventional Crank–Nicholson scheme.

Processing input data

The overall inputs to the building energy models come from two highly stochastic climate and user data sets. Other than for design loads where statistical climate extremes are required, climate data for modelling can be generated from two possible sources:

- 'Real' weather data records formatted into a typical meteorological year (TMY)
- Simulated weather data generated typically using autoregressive modelling methods

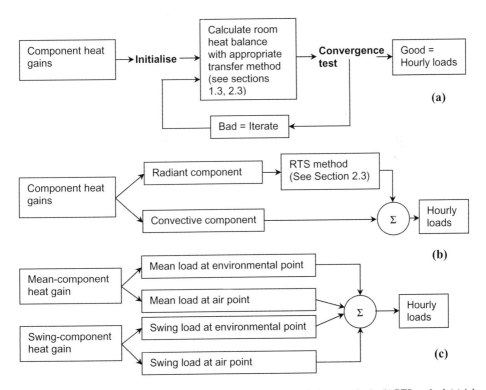

Fig. 6.2   Methods for daily periodic cooling load calculations. (a) Heat balance methods. (b) RTS method. (c) Admittance method.

For simulation modelling a maximum data time interval of one hour is usually specified (this is common for response factor conduction modelling methods, but a higher time resolution is preferable if available).

A typical climate data set based on hourly time series will usually include the following pattern of data:

- Station details: latitude and longitude
- Month number
- Day number
- Day hour
- Precipitation (mm)
- Atmospheric pressure (mbar)
- Wind speed ($m s^{-1}$)
- Wind direction (degrees from north)
- Dry bulb temperature (°C)
- Wet bulb temperature (°C) (or relative humidity %)
- Global horizontal solar radiation ($W m^{-2}$)

Though measurements of both direct-normal and diffuse components of solar radiation can be made, these tend to be restricted to experimental stations and a model will usually be required to break the global value into its component parts (see Section 2.2), in which case the station location angles are used. Weather data for energy modelling are formatted in a number of different ways (Table 6.2).

Fuller details can be found in Harriman *et al.* (1999), and Levermore & Doylend (2002) compare the North American WYEC, TMY and IWEC data (the abbreviations now in use are WYEC2 and TMY2 to distinguish current weather files from earlier versions) with European TRY and DRY data. Augustyn (1998) describes ASHRAE toolkit TC4.2 for the development and formatting of hourly weather files for energy calculations. Data for numerous North American (TMY) and other (IWEC) locations can be found in NREL (1995).

Table 6.2 Common weather data types.

Format	Description	Comment
TMY	Typical meteorological year	Hourly time series data for one complete typical year (North American locations)
TRY	Test reference year	Hourly time series forming 12 months of data drawn from different years to form a 'typical' year (European locations)
WYEC	Weather years for energy calculations	Hourly time series of typical meteorological data (North American locations)
IWEC	International weather for energy calculations	Hourly time series of typical meteorological data (locations outside North America)
DRY	Design reference year	Hourly time series of typical meteorological data (European locations)
BRY	Biomass reference year	Daily time-series of 12 months taken from different years (European locations)
EWY	Example weather year	Hourly time series of typical meteorological data (UK locations)

Whilst it is desirable to use site-specific measurement-based weather data for energy modelling there are instances in which this may not be possible – lack of data for a given site being the obvious one. Also, a growing number of emerging applications in, for example, embedded renewable energy applications (e.g. new technologies in wind concentrators – Campbell & Stankovic 1999) and building-integrated photovoltaics (e.g. Jones & Underwood 2001), as well as applications in control system simulation over short time horizons, demand much higher resolution such as minutely data for which there is a paucity of data based on measurements.

Methodologies for predicting weather data using models have been developed to deal with some of these cases. Most of these use stochastic methods based on auto-regressive models (examples on the use of these methods are given in Section 5.3) or the use of cross-correlation from other known weather variables. Examples include Knight *et al.* (1991) who dealt with annual hourly solar data prediction (with recommendations for procedures to model dry-bulb temperature and relative humidity); Gul *et al.* (1998) who cross-correlated solar data from other known data; Ren & Wright (2002) who proposed a hybrid deterministic-stochastic short time horizon predictor (24 hours) for solar and temperature data for use in, *inter alia*, adaptive controller development; van Paassen & Luo (2002) who develop a weather data prediction method for analysing climate change impact on buildings. Though most of these methods lead to hourly data, the methodologies used are in most cases equally applicable to higher time resolution predictions subject to available weather data sets.

Procedures to describe the highly stochastic inputs due to building users are less well developed. Various assumptions are made in practice-based patterns of discrete switching events. One possibility is to make use of monitoring data for lighting or small power demand as a proxy for mean user activity and, about which, a random component can be imposed (e.g. Jones & Underwood 2003). User activity will usually exhibit strong repeating sequence patterns defined by three timescales; daily per week, weekly per month and monthly per year.

Data for modelling heat gains from equipment sources have been reported in several sources (Parsloe & Hejab 1992; Wilkins & Hosni 2000; Jones & Underwood 2003). Most of what is reported centres on 'nameplate ratio' (i.e. actual power drawn by equipment divided by the nameplate power rating) comparisons of existing equipment and utilisation of equipment (i.e. diversity). Typical nameplate ratios were found to range from 25 to 50% and diversities from 37 to 78% with equipment loads of less than $20\,W\,m^{-2}$ in many instances, and often considerably less than this. Jones & Underwood (2003) additionally recommended procedures for profiling the variation in power utilisation over typical seasonal days based on second-order least squares regression fitting to half-day data sets. By introducing a random component whose amplitude is fixed by simulated maximum and minimum bounds on the data, a daily power profile emerges (e.g. Fig. 6.3).

Regarding the fraction of equipment heat gain attributable to either radiant or convective emission (the so-called 'radiant-convective split'), Hosni *et al.* (1998) evaluated this for a range of office and laboratory equipment, concluding that the fractions can vary considerably (11–61% radiant); and Liesen *et al.* (1998) reported on the sensitivity of cooling load calculations to its variation.

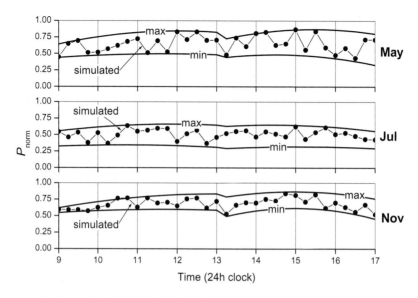

Fig. 6.3   Typical model-based power profiling (Jones & Underwood 2003).

6.2 Internal ventilation problems

Rapid expansion in the development of computational fluid dynamics (CFD) modelling techniques through the 1990s has seen this area become one of the fastest growing for modelling
building fluid flow problems in recent years. However, the high computational cost associated with the essentially iterative solution schemes used in steady-state CFD analysis
(much less dynamic-state CFD analysis) has meant that the integration of CFD into substantive building energy programs and codes has been restricted to a few bespoke cases. These
cases will be considered in more detail in Section 7.1.

Most of the applications to building fluid flow (whether as conjugate problems with heat
transfer or isothermal problems) have therefore tended to make use of specialist CFD software, the major determinants of a successful outcome mainly resting on grid-mesh setting
and the choice of turbulence model (see Section 3.2). By far the most popular choice for the
latter has continued to rest with the k–ε model, but as has been discussed at length in Section
3.2 there is now a far greater choice available for this. Russell & Surendran (2000) give a comprehensive review of these applications for internal ventilation flow problems, and Chow
(2001) discusses applications to a wider range of building-related flow problems. In this section some of the recent developments in modelling methodologies of relevance to building-
related ventilation problems are highlighted and reviewed.

Application examples using CFD

Allocca *et al.* (2003) studied ventilation flows for the 'difficult' ventilation regime that exists
with rooms having windows on one side only. They stacked spaces within an ambient domain
so that both wind driven and buoyancy driven flows could be considered (see Fig. 6.4 for
typical results) though they concluded that the CFD results under-predicted combined flows

when compared with an analytical verification test. Papakonstantinou *et al.* (2000) also considered single-sided naturally ventilated buildings, comparing modelling results with experimental data from a test cell and concluding good agreement between the two.

Investigations into the induced natural ventilation created by a solar-heated construction element, such as a roof channel or chimney, have been investigated by Ziskind *et al.* (2002) for a single storey building, and Letan *et al.* (2003) for the multi-storey case. Good results were obtained using this method even at low solar irradiances, especially with ventilation inlet at mid-height – see Figs 6.5a and 6.5b.

In an application concerning a smaller scale buoyancy problem related to indoor air quality (IAQ), Murakami *et al.* (1998) used a CFD model with an adapted low-Re $k–\varepsilon$ turbulence model to quantify the quality of breathing air due to rising vapour over a body carrying contaminants from the floor.

In applications related to equipment, Karki *et al.* (2003) used CFD modelling to determine the flow rates of air through perforated floor tiles in a ventilated floor application demonstrating good model agreement with experimental data. Bojic *et al.* (2003) studied the influence of baffling plates in the recesses of high rise buildings in Hong Kong in reducing the high temperatures at condensing units towards the top of the building caused by the buoyant plume induced by condenser heat rejection (see Fig. 6.6). It was concluded that the insertion of one

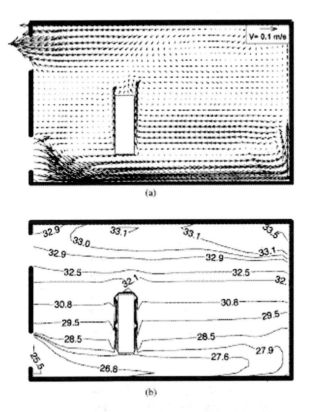

Fig. 6.4   Typical side-ventilation results: (a) Velocity. (b) Temperature. (Reprinted from *Energy and Buildings*, 35, Allocca, C., Chen, Q. & Glicksman, L.R., Design analysis of single-sided natural ventilation, 785–95, © (2003), with permission from Elsevier.)

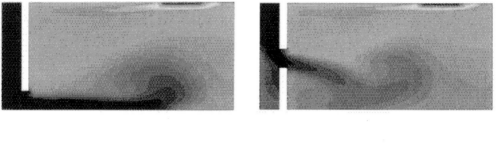

Fig. 6.5 Solar-induced ventilation modelling results with alternative inlet positions. Top: temperature. Bottom: velocity. (Reprinted from *Energy and Buildings*, 34, Ziskind, G., Dubovsky, V. & Letan, R., Ventilation by natural convection of a one-story building, 91–102, © (2002), with permission from Elsevier.)

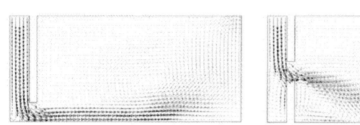

(a)   **(b)**   **(c)**

Fig. 6.6 Flow baffling for condenser-induced plumes. (a) Typical floor plan. (b) Meshing. (c) Typical results (velocity). (Reprinted from *Building and Environment*, 38, Bojic, M., Yik, F. & Lee, M., Influence of plates on flow inside a recessed space generated by heat rejection, 593–604, © (2003), with permission from Elsevier.)

baffling plate could reduce air temperatures at upper-most condensers in a 30 storey building by 30%.

Turbulence models for internal building flows

More development attention has been given to turbulence modelling than any other area of CFD analysis. Indeed building fluid flow problems represent one of the widest range of contrasting turbulence modelling requirements. Table 6.3 contrasts some of the possibilities (see Section 3.2 for details of methods and terminology).

Table 6.3 Choices of turbulence model.

Application	Turbulence Model
Wind flow around buildings	Two-layer $k-\varepsilon$; RNG $k-\varepsilon$; LES
Flow in ducts and pipes	PMLH; $k-\varepsilon$
Room air movement (isothermal or moderate ΔT)	Low-Re $k-\varepsilon$
Room air movement with surface heat transfers	Low-Re $k-\varepsilon$; Two-layer low-Re $k-\varepsilon$

Xu & Chen (1998) have compared the 'standard' $k-\varepsilon$ turbulence model with two adapted low-Re $k-\varepsilon$ turbulence models for modelling buoyancy-driven flow in a room with two differentially heated vertical walls, concluding that improved results were obtained from the two low-Re models.

Cook & Lomas (1998) have compared the $k-\varepsilon$ with an alternative RNG $k-\varepsilon$ model in which the turbulence model constants are calculated using renormalisation group theory. They applied both models to a problem involving buoyancy-driven displacement ventilation flows in a room and concluded that the $k-\varepsilon$ model failed to correctly predict entrainment flow into the displacement plume, whereas the RNG $k-\varepsilon$ provided better results in this region. Nielsen (1998) demonstrated applications for a zero-equation turbulence model, conventional two-layer $k-\varepsilon$ model, low-Re $k-\varepsilon$ model and large eddy simulation respectively applied to a low turbulence room air application; displacement ventilation; an application involving evaporative emissions and a general room air flow problem. Srebic *et al.* (1999) have tested the application of a zero-equation turbulence model of the PMLH type to a more complex indoor air flow problem involving displacement ventilation with infiltration and forced convection heating, concluding that the results agree well with experimental data.

Jiang & Chen (2001) have compared two types of large eddy simulation models of turbulence applied to cross natural ventilation of a space with both ambient and internal zones within the domain. The two models use conventional Smagorinsky and filtered sub-grid scale procedures respectively. The latter is found to give better results in near-wall regions and where the flow has both laminar and turbulent features. These methods also enable both mean and instantaneous flow predictions (Fig. 6.7), which is not possible using conventional Reynolds-averaged Navier–Stokes methods.

In an application related to ventilating plant, Djunaedy & Cheong (2002) developed a model of a square four-way ceiling diffuser; and went on to compare results of measured centreline velocity with both a conventional $k-\varepsilon$ turbulence model and a RNG $k-\varepsilon$ turbulence model. Results favour the latter method, especially in the terminal-decay region of the flow trajectory (Fig. 6.8).

Zonal and network ventilation modelling

The zonal and network modelling methods have generally developed as lower 'cost' alternatives to CFD modelling for two groups of applications – temperature stratification in rooms and indoor air quality (IAQ). In the former case, several applications involving a multi-cell or zonal approach by splitting up each zone into a number of sub-zones have been reported.

Voeltzel *et al.* (2001) used a multi-cell zonal ventilation model approach (see Fig. 6.9) to improve the prediction of temperature distribution and air movement in large spaces with high quantities of glazing, such as atria. Results show improvements in predictive accuracy against experimental data when compared with results from a mixed one-node approach.

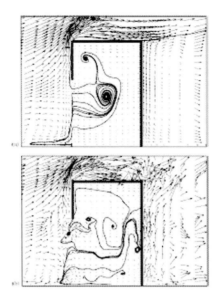

Fig. 6.7 Large eddy simulation applied to single-sided ventilation with a filtered sub-grid scale. (a) Mean flow stream. (b) Instantaneous flow stream. (Reprinted from *J. Wind End. Ind. Aerodynamics*, 89, Jiang, Y. & Chen, Q., Study of natural ventilation in buildings by large eddy simulation, 1155–78, © (2001), with permission from Elsevier.)

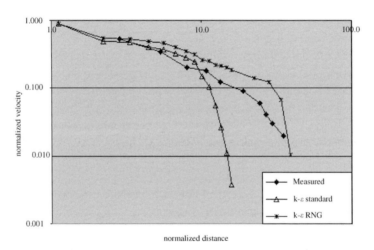

Fig. 6.8 Centreline velocity predictions from a ceiling diffuser. (Reprinted from *Building and Environment*, 37, Djunaedy, E. & Cheong, K.W.D., Development of a simplified technique of modelling four-way ceiling diffusers, 393–403, © (2003), with permission from Elsevier.)

Wurtz *et al.* (1999) compared results of two-dimensional and three-dimensional room descriptions from CFD modelling, experimental data and a zone model. Ren & Stewart (2003) again applied multi-celling of single zones and compare results with those from a CFD model concluding that, with reasonably refined cell divisions, a multi-cell zonal model can give results that compare well with CFD results but with much lower computational demand.

Cell grouping in zonal modelling is not straightforward because recognition needs to be made about the varying characteristics of flow within a zone. Hence Musy *et al.* (2002) have

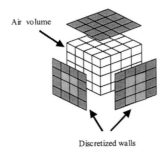

Air volume

Discretized walls

Fig. 6.9 Multi-celling a single zone for improved predictive accuracy. (Reprinted from *Energy and Buildings*, 33, Voeltzel, A., Carrie, F.R. & Guarracino, G., Thermal and ventilation modeling of large highly-glazed spaces, 121–32, © (2001), with permission from Elsevier.)

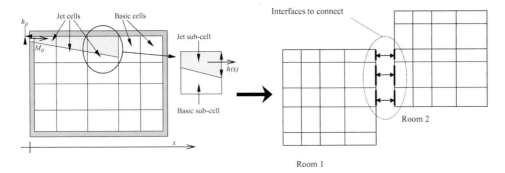

Fig. 6.10 Automated cell-setting in zonal modelling. (Reprinted from *Building and Environment*, 37, Musy, M., Winklemann, F., Wurtz, E. & Sergent, A., Automatically generated zonal models for building air flow simulation: Principles and applications, 873–81, © (2002), with permission from Elsevier.)

described a method for automatically generating cell groups for zonal air flow modelling applications which accounts for local air flow characteristics and cell-zone interfacing (Fig. 6.10). Recognising that momentum details are neglected in such models and that non-physical cell-to-cell pressure losses can arise, Axley (2001) has recommended improvements in cell-to-cell viscous loss descriptions using surface drag relations; and Griffith & Chen (2003) have introduced a method to account for zonal momentum. These developments have been shown to improve near-wall flow behaviour in zonal models as well as reducing the need to set special cell types to account for variations in flow character.

In recent applications of zonal and network ventilation modelling to IAQ problems, Haas *et al.* (2002) have described the application of the multi-zone modelling program COMIS for dealing with problems in pollutant generation (in this case occupancy-generated CO_2 – see Fig. 6.11).

Stewart & Ren (2003) have reported on the nesting of a multi-cell zonal model (again within COMIS) to predict pollutant dispersal (see Fig. 6.12); and Yik & Powell (2002) have used a network modelling approach to analyse pressure gradients and leakage flow rates around hospital isolation wards. For these problems, additional species concentration–continuity equations are needed (for details of the equations required for dealing with the transport of gaseous species, see Blondeau *et al.* 1998); but in some instances the treatment of energy calculations (unless buoyancy flows are likely to be significant) might be suppressed.

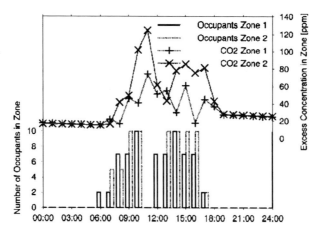

Fig. 6.11 Typical simulation results of pollutant generation. (Reprinted from *Energy and Buildings*, 34, Haas, A., Weber, A., Dorer, V., Keilholz, W. & Pelletret, R., COMIS v3.1 simulation environment for multizone air flow and pollutant transport modelling, 873–82, © (2002), with permission from Elsevier.)

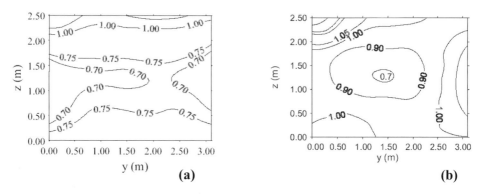

Fig. 6.12 Room sections showing simulated ventilation efficiency contours. (a) Supply top left; extract top right (contaminant centre). (b) Supply top left; extract bottom right (contaminant centre). (Reprinted from *Building and Environment*, 38, Stewart, J. & Ren, Z., Prediction of indoor gaseous pollutant dispersion by nesting sub-zones within a multizone model, 635–43, © (2003), with permission from Elsevier.)

6.3 Wind flow around buildings

Special consideration is given to progress in modelling wind flow around buildings (WFAB) because of the growing interest in the influence of wind flow behaviour over the following design issues:

- Improving the robustness of assessments of natural ventilation and infiltration rates
- Investigations of the impact of building clusters on local ambient air quality in pedestrian areas
- Quantifying the impact of local ambient air quality on the quality of natural ventilation air

Modelling WFAB has received considerable attention recently with most of the literature focusing on two themes, wind loading for structural design and pollutant dispersion. Also, most sources involve model validation using boundary layer wind tunnel (BLWT) tests. Though most of the BLWT test results are generated solely for the purpose of validating a particular model, experimental data for general WFAB model validation has been generated by Minson *et al.* (1995) using laser-Doppler anemometry for highly turbulent wind flows near buildings. Wong & Chin (2002) have developed a procedure for modelling wind pressure coefficients used in many branches of ventilation modelling (see Section 3.3) based on the statistical regression analysis of wind tunnel data.

Oliveira & Younis (2000) have assessed how well data in the wind loading of structures can be predicted by CFD making use of full-scale measurements around a single-span high-eaves glasshouse. They found that the use of a 2-D modelling strategy leads to a serious overestimation of roof suction loads, and that the standard k–ε turbulence model fails to predict flow separation at the roof corner. The roof corner vortex has been studied in detail by other workers (He & Song 1997); in this case using LES with a Smagorinsky sub-grid scale model for small scale turbulence and a log-law of the wall approach at surface boundaries. They point out that the roof corner detail generates two counter-rotating vortices when receiving a quartering wind with resulting high roof suction pressures. Results are compared with wind tunnel tests for three mesh sizes, concluding that if the mesh is fine enough, all scale levels of the fluctuating eddies can be captured.

Prediction of the stable atmospheric boundary layer over both flat terrain and terrain with topology has been studied by Huser *et al.* (1997). They used the commercial program CFX with the k–ε turbulence model but varied the four standard constants used in the latter. The main focus of the study was pollution dispersion over the terrain boundary layer case and results are presented as normalised eddy viscosity ($C_\mu k^2/\varepsilon$) profiles. They conclude that when the surface roughness is high the model tends to overpredict the eddy viscosity intensity.

Baskaran & Kashef (1996) considered wind flow around three cases – a single building, two parallel buildings and a multiple building configuration. They developed a bespoke simplified code called 'TWIST' for flow around the simple geometries and used PHOENICS for the multiple building configuration. Results are compared with BLWT test data from ten sources in the literature concerned with the evaluation of pedestrian wind environment conditions, although most of these are for the single building case.

Meroney *et al.* (1999) looked at the flow and dispersion of pollutants emitted by sources near buildings using FLUENT and compared several methods of turbulence modelling in the process (k–ε; RNG k–ε; Reynolds–stress (RSM) closure approximations). They also summarised the findings of a substantial review of the experimental and modelling analysis of wind effects (Architectural Institute of Japan 1998). Results from their modelling are compared with four wind tunnel test cases, three of which are described by cubes or rectangular prisms, and the fourth is a street canyon. They conclude that pollutant concentrations are overpredicted by Reynolds-averaged turbulence models though predicted mean wind pressure coefficients were found to be reasonably accurate (and not particularly sensitive to grid adaptation or the turbulence model used). The RSM model was generally found to produce more realistic results than either k–ε or RNG k–ε. A further finding was that grid adaptation (rather than grid density) is a convenient way of improving the results of flow separation and reattachment.

Yik *et al.* (2003) generated wind data for natural ventilation design in Hong Kong using a statistical analysis of 15 years of wind data records, including an assessment of the frequency and periods of stagnation (see Fig. 6.13).

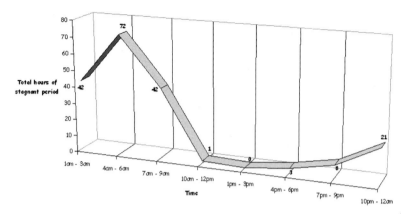

Fig. 6.13   Total daily hours of wind stagnation in Hong Kong (1984–98) (Yik *et al*. 2003).

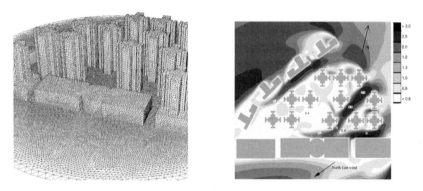

Fig. 6.14   Unstructured grid mesh (left) and simulated wind amplification factors around a high-rise residential building cluster in Hong Kong (Bojic *et al*. 2002).

In related work they used the data to investigate the modelling of wind flow around high rise residential buildings in Hong Kong using CFD modelling (Bojic *et al*. 2002) – Fig. 6.14.

Turbulence models for WFAB flows

Most of the available sources highlight shortcomings in the classical k–ε turbulence model for urban wind modelling problems and, correspondingly, seek to find better alternatives in some cases through sub-grid scale methods such as large eddy simulations (LES). Again, most of the reported work tends to rely on experimental data from BLWT tests for model validation.

Some workers have addressed the failings of k–ε turbulence modelling by developing improvements to it. Zhou & Stathopoulos (1997) noted that the k–ε model is especially flawed when trying to predict the flow behaviour after separation over roof structures and sides walls of a building. They suggest that a turbulence modelling method is needed, with the computational cost of k–ε but the accuracy of LES.

Since k–ε is known to predict generally mainstream conditions of flow around buildings quite well but tends to break down in the near-wall regions, Zhou & Stathopoulos considered a two-layer method. In this, the zones of the computational domain are divided into two layers or regions – inner regions and external regions. Flow in the external regions uses the standard k–ε model whereas flow in the inner regions use a one-equation k-model approach (for details see Norris & Reynolds 1975). Results comparing this method with a conventional k–ε approach for both surface-normal and quartering winds are compared with experimental data sourced from the literature. They conclude that flow separation is correctly predicted by the two-layer model but not by the standard k–ε model. The latter also overpredicts the suction pressure across the roof edge.

Ehrhard & Moussiopoulos (2000) argue that much of the work that has concentrated on refining and improving k–ε for WFAB problems has failed to make a significant impact on accuracy in all regions of typical wind flow fields. They go on to develop a new non-linear turbulence model, the premise of which is to retain terms up to cubic level (the majority of conventional turbulence models being based on linear or quadratic stress–strain relationships). Results generated with experimental comparisons from a BLWT for flow over a surface mounted cube and an asymmetric street canyon both show an improvement over the conventional k–ε approach.

Kawamoto (1997) has compared two types of k–ε–ϕ model with standard k–ε for the prediction of wind pressure coefficients around buildings. In this approach, ϕ is an anisotropy parameter of Reynolds stresses and is set to decrease with an adverse pressure gradient and to increase with a favourable pressure gradient. Two model structures are developed – one in which surface-impingement and adverse pressure gradients are treated separately, and a simpler model structure in which adverse and impingement pressure gradients are combined. It is concluded that both model structures give significant improvement over standard k–ε for predicting pressure distributions over buildings.

Significant work is reported in the literature on the application of LES for WFAB problems. Murakami (1998) has reviewed the available turbulence models for computational wind engineering problems, identifying special modelling problems for flow around buildings due to impingement, bluff body sharp edges and inflow/outflow boundary condition specification. Approaches based on direct numerical simulation (DNS), LES and Reynolds-averaged (RANS) methods are focused on, as are modifications to the standard k–ε method to minimise its classical problem of over-generating k in the impingement zone.

Tutar & Oguz (2002) also conducted a comparative assessment of different methods focusing on k–ε, RNG k–ε, RNG sub-grid scale and Smagorinsky sub-grid scale methods. They looked at turbulent wind flow around two parallel buildings with different wind directions and geometries. Results compared with experimental data suggest that the RNG sub-grid scale model gives better results than other turbulence models. Nozawa & Tamura (2002) considered the problem of specifying boundary inlet conditions in WFAB domains with turbulent approaching flows. They developed patterns of inlet boundary conditions for sites with both smooth and rough surfaces, concluding that results strongly depend on the turbulence intensity of the approaching flow. Thomas & Williams (1999) also noted the importance of giving an accurate specification of boundary flow characteristics and inflow turbulence in BLWT applications, and investigated a method for developing this specification specifically for LES applications. Their method is based on applying a separate and independent simulation of the flow stream upstream of a bluff body, and though computationally expensive, the results can be reused for a wide range of modelling applications.

Jiang & Chen (2002) argue that ventilation patterns (specifically cross-flow ventilation) in buildings are difficult to investigate using BLWT techniques because of changes in wind direction. They also point out deficiencies in DNS and RANS modelling and identify LES methods as the most promising for this application. Results using LES with a Smagorinsky sub-grid scale model are compared with wind tunnel data with reasonable agreement.

Street canyons

A significant number of sources deal with the specific issues of wind flow in street canyons. Most of this is focused on pollutant dispersion problems and, again, most of these sources have both CFD modelling and experimental components to them (the latter mostly through scaled BLWT tests). Early, purely experimental, work on this problem was reported by De-Paul & Sheih (1986) and Hoydysh & Dabberdt (1988). In the former, tracer balloons were used to measure cross-wind velocities in a deep street canyon in Chicago. A dominant vortex cell was identified in the canyon consistent with later modelling work (e.g. Ca *et al.* 1995; Chan *et al.* 2002 – see Fig. 6.15). They also concluded that traffic in street canyons can have a significant impact on distribution up to heights of 7 m.

The work reported by Hoydysh & Dabberdt (1988) investigated flow behaviour in an asymmetric street canyon using tracer gas and bubble jet visualisation methods in a BLWT. This gives a useful source of validation data for CFD modelling studies related to pollutant concentrations. In other recent work, Chan *et al.* (2003) investigated the impacts of canopies on air quality in canyons; Sagrado *et al.* (2002) considered the impact of the downstream canyon buildings on wind flow; Chang & Meroney (2001) analysed chemical and biological agent dispersal in urban street canyons with both modelling and experimental approaches; Spadaro & Rabl (2001) investigated wider canyon groupings with a focus on the cost due to damage by traffic air pollution (Fig. 6.16). A wider review of the work on street canyon air quality modelling can be found in Vardoulakis *et al.* (2003).

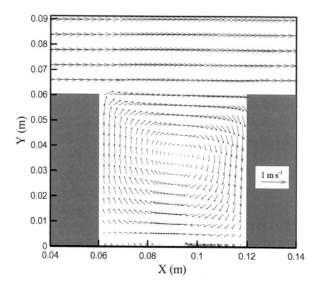

Fig. 6.15 Street canyon vortex. (Reprinted from *Atmospheric Environment*, 36, Chan, T.L., Dong, G., Leung, C.W., Cheung, C.S. & Hung, W.T., Validation of a two-dimensional pollutant dispersion model in an isolated street canyon, 861–72, © (2002), with permission from Elsevier.)

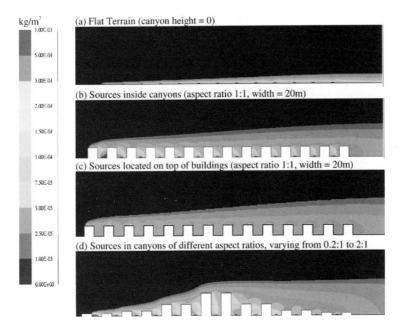

Fig. 6.16 Pollutant concentration patterns around multiple street canyons with pollutant source at the centre of each canyon. (Reprinted from *Atmospheric Environment*, 35, Spadaro, J.V. & Rabl, A., Damage costs due to automotive air pollution and the influence of street canyons, 4763–75, © (2001), with permission from Elsevier.)

A number of modelling studies on street canyon wind patterns have been reported. Ca *et al.* (1995) considered both heat and mass transfer in a north-south street canyon with cross flow using a 2-D model with LES. The focus was to analyse heat transfer from (potentially wetted) buildings and road/pavement surfaces, including radiation.

There was no experimental validation of the modelling and it was found that calculations required to be updated at, typically, one-hour simulation time intervals to account for the surface thermal capacities. Hassan & Crowther (1998) used the 2-D form of PHOENICS with a standard k–ε turbulence model to study the cross wind behaviour and pollutant dispersion in a street canyon. Again, they confirm the existence of a dominant vortex in the canyon. Results were generated and compared with measurements of carbon monoxide concentrations in a Glasgow shopping street. Scarperdas & Colville (1999) used StarCD with the standard k–ε model to investigate problems associated with the measurement of air quality measurements in city centres. Since many European cities routinely conduct air quality measurements at street intersections, guidance is needed on the siting and location of the monitoring sensors. They concluded that measured carbon monoxide at street intersections is significantly affected by wind direction, and noted that CFD results can help to establish dominant flow patterns.

Walton *et al.* (2002) looked at the turbulence structure and physical dispersion of pollutants in an urban street canyon using CFX with both LES and k–ε. They first validated the model using experimental data from a canyon-like roof garden using full scale sonic anemometry. They concluded that the k–ε model predicts a weaker canyon vortex whereas the LES approach gives results nearer the experiment. In a linked paper, Walton & Cheng (2002) used the LES-based model validated previously for the prediction of flow behaviour in an idealised street canyon, and further concluded that the LES method gives higher turbulent intensities than k–ε. In a rare look at the thermodynamics of street canyons, Papadopoulos (2001) considered the impact of local heat transfers (including heat rejection from the condensing units of air conditioners), and recommended a procedure for flow and temperature prediction.

6.4 Applications to plant

Considerable progress has been made on the modelling of plant for energy in buildings, though this has tended to lag behind the earlier developments of methods and practices for modelling building envelopes. Building energy plant can be classified into two groups:

- Secondary plant
- Primary plant

Secondary plant is generally taken to mean equipment at room and, in some cases, distribution level. Room emitters, chilled ceiling and beams, coils, pumps and fans, conduits are examples of equipment in this category. Primary plant is associated with centralised equipment for the generation (or dissipation) of energy. Examples include boilers, chillers and combined heat and power plant. Probably the most comprehensive treatments to date of models for equipment in both categories (including recommended Fortran code listings) can be found in Brandemeuhl (1993) (secondary systems) and LeBrun *et al.* (1999) (primary plant).

A further classification related to modelling concerns the extent to which detail is attached to the description of plant. Fundamentally, there are three possibilities as far as plant descriptions integrated within building energy models are concerned as summarised in Table 6.4.

A review and examples of some the recent progress in plant modelling, with particular reference to steady-state applications, now follows. Section 6.5 then concentrates on some of the recent developments in dynamic-state applications with particular reference to control problems.

Table 6.4 Plant descriptions in building energy models.

Method	Comment	Typical Applications
Control functions	Plant not modelled Perfect set point tracking assumed Low computational cost	Seasonal energy predictions Outline control strategy analysis

Steady-state	Plant modelled using steady-state
	    energy and continuity balances
	'Quasi-steady-state' model when
	    building model is fully dynamic
	Low/moderate computational cost
		'Accurate' seasonal energy predictions
		Plant sizing and selection analyses
		Control strategy design and some areas
		    of fault detection and diagnosis

Dynamic-state	Nonlinear (time-variant) methods for greatest utility High/very high computational cost	Control Thermal storage Plant 'troubleshooting' Fault detection and diagnosis

Secondary plant

Considerable attention has been given to the treatment of cooling coils mostly using steady-state theoretical methods (McQuiston 1975; Mirth & Ramadhyani 1993a; Mirth & Ramadhyani 1993b; Yik *et al*. 1997; Morisot *et al*. 2002). A rigorous treatment of the subject with heat and mass transfer can be relaxed to a simple case involving sensible heat exchange only where needed (see Section 5.1) enabling a wide range of problem areas to be captured through an understanding of this one particular case. Recently, there has been some interest in using manufacturers' catalogue data for constructing and verifying models of cooling coils and other items of related plant (e.g. Rabehl *et al*. 1999). An example of results of cooling coil modelling verified using catalogue data is shown in Fig. 6.17.

A number of applications have been reported concerning modelling treatment of chilled ceilings, essentially, as sensible heat exchange applications within room spaces. Niu *et al*. (1997) modelled a chilled ceiling application dynamically including a treatment of radiant effects. Miriel *et al*. (2002) modelled ceiling panels for both winter heating and summer cooling with results compared with experimental data. Laurenti *et al*. (2002) developed a finite element model description of a panel heating and cooling system that has structural similarities with the transfer function method for building conduction calculations (see Section 1.3). Jeong & Mumma (2003) modelled the impact of mixed convection on the outputs of chilled ceiling panels with local primary air supply from diffusers, concluding previous assumptions of cooling capacity in these applications by up to 35%. Fig. 6.18 illustrates the total panel capacities with the radiant component at a variety of mechanical air supply velocities.

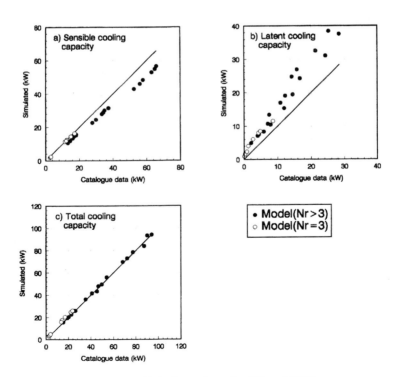

Fig. 6.17   Cooling coil model results verified using catalogue data (Yik *et al*. 1997).

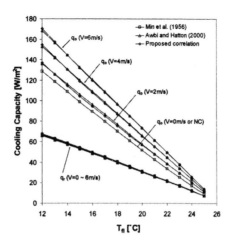

Fig. 6.18 Chilled ceiling radiant and total sensible cooling capacities with mechanical air supply at various veloc-ities. (ASHRAE Int. J. HVAC & R Res. 9(3) 251–7 2003. © ASHRAE, www. ashrae. org.)

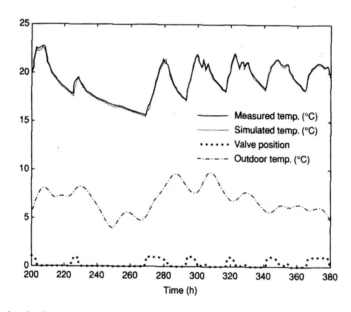

Fig. 6.19 Reduced order space heating model – temperature results compared with field monitoring data (Gouda *et al*. 2000).

In other secondary HVAC plant modelling applications, Khoo *et al*. have proposed steady-state (Khoo *et al*. 1998a) and dynamic-state (Khoo *et al*. 1998b) models of variable-air-volume terminal units using curve fits to experimental data. Gouda *et al*. (2000) have demon-strated that a first-order dynamic model of a heating system coupled to a low-order model of a space can give excellent tracking results over relatively short time horizons as shown in Fig. 6.19. Musy *et al*. (2001) have considered the detailed steady-state response of radiant/convec-tive heating in a space by linking an emitter model to the multi-cell zonal modelling approach discussed in Section 6.3. In applications related to distribution, Hanby *et al*. (2002) proposed a dynamic model of a general flow conduit based on its discretisation into a sequence of well mixed zones. They concluded that optimality arises with 46 zone nodes but satisfactory results for most cases will be obtained with 20 nodes, the results being largely independent of the length-to-diameter ratio of the conduit.

Yik (2001) modelled the performance of distributed systems serviced by multiple pumps and chillers and identified a problem due to flow starvation when pump and chiller switching sequences take effect – the resulting capacity reduction is shown in Fig. 6.20. It was concluded that two-port valves should be used with a primary-circuit differential pressure regulator on secondary loads such as cooling coils. Olsen & Chen (2003) recognised that the particularly mild climate conditions of the UK require attention to be placed on hybrid solutions that maximise passive opportunities, such as natural ventilation and night cooling, but also permit the use of active system intervention such as chilled ceilings and variable-air-volume systems when necessary. Example results using the integrated modelling program EnergyPlus which includes steady-state descriptions of plant are shown in Fig. 6.21.

Primary plant

Numerous examples of chiller models and their applications can be found in the literature (Braun *et al.* 1987; Wang *et al.* 2000; Yik & Lam 1998). Wang *et al.* deal with the dynamic modelling of centrifugal chillers using a theoretical basis but 'tuned' with field measurement data. Yik & Lam deal with centrifugal, reciprocating and screw compressor-driven chillers for both air- and water-cooled condensers based on curve-fitting to manufacturers' data. In later work, these models are used to predict annual performance profiles using alternative methods of condenser cooling (Yik *et al.* 2001). Figure 6.22 shows results based

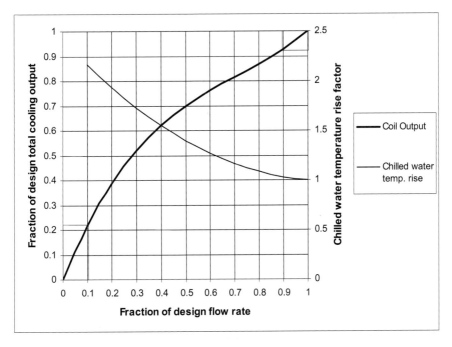

Fig. 6.20   Capacity reduction in cooling coils with reducing chilled water flow rate (Yik 2001).

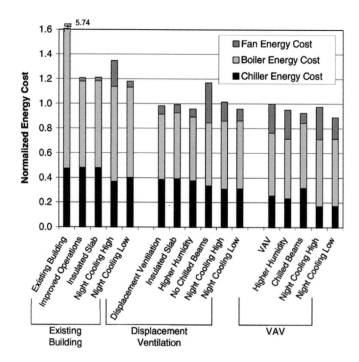

Fig. 6.21 Energy use by various 'hybrid' system solutions in a UK application. (Reprinted from *Energy and Buildings*, 35, Olsen, E.L. & Chen, C., Energy consumption and comfort analysis for different low-energy cooling systems in a mild climate, 561–71, © (2003), with permission from Elsevier.)

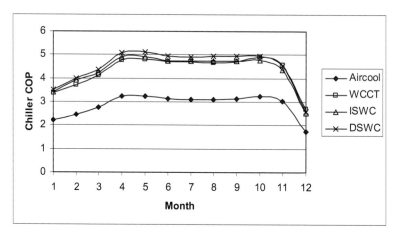

Fig. 6.22 Predicted chiller performance using alternative methods of heat rejection. (Reprinted from *Energy and Buildings*, 33, Yik, F.W.H., Burnett, J. & Prescott, I., Predicting air conditioning energy consumption of a group of buildings using different heat rejection methods, 151–66, © (2001), with permission from Elsevier.)

on water-cooled, air-cooled, indirect sea-water and direct sea-water methods which show that, in spite of low seasonal sink temperatures, reduced load operation during winter months results in reduced performance.

Another area of modelling interest related to refrigeration plant lies in ice storage plant (Arnold 1994; Zhu & Zhang 2000). The high thermal capacity of these thermal storage sys-.

tems coupled with the need to investigate variably switched loads during charging and discharging transients requires that modelling be carried out in the dynamic state.

Progress in boiler and combustion plant modelling has been more generic (other than for general energy prediction and part-load performance models; e.g. LeBrun *et al.* 1999) concentrating mainly on larger scale plants with an emphasis on gaining improved understanding on the mechanisms of, and remedies for, emission species. Interest in particular in CFD applications to these problems has emerged – Eaton *et al.* (1999).

Models of cogeneration or combined heat and power (CHP) plant for buildings have also evolved. The earliest models consisted of spreadsheet-based energy profilers for matching the steady-state energy balances of plants at rated and possibly turndown operating conditions to prevailing energy demand patterns (e.g. Williams *et al.* 1996). Some of the more rigorous modelling approaches using steady- or quasi-steady-state and dynamic-state modelling of the plants themselves have also been reported for applications involving integrated and hybrid plants, control problems and energy storage, etc.. van Schijndel (2002) has reported on the development of a quasi-steady-state model for of CHP plant for year-round simulations of control strategy evaluation, and Lin & Yi (2000) used a similar approach for a very large scale steam turbine-based application to district heating. McCrorie *et al.* (1996) developed a dynamic spark-ignition model for small scale CHP plant analysis. In an application that is developing growing interest in co-generation due to the stabilisation of year-round heat utilisation, Shimazaki (2003) has modelled the concept of heat-cascading (Fig. 6.23) in order to target waste heat on potential consumers such as absorption cycle refrigeration; and Bruno *et al.* (1999) have modelled a combined-cycle gas turbine CHP plant coupled to an absorption cycle chiller which uses the power plant waste heat streams (Fig. 6.24). Further perspectives on co-generation appear in the next section.

Renewable and low carbon systems

There has been rapidly growing interest in the treatment of a new class of embedded generation and low energy use technologies, specifically related to managing the reduction in

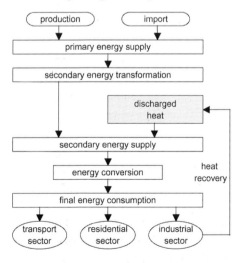

Fig. 6.23 Heat-recovery cascading model. (Reprinted from *Energy Policy*, 31, Shimazaki, Y., Evaluation of refrigeration and air-conditioning technologies in heat cascading systems under the carbon dioxide emissions constraint: The proposal of the energy cascade balance table, 1685–97, © (2003), with permission from Elsevier.)

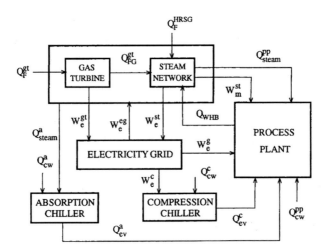

Fig. 6.24   Absorption chiller-coupled CHP plant. (Reprinted from *Appl. Thermal Eng.*, 19, Bruno, J.C., Miquel, J. & Castells, F., Modeling of ammonia absorption chillers integration in energy systems of process plants, 1297–328, © (1999), with permission from Elsevier.)

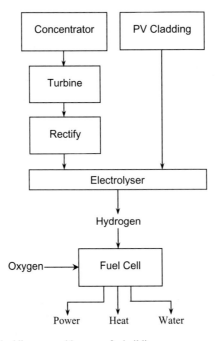

Fig. 6.25   Illustration of an embedding renewable energy for buildings.

hydrocarbon fuel dependency as well as fulfilling various countries' objectives for carbon reduction, under the terms of the various earth summits. It is likely, as world communities manage their reducing dependency on hydrocarbon fuels, that new carbon-free alternatives will emerge, some if not all of which will depend on hydrogen as a means of storage and distribution. As an illustration, consider the use of photovoltaic (PV) rainscreen cladding on a building supplemented by a wind concentrator containing a wind turbine. Both power sources generate temporally according to the availability of driving inputs and, when generating, power outputs may not be in phase with matching building demands. Figure 6.25 illustrates how this situation can be managed through the use of a fuel cell. All three technologies are expected to have potential for embedded application in buildings of the future.

A brief review of approaches to modelling these and other related 'low carbon' technologies follows.

Considerable attention has been given to the development of ground-coupled heat pump systems as a class of low-carbon system. The mild and stable temperatures of earth combine to provide higher heat pump evaporating temperatures (and, therefore, higher coefficient of performance) than is the case with conventional ambient air source applications. Like Rabehl *et al.* (1999), Jin & Spitler (2002) use curve fitting to catalogue data to develop a steady-state model of the typical water-to-water compression heat pump that can be used in quasi-steady-state modelling of ground-coupled applications.

Tarnawski & Leong (1993) deal with the design and modelling of horizontal-type ground heat exchangers, and the effect of soil condition on ground-coupled heat pump performance has been investigated by Leong *et al.* (1998). To deal with the problem of earth temperature saturation after several years of heat extraction, Spitler & Underwood (2003) modelled scenarios for sustainable earth heat exchange in vertical borehole heat exchangers based on rejecting heat in summer from chilled ceiling plant. Figure 6.26 shows that, after ten years of heat transfer cycles, there is minimal overall change in exiting water temperature from the ground loop (and, hence, earth temperature) for typical UK commercial building applications. A comprehensive review of the technology, design and installation of ground coupled systems from a UK perspective can be found in Rawlings (1999).

Interest in fuel cells has, until recently, been restricted to basic research and development. Recently, however, as experience has grown with the demonstration of this technology in a variety of applications, interest has grown in the potential role of fuel cells in embedded power generation applications in buildings. Ellis (2002) provides a comprehensive review of the potential for building applications of fuel cells, and a treatment of the fundamentals can be found in Blomen & Mugerwa (1993).

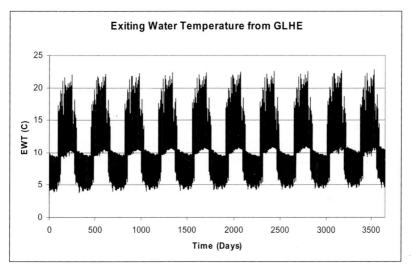

Fig. 6.26 Sustainable earth exchange (Spitler & Underwood 2003).

Many current applications for co-generation centre on methane-fuelled systems with steam internal reforming to break the methane down into hydrogen and carbon dioxide. As dependency on hydrocarbon fuels diminishes, a role for the fuel cell that is more in-keeping with the renewable scheme of Fig, 6.25 is more likely to prevail. A general method for modelling stacked fuel cells can be found in Lee & Lalk (1998). The most promising options for building co-generation systems, at least in the short-to-medium term, are (classified by electrolyte) the solid oxide (SOFC) and molten carbonate (MCFC) fuel cells. These types operate at temperatures of 1000°C and 650°C respectively making them suitable for high grade heat recovery as well as the potential for steam raising for internal reforming if needed. Simple flowsheet modelling approaches for a SOFC can be found in Riensche *et al.* (1998) and, for the MCFC, in Au *et al.* (2003).

Interest in wind turbine applications on buildings is based on the idea of accelerating or concentrating the dominant wind site-trajectory onto a wind turbine. Horizontal axis wind turbines (HAWT) are of greatest practical interest due to the limited commercial availability of vertical axis turbines. Campbell & Stankovic (1999) present prototype designs for a variety of concentrator possibilities based on a number of built forms such as twin towers and building infill sections. Modelling of wind turbines has been well documented in the literature (e.g. Nathan 1980; Welfonder *et al.* 1997; Maalawi & Badawy 2001), the simple curve-fit based model of Welfoner *et al.* being particularly appealing for integration within building energy models:

$$P_{\text{turbine}} = \frac{\rho}{2} \times C_{\text{P}} A u_{\text{w}}^3 \qquad (6.1)$$

in which:

P_{turbine} = turbine power extracted from wind (W)
ρ = air density at turbine site (kg m^{-3})
C_{P} = power coefficient
A = rotor swept area (m^2)
u_{w} = axis-corrected wind speed (m s^{-1})

The power coefficient, C_{P}, can be fitted from test data (Welfonder *et al.* 1997).

Most photovoltaic (PV) modelling applications for building energy problems can quite adequately be represented at module detail rather than at cell detail where a very good correlation between in-plane irradiance and power yield can be expected over the uniformly insolated surface. However, it is desirable to model the problem with the inclusion of module temperature since this will vary according to the instantaneous module heat balance as well as across the seasons. A method for modelling the common monocrystalline-based PV module in this way as part of a whole building has been reported by Jones & Underwood (2002) together with allowances for module temperature variation (Jones & Underwood 2001). Their model is based on a justification for a constant fill factor of a PV module – the maximum power point at the 'knee' of the current–voltage characteristic divided by the product of the short-circuit current and open-circuit voltage (see Fig. 6.27).

The power model arising from this is as follows:

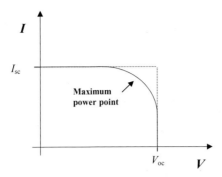

Fig. 6.27   PV current–voltage curve for a photovoltaic module.

$$P_{\text{array}} = \frac{1}{T_{\text{m}}} \times C_{\text{FF}} N_{\text{m}} \eta_{\text{inv}} I_{\text{s}} \ln(I_{\text{s}} \times 10^{6})$$ (6.2)

in which:

P_{array} = array output power (W)
C_{FF} = fill factor model constant (K m^2)
N_{m} = number of modules forming the array
η_{inv} = inverter efficiency (fraction)
I_{s} = total solar irradiance normal to module surface (W m^{-2})
T_{m} = module temperature (K)

C_{FF} can be calculated from reference data (Jones & Underwood 2002) with typical values around 1.22 K m^2. Allowances for inverter losses are also discussed.

Examples of more detailed models for PV systems are as follows: Sukamongkol *et al.* (2002) develop the conventional 'diode' type theoretical model; Durisch *et al.* (2000) report on detailed curve-fit types models; and Armenta-Deu (2003) investigated the performance of battery-coupled PV systems in a stand-alone context.

More recently, attention has begun to turn to the modelling of coupled or hybrid renewable energy systems for buildings. Clarke & Kelly (2001) describe the development of a generic modelling approach for integrating power sources into building energy models. Lund & Munster (2003) consider the large-scale integration of wind power and CHP in Denmark; and Iqbal (2003) has developed a model of a wind turbine-coupled fuel cell system for small-scale applications.

6.5 Applications to control and fault detection

Modelling applications in control and plant response have continued to focus on the development of adaptive and 'smart' control algorithms throughout most of the 1990s and beyond. General reviews of these applications can be found in Underwood (1999). Recently, interest has turned to other problems in plant response, the most prominent of which has been in the area of fault detection and diagnosis (FDD).

Interest in variable air volume air (VAV) conditioning has continued to be prominent. Mei & Levermore (2002) have synthesised a VAV terminal box model (Khoo *et al.* 1998a) with controls and an artificial neural network-based fan model for control system simulation applications. Haves *et al.* (1998) have developed a set of HVAC plant component models for use in the testing and evaluation of interaction in multi-loop control strategies with particular emphasis on VAV systems; and Wang (1999) has implemented a VAV system model on-line in a building management system to assist in the supervisory control set point scheduling. Other on-line applications with building management systems have been reported by Salsbury & Diamond (2000) and Clarke *et al.* (2002); the first of these uses a model to generate performance targets for plant response that can be compared with monitored system outputs for performance validation and energy usage analysis, and the latter uses an embedded model in a building management system for predictive control (Fig. 6.28).

Several other applications use simulation models of plant and control to deal with the problems associated with control loop interaction in complete assemblies of plant typically found in air handling units (Jetté *et al.* 1998; Mathews *et al.* 2001; Seem *et al.* 1999; Salsbury 1998).

Seem *et al.* (1999) introduce a novel 'finite state machine' strategy as an alternative to the conventional sequenced control of heating, free-cooling and refrigeration-based cooling in which the transfer between functions is governed by a state transition delay (see Fig. 6.29). Salsbury (1998) found that the introduction of feedforward control based on inverse plant models improves stability by reducing reliance on the main feedback loops (design methods for feedforward control can be found in Underwood 1999).

Modelling applications to control at higher levels of HVAC plant detail than those described above have been reported. Riederer *et al.* (2002) have developed a system model based around a zonal modelling approach (see Section 6.2) enabling the space to be split into a number of air zones (three in the vertical plane in this case) so that an analysis of sensor position could be made. Underwood (2001) has developed a detailed dynamic model of a vapour-compression refrigeration plant in order to investigate coupling between the two rival control loops – that acting on the compressor for capacity control and that acting on the ex-

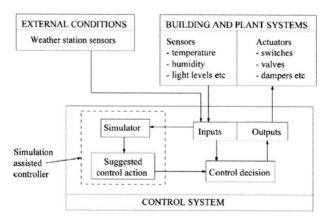

Fig. 6.28 Simulation-assisted control example. (Reprinted from *Energy and Buildings*, 34, Clarke, J.A., Cockcroft, J., Conner, S., Hand, J. W., Kelly, N.J., Moore, R., O'Brien, T. & Strachan, P. Similation-assisted control in building energy management systems, 933–40, © (2002), with permission from Elsevier.)

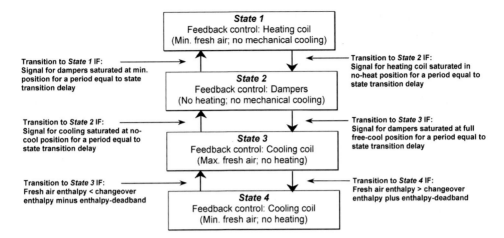

Fig. 6.29 Finite state machine control of sequential processes. (ASHRAE Int. J. HVAC & R Res 5(1) 35–8 1999. © ASHRAE, www. ashrae. org.)

pansion device for evaporator refrigerant outlet state control. Deng & Missenden (1999) have developed two types of model for a direct expansion air conditioning plant – a detailed non-linear model and a linear transfer function-based model of the plant. They use the latter for verification and simplification of the non-linear model. Underwood (2000) has also concluded that detailed non-linear models of HVAC plants can be alternatively expressed as a simple linear model with an uncertainty specification. The first-order model with delay time (Equation 6.3):

$$G(s) = k_p \times \frac{\exp(-d_p s)}{(\tau_p s + 1)} \tag{6.3}$$

can be used to model many non-linear processes subject to a parametric uncertainty envelope:

$$k_p \in k_p^-, k_p^+; \quad d_p \in d_p^-, d_p^+; \quad \tau_p \in \tau_p^-, \tau_p^+$$

where k_p, d_p, τ_p are the mean process gain, delay-time and time constant respectively and k_p^-, k_p^+ represent minimum and maximum bounds on the process gain respectively (likewise for the other process parameters). This specification has been used to develop a particular class of fixed-parameter 'robust' controllers for HVAC plant (Underwood 2000; Noda *et al.* 2003).

Examples of applications to the highly damped problem of space heating control are found in Fraisse *et al.* (1999) and Cho & Zaheer-uddin (2003), in the latter case dealing with predictive control of intermittently operating under-floor heating. Fraisse *et al.* (1999) compare several methods of heating start-up control, including two new methods, one of which uses a fuzzy logic control algorithm.

Fuzzy control continues to excite some interest in HVAC applications due to the wide fund of experiential data available to inform them (gaps in which can easily be plugged by models) together with the avoidance of the need for inconvenient (and frequently intractable) commissioning and tuning. Gouda *et al.* (2001) also explored this application for a quasi-adaptive heating control in which look-ahead comfort predictions from an ANN-based model form one of several fuzzified inputs to the controller. Results from this work are given in Fig. 6.30 showing that afternoon overheating in buildings with passive solar heat gain can be virtually eliminated with this method. Another example is reported by Ghiaus (2001) dealing with the development of an adaptive fuzzy controller for a fan coil unit application using a more computationally efficient Sugeno inference method instead of the conventional Mamdani method (see Section 5.3).

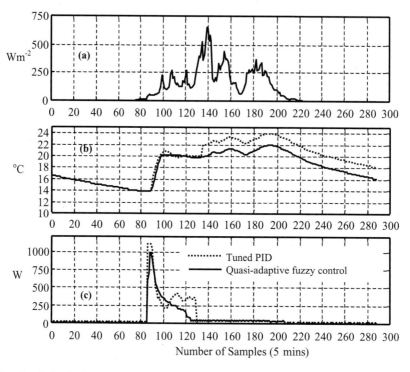

Fig. 6.30   Quasi-adaptive fuzzy control of heating. (a) Solar irradiance on window. (b) Internal air temperature. (c) Heater load.

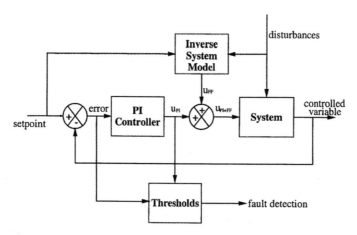

Fig. 6.31   Fault detection in combined feedback/feedforward schemes. (Reprinted from *Energy and Buildings*, 33, Salsbury, T.J. & Diamond, R., Fault detection in HVAC systems using model-based feedforward control, 403–15, © (2001), with permission from Elsevier.)

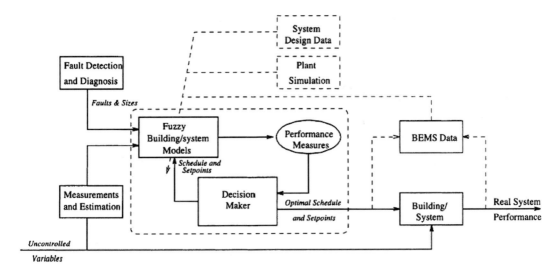

Fig. 6.32   Fault-tolerant control scheme. (Reprinted from *Energy and Buildings*, 33, Liu, X.-F. & Dexter, A., Fault-tolerant supervisory control of VAV air-conditioning systems, 379–89, © (2001), with permission from Elsevier.)

The most recently developed area work involving the use of plant response models relates to the field-operational problem of fault detection and diagnosis (FDD) – for examples see Ngo & Dexter (1999); Salsbury (1999); Salsbury & Diamond (2001); Liu & Dexter (2001); Buswell *et al.* (2003).

Most of this early work has confined itself to problems in air handling plants. Ngo & Dexter developed a model-based approach to try to improve fault detection sensitivity, given the poor levels of measurement precision and accuracy frequently found in practice. Salsbury & Diamond recognise a further merit in the feedforward control scheme mentioned earlier (Salsbury 1998) in that fault levels can be inferred from increased activity in the feedback loop – see Fig. 6.31. Liu & Dexter attempted to incorporate fault tolerance into plant, using fuzzy plant models to predict plant control response which informs a control algorithm that adapts to degradation faults such as fouling – see Fig. 6.32.

References

Adjali, M.H., Davies, M. & Littler, J. (1998) Earth-contact heat flows: Review and application of design guidance. *Building Services Engineering Research & Technology* 19(3) pp.111–21.

Allocca, C., Chen, Q. & Glicksman, L.R. (2003) Design analysis of single-sided natural ventilation. *Energy and Buildings* 35 pp.785–95.

Architectural Institute of Japan (1998) *Numerical Prediction of Wind Loading on Buildings and Structures.* Sub-committee on Wind Engineering Data for Structural Design, Architectural Institute of Japan, Tokyo.

Armenta-Deu, C. (2003) Prediction of battery behaviour in SAPV applications. *Renewable Energy* 28 pp.1671–84.

Arnold, D. (1994) Dynamic simulation of encapsulated ice tanks: Part II – model development and validation. *ASHRAE Transactions* 100(1) pp.1245–54.

Au, S.F., McPhail, S.J., Woudstra, N. & Hemmes, K. (2003) The influence of operating temperature on the efficiency of a combined heat and power fuel cell plant. *Journal of Power Sources* 122 pp.37–46

Augustyn, J. (1998) WYEC2 User's Manual and software toolkit. *ASHRAE Transactions* 104(1A) pp.32–40.

Axley, J.W. (2001) Surface-drag flow relations for zonal modelling. *Building and Environment* 36 pp.843–50.

Baskaran, A. & Kashef, A. (1996) Investigation of air flow around buildings using computational fluid dynamics techniques *Engineering Structures* 18(11) pp.861–75.

Beausoleil-Morrison, I. (2001) An algorithm for calculating convection coefficients for internal building surfaces for the case of mixed flow in rooms. *Energy and Buildings* 33 pp.351–61.

Beausoleil-Morrison, I. (2002) The adaptive simulation of convective heat transfer at internal building surfaces. *Building and Environment* 37 pp.791–806.

Blomen, L.J.M.J. & Mugerwa, M.N.(eds) (1993) *Fuel Cell Systems.* Plenum, New York.

Blondeau, P., Sperandio, M., Allard, F. & Haghighat, F. (1998) Detailed modelling of gas phase chemistry mechanisms in IAQ simulation. *ASHRAE Transactions* 104(2) pp.1309–17.

Bojic, M., Yik, F.W.H. & Burnett, J. (2002) The air flow around and inside high-rise residential buildings in Hong Kong. *Proceedings, Advances in Building Technology – ABT 2002, Hong Kong* 2 pp.331–8.

Bojic, M., Yik, F. & Lee, M. (2003) Influence of plates on flow inside a recessed space generated by heat rejection. *Building and Environment* 38 pp.593–604.

Brandemuehl, M. (1993) *HVAC 2 Toolkit: Algorithms and Subroutines for Secondary HVAC System Energy Calculations.* American Society of Heating, Refrigerating and Air Conditioning Engineers Inc. Atlanta, GA.

Braun, J.E. & Chaturvedi, N. (2002) An inverse gray-box model for transient building load prediction. *International Journal of Heating, Ventilating, Air-conditioning and Refrigerating Research* 8(1) pp.73–99.

Braun, J., Mitchell, J.E., Klein, S.A. & Beckman, W.A. (1987) Models for variable-speed centrifugal chillers. *ASHRAE Transactions* 93(1) pp. 1794–1813.

Braun, J.E., Montgomery, K.W. & Chaturvedi, N. (2001) Evaluating the performance of building thermal mass control strategies. *International Journal of Heating, Ventilating, Air-conditioning and Refrigerating Research* 7(4) pp.403–28.

Bruno, J.C., Miquel, J. & Castells, F. (1999) Modelling of ammonia absorption chillers integration in energy systems of process plants. *Applied Thermal Engineering* 19 pp.1297–328.

Buswell, R.A., Haves, P. & Wright, J.A. (2003) Model-based condition monitoring of a HVAC cooling coil sub-system in a real building. *Building Services Engineering Research & Technology* 24(2) pp.103–6.

Ca, V.T., Asaeda, T., Ito, M. & Armfield, S. (1995) Characteristics of wind field in a street canyon. *Journal of Wind Engineering and Industrial Aerodynamics* 57 pp.63–80.

Campbell, N.S. & Stankovic, S. (1999) *Wind Energy for the Built Environment (Project WEB) – Final Report of EU Contract JOR3-CT98–0279 – Joule III Programme.* European Commission, Brussels.

Carpenter, S.C., Kosny, J. & Kossecka, E. (2003) Modelling transient performance of two-dimensional and three-dimensional building assemblies. *ASHRAE Transactions* 109(1) pp.566–71.

Chan, A.T., Au, W.T.W. & So, E.S.P. (2003) Strategic guidelines for street canyon geometry to achieve sustainable street air quality – part II: Multiple canopies and canyons. *Atmospheric Environment* 37 pp.2761–72.

Chan, T.L., Dong, G., Leung, C.W., Cheung, C.S. & Hung, W.T. (2002) Validation of a two-dimensional pollutant dispersion model in an isolated street canyon. *Atmospheric Environment* 36 pp.861–72.

Cho, S.H. & Zaheer-uddin, M. (2003) Predictive control of intermittently operated radiant floor heating system. *Energy Conversion and Management* 44 pp.1333–42.

Chow, W.K. (2001) Numerical studies of airflows induced by mechanical ventilation and air conditioning (MVAC) systems. *Applied Energy* 68 pp.135–59.

Chang, C-H. & Meroney, R.N. (2001) Numerical and physical modelling of bluff body flow and dispersion in urban street canyons. *Journal of Wind Engineering and Industrial Aerodynamics* 89 pp.1325–34.

Clarke, J.A., Cockroft, J., Conner, S., Hand, J.W., Kelly, N.J., Moore, R., O'Brien, T. & Strachan, P. (2002) Simulation-assisted control in building energy management systems. *Energy and Buildings* 34 pp.933–40.

Clarke, J.A., Janak, M. & Ruyssevelt, P. (1998) Assessing the overall performance of advanced glazing systems. *Solar Energy* 63(4) pp.231–41.

Clarke, J.A., Johnstone, C.M., Kelly, N.J., McLean, R.C. & Anderson, J.A. (1999) A technique for the prediction of the conditions leading to mould growth in buildings. *Building and Environment* 34 pp.515–21.

Clarke, J.A. & Kelly, N.J. (2001) Integrating power flow modelling with building simulation. *Energy and Buildings* 33 pp.333–40.

Cook, M.J. & Lomas, K.J. (1998) Buoyancy-driven displacement ventilation flows: Evaluation of two eddy viscosity turbulence models for prediction. *Building Services Engineering Research & Technology* 19(1) pp.15–21.

Crowley, M.E. & Hashmi, M.S.J. (2000) Improved direct solver for building energy simulation. *Building Services Engineering Research & Technology* 21(3) pp.167–73.

Davies, M.G. (1999) Current methods to handle wall conduction and room internal heat transfer. *ASHRAE Transactions* 105(2) pp.142–53.

Davies, M., Zoras, S. & Adjali, M.H. (2001) Improving the efficiency of the numerical modelling of built environment earth-contact heat transfers. *Applied Energy* 68 pp.31–42.

Deng, S-M. & Missenden, J.F. (1999) Applying classical control theory for validating and simplifying a dynamic mathematical model of an air-conditioning plant. *ASHRAE Transactions* 105(1) pp.140–8.

DePaul, F.T. & Sheih, C.M. (1986) Measurement of wind velocities in a street canyon. *Atmospheric Environment* 20(3) pp.455–59.

Déqué, F., Ollivier, F. & Poblador, A. (2000) Grey boxes used to represent buildings with a minimum number of geometric and thermal parameters. *Energy and Buildings* 31 pp.29–35.

Deru, M., Judkoff, R. & Neymark, J. (2003) Whole building energy simulation with a three-dimensional ground-coupled heat transfer model. *ASHRAE Transactions* 109(1) pp.557–65.

Djunaedy, E. & Cheong, K.W.D. (2002) Development of a simplified technique of modelling four-way ceiling diffusers. *Building and Environment* 37 pp.393–403.

Durisch, W., Tille, D., Wörz, A. & Plapp, W. (2000) Characterisation of photovoltaic generators. *Applied Energy* 65 pp.273–84.

Eaton, A.M., Smoot, L.D., Hill, S.C. & Eatough, C.N. (1999) Components, formulations, solutions, evaluation, and application of comprehensive combustion models. *Prog. Energy Combustion Science* 25(4) pp.387–436.

Ehrhard, J. & Moussiopoulos, N. (2000) On a new nonlinear turbulence model for simulating flows around building-shaped structures. *Journal of Wind Engineering and Industrial Aerodynamics* 88 pp.91–9.

Ellis, M.W. (2002) *Fuel Cells for Air Conditioning Applications.* American Society of Heating, Refrigerating and Air Conditioning Engineers, Inc., Atlanta, GA.

Fraisse, G., Viardot, C., Lafabrie, O. & Achard, G. (2002) Development of a simplified and accurate building model based on electrical analogy. *Energy and Buildings* 34 pp.1017–31.

Fraisse, G., Virgone, J. & Yezou, R. (1999) A numerical comparison of different methods for optimising heating-restart time in intermittently occupied buildings. *Applied Energy* 62 pp.125–40.

Ghiaus, C. (2001) Fuzzy model and control of a fan-coil. *Energy and Buildings* 33 pp.545–51.

Ghiaus, C. (2003) Free-running building temperature and HVAC climatic suitability. *Energy and Buildings* 35 pp.405–11.

Gouda, M.M., Danaher, S. & Underwood, C.P. (2000) Low order model for the simulation of a building and its heating system. *Building Services Engineering Research and Technology* 21(3) pp. 199–208.

Gouda, M.M., Danaher, S. & Underwood, C.P. (2001) Thermal comfort based fuzzy logic controller. *Building Services Engineering Research & Technology* 22(4) pp.237–53.

Gouda, M.M., Danaher, S. & Underwood, C.P. (2002) Application of an artificial neural network for modelling the thermal dynamics of a building's space and its heating system. *Mathematical and Computer Modelling of Dynamical Systems* 8(3) pp.333–44.

Griffith, B. & Chen, Y. (2003) A momentum-zonal model for predicting zone airflow and temperature distributions to enhance building load and energy simulations. *International Journal of Heating, Ventilating, Air-conditioning and Refrigerating Research* 9(3) pp.309–25.

Gul, M.S., Muneer, T. & Kambezidis, H.D. (1998) Models for obtaining solar radiation from other meteorological data. *Solar Energy* 64(1–3) pp.99–108.

Haas, A., Weber, A., Dorer, V., Keilholz, W. & Pelletret, R. (2002) COMIS v3.1 simulation environment for multizone air flow and pollutant transport modelling. *Energy and Buildings* 34 pp.873–82.

Hanby, V.I., Wright, J.A., Fletcher, D.W. & Jones, D.N.T. (2002) Modelling the dynamic response of conduits. *International Journal of Heating, Ventilating, Air-conditioning and Refrigerating Research.* 8(1) pp.1–12.

Harriman, L.G., Colliver, D.G. & Quinn-Hart, K. (1999) New weather data for energy calculations. *ASHRAE Journal* 41,(3) pp.31–8.

Hassan, A.A. & Crowther, J.M. (1998) Modelling of fluid flow and pollution dispersion in a street canyon. *Environmental Monitoring and Assessment* 52(1–2) pp.281–97.

Haves. P., Norford, L.K. & DeSimone, M. (1998) A standard simulation test bed for the evaluation of control algorithms and strategies. *ASHRAE Transactions* 104(1A) pp.460–73.

He, J. & Song, C.C.S. (1997) A numerical study of wind flow around the TTU building and the roof corner vortex. *Journal of Wind Engineering and Industrial Aerodynamics* 67 & 68 pp.547–8.

Hosni, M.H., Jones, B.W. & Sipes, J.M. (1998) Total heat gain and the split between radiant and convective heat gain from office and laboratory equipment in buildings. *ASHRAE Transactions* 104(1A) pp.356–65.

Hoydysh, W.G. & Dabberdt, W.F. (1988) Kinematics and dispersion characteristics of flows in asymmetric street canyons. *Atmospheric Environment* 22(12) pp.2677–89.

Huser A, Nilsen PJ and Skåtun H (1997) Application of k–ε model to the stable ABL: Pollution in complex terrain *Journal of Wind Engineering and Industrial Aerodynamics* 67 & 68 pp.425–36.

Iqbal, M.T. (2003) Simulation of a small wind fuel cell hybrid energy system. *Renewable Energy* 28 pp.511–22.

Jeong, J.W. & Mumma, S.A. (2003) Impact of mixed convection on ceiling cooling panel capacity. *International Journal of Heating, Ventilating, Air-conditioning and Refrigerating Research* 9(3) pp.251–7.

Jetté, I., Zaheer-uddin, M. & Fazio, P. (1998) PI-control of dual duct systems: Manual tuning and control loop interaction. *Energy Conversion and Management.* 39(14) pp.1471–82.

Jiang, Y. & Chen, Q. (2001) Study of natural ventilation in buildings by large eddy simulation. *Journal of Wind Engineering and Industrial Aerodynamics* 89 pp.1155–78.

Jiang, Y. & Chen, Q. (2002) Effect of fluctuating wind direction on cross natural ventilation in buildings from large eddy simulation. *Building and Environment* 37 pp.379–86.

Jin, H. & Spitler, J.D. (2002) A parameter estimation based model of water-to-water heat pumps for use in energy calculation programs. *ASHRAE Transactions* 108(1) pp.3–17.

Jones, A.D. & Underwood, C.P. (2001) A thermal model for photovoltaic systems. *Solar Energy* 70(4) pp.349–59.

Jones, A.D. & Underwood, C.P. (2002) A modelling method for building-integrated photovoltaic power supply. *Building Services Engineering Research & Technology* 23(3) pp.167–77.

Jones, A.D. & Underwood, C.P. (2003) A method for profiling lighting and small power demands in buildings. *Building Services Engineering Research & Technology* 24(1) pp.47–53.

Karki, K.C., Radmehr, A. & Patankar, S.V. (2003) Use of computational fluid dynamics for calculating flow rates through perforated tiles in raised-floor data centres. *International Journal of Heating, Ventilating, Air-conditioning and Refrigerating Research* 9(2) pp.153–66.

Karlsson, J. & Roos, A. (2000) Modelling the angular behaviour of the total solar energy transmittance of windows. *Solar Energy* 69(4) pp.321–9.

Kawamoto, S. (1997) Improved turbulence models for estimating wind loading. *Journal of Wind Engineering and Industrial Aerodynamics* 67 & 68 pp.589–99.

Khoo, I., Levermore, G.J. & Letherman, K.M. (1998a) Variable-air-volume terminal units I: Steady-state models *Building Services Engineering Research & Technology* 19(3) pp.155–62.

Khoo, I., Levermore, G.J. & Letherman, K.M. (1998b) Variable-air-volume terminal units II: Dynamic model. *Building Services Engineering Research & Technology* 19(3) pp.163–71.

Knight, K.M., Klein, S.A. & Duffie, J.A. (1991) A methodology for the synthesis of hourly weather data. *Solar Energy* 46(2) pp.109–20.

Laurenti, L., Marcotullio, F. & Zazzini, P. (2002) A proposal for the calculation of panel heating and cooling system based on the transfer function method. *ASHRAE Transactions* 108(1) pp.182–201.

LeBrun, J., Bourdouxhe, J-P. & Grodent, M. (1999) *HVAC 1 Toolkit: A Toolkit for Primary HVAC System Energy Calculation.* American Society of Heating, Refrigerating and Air Conditioning Engineers Inc., Atlanta, GA.

Lee, J.H. & Lalk, T.R. (1998) Modelling fuel cell stack systems. *Journal of Power Sources* 73 pp.229–41.

Leong, W.H., Tarnawski, V.R. & Aittomäki, A. (1998) Effect of soil type and moisture content on ground heat pump performance. *International Journal of Refrigeration* 21(8) pp.595–606.

Letan, R., Dubovsky, V. & Ziskind, G. (2003) Passive ventilation and heating by natural convection in a multi-storey building. *Building and Environment* 38 pp.197–208.

Levermore, G.J. & Doylend, N.O. (2002) North American and European hourly-based weather data and methods for HVAC building energy analyses and design by simulation. *ASHRAE Transactions* 108(2) pp.1053–62.

Liesen, R.J. & Pedersen, C.O. (1999) Modelling the energy effects of combined heat and mass transfer in building elements: Part I – theory. *ASHRAE Transactions* 105(2) pp.941–53.

Liesen, R.J., Strand, R.K. & Pedersen, C.O. (1998) The effect of varying the radiative/convective split for internal gains on cooling loads calculations: A case study for the Pentagon. *ASHRAE Transactions* 104(1B) pp.1790–802.

Lin, F. & Yi, J. (2000) Optimal operation of a CHP plant for space heating as a peak load regulating plant. *Energy* 25 pp.283–98.

Liu, X-F. & Dexter, A. (2001) Fault-tolerant supervisory control of VAV air-conditioning systems. *Energy and Buildings* 33 pp.379–89.

Lucas, F., Adelard, L., Garde, F. & Boyer, H. (2002) Study of moisture in buildings for hot humid climates. *Energy and Buildings* 34 pp.345–55.

Lund, H. & Münster, E. (2003) Modelling of energy systems with a high percentage of CHP and wind power. *Renewable Energy* 28 pp.2179–93.

Maalawi, K.Y. & Badawy, M.T.S. (2001) A direct method for evaluating performances of horizontal axis wind turbines. *Renewable and Sustainable Energy Reviews* 5 pp.175–90.

Manz, H. (2003) Numerical simulation of heat transfer by natural convection in cavities of façade elements. *Energy and Buildings* 35 pp.305–11.

Manzan, M. & Saro, O. (2002) Numerical analysis of heat and mass transfer in a passive building component cooled by water evaporation. *Energy and Buildings* 34 pp.369–75.

Mathews, E.H., Botha, C.P., Arndt, D.C. & Malan, A. (2001) HVAC control strategies to enhance comfort and minimise energy usage. *Energy and Buildings* 33 pp.853–63.

McCrorie, K.A.B., Underwood, C.P. & LeFeuvre, R.F. (1996) Small-scale combined heat and power simulations: Development of a spark-ignition engine model. *Building Services Engineering Research & Technology* 17(3) pp.153–60.

McQuiston, F.C. (1975) Fin efficiency with combined heat and mass transfer *ASHRAE Transactions* 81(1) pp.350–5.

Mei, L. & Levermore, G.J. (2002) Simulation of a VAV system with an ANN fan model and non-linear VAV box model. *Building and Environment* 37 pp.277–84

Mendes, N., Winklemann, F.C., Lamberts, R. & Philipi, P.C. (2003) Moisture effects on conduction loads. *Energy and Buildings* 35 pp.631–44.

Ménézo, C., Roux, J.J. & Virgone, J. (2002) Modelling heat transfers in building by coupling reduced-order models. *Building and Environment* 37 pp.133–44.

Meroney, R.N., Leitl, B.M., Rafailidis, S. & Schatzmann, M. (1999) Wind-tunnel and numerical modelling of flow and dispersion about several building shapes. *Journal of Wind Engineering and Industrial Aerodynamics* 81 pp.333–45.

Minson, A.J., Wood, C.J. & Belcher, R.E. (1995) Experimental velocity measurements for CFD validation. *Journal of Wind Engineering and Industrial Aerodynamics* 58 pp.205–15.

Miriel, J., Serres, L. & Trombe, A. (2002) Radiant panel heating-cooling systems: Experimental and simulated study of the performances, thermal comfort and energy consumptions. *Applied Thermal Engineering* 32 pp.1861–73.

Mirth, D.R. & Ramadhyani, S. (1993a) Prediction of cooling-coil performance under condensing conditions. *International Journal Heat Fluid Flow* 14(4) pp.391–400.

Mirth, D.R. & Ramadhyani, S. (1993b) Comparison of methods of modelling the air-side heat and mass transfer in chilled water cooling coils. *ASHRAE Transactions* 99(2) pp.285–9.

Morisot, O., Marchio, D. & Stabat, P. (2002) Simplified model for the operation of chilled water cooling coils under non-nominal conditions. *International Journal of Heating, Ventilating, Air-conditioning and Refrigerating Research* 8(2) pp.135–58.

Murakami, S. (1998) Overview of turbulence models applied in CWE – 1997. *Journal of Wind Engineering and Industrial Aerodynamics* 74–76 pp.1–24.

Murakami, S., Kato, S. & Zeng, J. (1998) Numerical simulation of contaminant distribution around a modelled human body: CFD study on computational thermal manikin – Part II. *ASHRAE Transactions* 104(2) pp.226–33.

Musy, M., Winklemann, F., Wurtz, E. & Sergent, A. (2002) Automatically generated zonal models for building air flow simulation: Principles and applications. *Building and Environment* 37 pp.873–81.

Musy, M., Wurtz, E., Winklemann, F. & Allard, F. (2001) Generation of a zonal model to simulate natural convection in a room with a radiative/convective heater. *Building and Environment* 36 pp.589–96

Nathan, G.K. (1980) A simplified design method and wind tunnel study of horizontal-axis windmills. *Journal of Wind Engineering and Industrial Aerodynamics* 6 pp.189–205.

Ngo, D. & Dexter, A. (1999) A robust model-based approach to diagnosing faults in air-handling units. *ASHRAE Transactions* 105(1) pp.1078–86

Nielsen, P.V. (1998) The selection of turbulence models for prediction of room airflow. *ASHRAE Transactions* 104(1B) pp.1119–27.

Niu, J., v.d. Kooi, J. & v.d. Ree, H. (1997) Cooling load dynamics of rooms with cooled ceilings. *Building Services Engineering Research & Technology* 18(4) pp.210–7.

Noda, Y., Yamazaki, T., Matsuba, T., Kamimura, K., Kurosu, S. (2003) Comparison in control performance between PID and H_∞ controllers for HVAC control. *ASHRAE Transactions* 109(1) pp.3–11.

Norris, L.H. & Reynolds, W.C. (1975) *Report No. FM-10*. Stanford University Department of Mechanical Engineering, Stanford, CA.

Nozawa, K. & Tamura, T. (2002) Large eddy simulation of the flow around a low-rise building immersed in a rough wall turbulent boundary layer. *Journal of Wind Engineering and Industrial Aerodynamics* 90 pp.1151–62.

NREL (1995) *Report Rep. NREL/SP-463–7668 – TMY2 User's Manual National Report*. National Renewable Energy Laboratory, Golden, CO.

Oliveira, P.J. & Younis, B.A. (2000) On the prediction of turbulent flows around full-scale buildings. *Journal of Wind Engineering and Industrial Aerodynamics* 86 pp.203–2.

Olsen, E.L. & Chen, Q. (2003) Energy consumption and comfort analysis for different low-energy cooling systems in a mild climate. *Energy and Buildings* 35 pp.561–71.

van Paassen, D.H.C. & Luo, Q.X. (2002) Weather data generator to study climate change on buildings. *Building Services Engineering Research & Technology* 23(4) pp.251–8.

Papadopoulos, A.M. (2001) The influence of street canyons on the cooling loads of buildings and the performance of air conditioning loads. *Energy and Buildings* 33 pp.601–7.

Papakonstantinou, K.A., Kiranoudis, C.T. & Markatos, N.C. (2000) Numerical simulation of air flow field in single-sided ventilated buildings. *Energy and Buildings* 33 pp.41–8.

Parsloe, C. & Hejab, M. (1992) Small power loads. *BSRIA Technical Note TN8/92*, Building Services Research and Information Association, Bracknell.

Pedersen, C.O., Fisher, D.E., Liesen, R.J. & Strand, R.K. (2003) ASHRAE Toolkit for building load calculations. *ASHRAE Transactions* 109(1) pp.583–9.

Phillips, J., Naylor, D., Oosthuizen, P.H. & Harrison, S.J. (2001) Numerical study of convective and radiative heat transfer from a window glazing with a Venetian blind. *International Journal of Heating, Ventilating, Air-conditioning and Refrigerating Research* 7(4) pp.383–402.

Pinney, A.A. & Parand, F. (1991) The effect of computational parameters on the accuracy of results from detailed thermal simulation programs. *Proceedings, Building Environmental Performance – BEP '91, Canterbury* pp.207–22.

Rabehl, R.J., Mitchell, J.W. & Beckman, W.A. (1999) Parameter estimation and the use of catalog data in modelling heat exchangers and coils. *International Journal of Heating, Ventilating, Air-conditioning and Refrigerating Research* 5(1) pp.3–17.

Rawlings, R.H.D. (1999) Ground Source Heat Pumps – A Technology Review. *BSRIA TN 18/99* Building Services Research and Information Association, Bracknell.

Ren, Z. & Stewart, J. (2003) Simulating air flow and temperature distribution inside buildings using a modified version of COMIS with sub-zonal divisions. *Energy and Buildings* 35 pp.257–71.

Ren, M.J. & Wright, J.A. (2002) Adaptive diurnal prediction of ambient dry-bulb temperature and solar radiation. *International Journal of Heating, Ventilating, Air-conditioning and Refrigerating Research* 8(4) pp.383–401.

Riederer, P., Marchio, D., Visier, J.C., Hussaundee, A. & Lahrech, R. (2002) Room thermal modelling adapted to the test of HVAC control systems. *Building and Environment* 37 pp.777–90.

Riensche, E., Stimming, U. & Unverzagt, G. (1998) Optimization of a 200 kW SOFC cogeneration plant Part I: Variation of process parameters. *Journal of Power Sources* 73 pp.251–6.

Rode, C. & Grau, K. (2003) Whole building hygrothermal simulation model. *ASHRAE Transactions* 109(1) pp.572–82.

Russell, M.B. & Surendran, P.N. (2001) Use of computational fluid dynamics to aid studies of room air distribution: A review of some recent work. *Building Services Engineering Research & Technology* 21(4) pp.241–7.

Sagrado, A.P.G., van Beeck, J., Rambaud, P. & Olivari, D. (2002) Numerical and experimental modelling of pollutant dispersion in a street canyon. *Journal of Wind Engineering and Industrial Aerodynamics* 90 pp.321–39.

Salsbury, T.I. (1998) A temperature controller for VAV air-handling units based on simplified physical models. *International Journal of Heating, Ventilating, Air-conditioning and Refrigerating Research* 4(3) pp. 265–79

Salsbury, T.I. (1999) A practical algorithm for diagnosing control loop problems. *Energy and Buildings* 29 pp.217–27.

Salsbury, T.I. & Diamond, R. (2000) Performance validation and energy analysis of HVAC systems using simulation. *Energy and Buildings* 32 pp.5–17.

Salsbury, T.I. & Diamond, R. (2001) Fault detection in HVAC systems using model-based feedforward control. *Energy and Buildings* 33 pp.403–15.

Scarperdas, A. & Colville, R.N. (1999) Assessing the representativeness of monitoring data from an urban intersection site in central London, UK. *Atmospheric Environment* 33 pp.661–74.

van Schijndel, A.W.M. (2002) Optimal operation of a hospital power plant. *Energy and Buildings* 34 pp.1055–65.

Seem, J.E., Park, C. & House, J.M. (1999) A new sequencing control strategy for air-handling units. *International Journal of Heating, Ventilating, Air-conditioning and Refrigerating Research* 5(1) pp.35–58.

Shimazaki, Y. (2003) Evaluation of refrigeration and air-conditioning technologies in heat cascading systems under the carbon dioxide emissions constraint: The proposal of the energy cascade balance table. *Energy Policy* 31 pp.1685–97.

Spadaro, J.V. & Rabl, A. (2001) Damage costs due to automotive air pollution and the influence of street canyons. *Atmospheric Environment* 35 pp.4763–75.

Spitler, J.D. & Fisher, D.E. (1999a) On the relationship between the radiant time series and transfer function methods for design cooling load calculations. *International Journal of Heating, Ventilating, Air-conditioning and Refrigerating Research* 5(2) pp.125–38.

Spitler, J.D. & Fisher, D.E. (1999b) Development of periodic response factors for use with the radiant time series method. *ASHRAE Transactions* 105(2) pp.491–509.

Spitler, J.D., Fisher, D.E. & Pedersen, C.O. (1997) The radiant time series cooling load calculation procedure. *ASHRAE Transactions* 103(2) pp.503–15.

Spitler, J.D. & Underwood, C.P. (2003) UK application of direct cooling ground source heat pump systems. *Proceedings, CIBSE/ASHRAE Conference, Edinburgh.*

Srebric, J.D., Chen, Q., Glicksman, L.R. (1999) Validation of a zero-equation turbulence model for complex indoor airflow simulation. *ASHRAE Transactions* 105(2) pp. 414–27.

Stewart, J. & Ren, Z. (2003) Prediction of indoor gaseous pollutant dispersion by nesting sub-zones within a multizone model. *Building and Environment* 38 pp.635–43.

Sukamongkol, Y., Chungpaibulpatana, S. & Ongsakul, W. (2002) A simulation model for predicting the performance of a solar photovoltaic system with alternating current loads. *Renewable Energy* 27 pp.237–58.

Tarnawski, V.R. & Leong, W.H. (1993) Computer analysis, design and simulation of horizontal ground heat exchangers. *International Journal of Energy Research* 17 pp.467–77.

Thomas, T.G. & Williams, J.J.R. (1999) Generating a wind environment for large eddy simulation of bluff body flows. *Journal of Wind Engineering and Industrial Aerodynamics* 82 pp.189–208.

Tutar, M. & Oguz, G. (2002) Large eddy simulation of wind flow around parallel buildings with varying configurations. *Fluid Dynamics Research* 31(5–6) pp.289–315.

Underwood, C.P. (1999) *HVAC Control: Modelling, Analysis and Design.* E. & F.N. Spon, London.

Underwood, C.P. (2000) Robust control of HVAC plant II: Controller design. *Building Services Engineering Research & Technology* 21(1) pp.63–71

Underwood, C.P. (2001) Analysing multivariable control of refrigeration plant using Matlab/Simulink *Proceedings, 7th IBPSA Conference, Rio* 1 pp.287–94.

Vardoulakis, S., Fisher, B.E.A., Pericleous, K. & Gonzalez-Flesca, N. (2003) Modelling air quality in street canyons: A review. *Atmospheric Environment* 37 pp.155–82.

Voeltzel, A., Carrie, F.R. & Guarracino, G. (2001) Thermal and ventilation modelling of large highly-glazed spaces. *Energy and Buildings* 33 pp.121–32.

Walton, A. & Cheng, A.Y.S. (2002) Large eddy simulation of pollution dispersion in an urban street canyon – Part II: Idealised canyon simulation. *Atmospheric Environment* 36 pp.3615–27.

Walton, A., Cheng, A.Y.S. & Yeung, W.C. (2002) Large eddy simulation of pollution dispersion in an urban street canyon – Part I: Comparison with field data. *Atmospheric Environment* 36 pp. 3601–13.

Wang, S. (1999) Dynamic simulation of building VAV air-conditioning system and evaluation of EMCS on-line control strategies. *Building and Environment* 34 pp.681–705.

Wang, S. & Chen, Y. (2001) A novel simple building load calculation model for building and system dynamic simulation. *Applied Thermal Engineering* 21 pp.683–702.

Wang, S., Wang, J. & Burnett, J. (2000) Mechanistic model of centrifugal chillers for HVAC system dynamics simulation *Building Services Engineering Research & Technology* 21(2) pp.73–83.

Welfonder, E., Neifer, R. & Spanner, M. (1997) Development and experimental identification of dynamic models for wind turbines. *Control Engineering Practice* 5(1) pp.63–73.

Wilkins, C.K. & Hosni, M.H. (2000) Heat gain from office equipment. *ASHRAE Journal* 42 pp.33–44.

Williams, J.M., Griffiths, A.J. & Knight, I.P. (1996) Combined heat and power plant: Sizing for new hospitals. *Building Services Engineering Research & Technology* 17(3) pp.147–152.

Wong, N.H. & Chin, H.K. (2002) An evaluation exercise of a wind pressure distribution model. *Energy and Buildings* 34 pp.291–309.

Wurtz, E., Nataf, J-M. & Winklemann, F. (1999) Two- and three-dimensional natural and mixed convection simulation using modular zonal models in buildings. *International Journal of Heat Mass Transfer* 42 pp.923–40.

Xu, W. & Chen, Q. (1998) Numerical simulation of airflow in a room with differentially heated vertical wall. *ASHRAE Transactions* 104(1A) pp.168–75.

Yik, F.W.H. (2001) Analyses of operations of control valves in air conditioning systems comprising multiple chillers and chilled water pumps. *Transactions of the Hong Kong Institute of Engineers* 8(3) pp.51–7.

Yik, F.W.H., Burnett, J. & Prescott, I. (2001) Predicting air conditioning energy consumption of a group of buildings using different heat rejection methods. *Energy and Buildings* 33(2) pp.151–66.

Yik, F.W.H. & Lam, V.K.C. (1998) Chiller models for plant design studies. *Building Services Engineering Research & Technology* 19(4) pp.233–42.

Yik, F.W.H., Lo, T.Y. & Burnett, J. (2003) Wind data for natural ventilation design in Hong Kong. *Building Services Engineering Research & Technology* 24(2) pp.125–36.

Yik, F.W.H. & Powell, G. (2002) Review of isolation ward ventilation design and evaluation by simulation. *Proceedings, International Conference Indoor Environmental Quality in Hospitals, Prague.*

Yik, F.W.H., Underwood, C.P. & Chow, W.K. (1997) Chilled-water cooling and dehumidifying coils with corrugated plate fins: Modelling method. *Building Services Engineering Research & Technology* 18(1) pp.47–58.

Zhou, Y. & Stathopoulos, T. (1997) A new technique for the numerical simulation of wind flow around buildings. *Journal of Wind Engineering and Industrial Aerodynamics* 72 pp.137–47.

Zhu, Y. & Zhang, Y. (2000) Dynamic modelling of encapsulated ice tank for HVAC system simulation. *International Journal of Heating, Ventilating, Air-conditioning and Refrigerating Research* 6(3) pp.213–28.

Ziskind, G., Dubovsky, V. & Letan, R. (2002) Ventilation by natural convection of a one-story building. *Energy and Buildings* 34 pp.91–102.

Zmeureanu, R., Iliescu, D., Dauce, D. & Yacob, Y. (2003) Radiation from cold or warm windows: Computer model development and experimental evaluation. *Building and Environment* 38 pp.427–34.

Chapter 7
Modelling in Practice II

Modelling for energy use in buildings has historically developed as a hierarchy of methodologies, each level developed for the specific purpose of solving a particular problem. Individually, many of these methodologies have been dealt with in detail in the preceding chapters. Building energy problems involve a complex set of interactions. Tackling problems on a 'bite-size' basis requires boundary conditions to be defined at inappropriate parts of the problem space in order to effect a solution. Often these boundary conditions will be uncertain and, inevitably, the outcome is a compromise often amounting to an extreme and, therefore, unrealistic solution.

For example, modelling heat and mass transfer in a construction element as an isolated problem requires assumptions to be placed on internal and (possibly) external surface boundary conditions whereas, in practice, the inclusion of the bounding zone air state at least should be permitted. If the bounding zone air state is allowed to participate, then assumptions are needed regarding the interaction between this zone and its neighbours. Then there is the role of the plant. Practical modelling therefore requires the correct interactions between all variables in the hierarchy of methodologies to be defined and made active. Programs have been developed to address the entire problem space of building energy interactions to a greater or lesser extent. The purpose of this chapter is to review some of the progress with these developments and to consider the quality and reliability of these tools.

7.1 Interrelationships between methodologies

Figure 7.1 illustrates energy-elemental interactions at zone level.

The zone can now be contextualised within the building as shown in Fig. 7.2 to express the various interrelationships between methods. The numbers in brackets refer to the sections in previous chapters where further details of each method can be sought.

Many programs are available or are developing to address parts of the integrated problem depicted by Fig. 7.2 and some would claim to address all of it and more besides. It is not the intention to compare and evaluate these but instead to highlight the extent to which they address the collective modelling objectives depicted typically by Fig. 7.2. Thus in Section 7.2 follows a review of how some of the available programs have evolved as integrated energy design and analysis tools, whereas Section 7.3 deals with the development of procedures for evaluating the reliability, robustness and quality of modelling tools and methods.

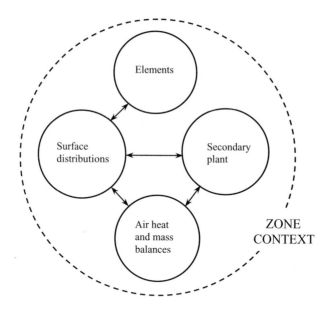

Fig. 7.1   Energy-elemental zone interactions.

7.2 Tools and their integration

Table 7.1 recasts Table 6.1 with additional perspectives on the various methods that have been explored in detail in previous chapters and which find themselves embedded, to a greater or lesser extent, in currently available programs and codes.

Methods are classified in terms of:

* Utility – the range of applications to which a particular method can be put
* Rigour – the level of detail required
* Cost – relative computational power required

It is tempting to use rigour as a proxy for accuracy. Whilst this may have some basis, there is, as yet, no robust and comprehensive method or set of methods for assessing the accuracy of modelling outcomes, although it is acknowledged that some progress towards this goal has been made (see Section 7.3). As there is no 'Gold Standard' for these criteria, they remain qualitative and relative. Table 7.1 is therefore suggested as a means to assess *relative* strengths of the various approaches available for tackling a variety of modelling problems.

Integration of methods – needs and trends

de Wilde *et al.* 2002 contextualise the locations of energy and related modelling within the whole building design process expressed as an 'objective tree', as shown in Fig. 7.3.

Energy and fluid flow problems in buildings are in any case interactive but single program codes that treat all areas in an integrated manner are still under development. Practical tools for the thermal comfort, air quality, energy use (and emissions) objectives referred to in de Wilde *et al.*'s objective tree are encompassed variously by three groups of software at present:

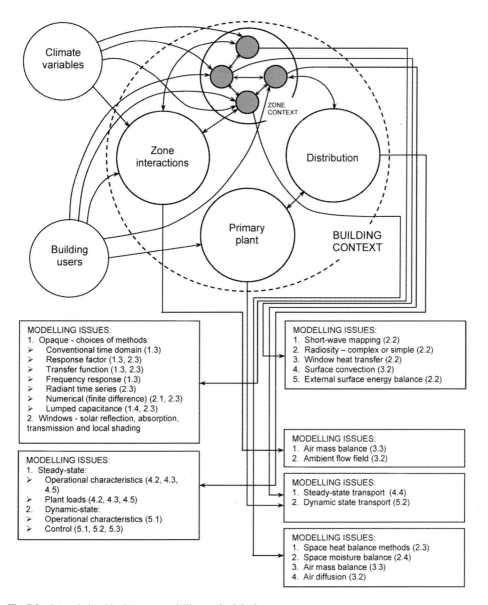

Fig. 7.2   Interrelationships between modelling methodologies.

Group 1:   Building energy and environment
Group 2:   Plant and control
Group 3:   Zonal ventilation and computational fluid dynamics

Though fully integrated programs encompassing all groups are possible in principle, the current state-of-the-art remains somewhat more fragmented, as shown in Fig. 7.4. Several programs provide full treatments of plant and controls (whilst others do not) but integration with the two main types of ventilation model remains to be effected through a linkage mechanism in the few cases that address this at present.

Table 7.1 Typical features of modelling methods.

Treatment	Options	Utility	Rigour	Cost
Conduction heat transfer	Conventional time domain	Limited	Restricted	Low
	Response factor	Moderate	Moderate	Low
	Transfer function	Moderate	Moderate	Low
	Radiant time series	Limited	Moderate	Low
	Frequency response	Limited	Moderate	Low
	Numerical – finite difference	High	Potentially high	Moderate/high
	Lumped capacitance	Moderate	Restricted	Low
Window modelling	Simple – fixed parameter	Low	Low	Low
	Optical properties	Moderate	Moderate	Low
Short-wave radiation dist.	Simplified surface-targeted	Moderate	Low	Low
	Mapped	High	High	Moderate
Long-wave radiation dist.	Lumped with convection	Low	Low	Low
	Simplified surface fractions	Moderate	Moderate	Low
	Rigorous	High	High	Moderate
Surface convection	Constant coefficients	Low	Low	Low
	Modelled coefficients	High	Moderate	Moderate
Moisture mass transfer	Zone moisture balance only	Moderate	Low	Low
	Fabric moisture sorption	High	High	Moderate/high
Air mass balance	Constant air change rate	Low	Low	Low
	Zonal ventilation model	High	High	Moderate
Air diffusion	Perfectly mixed	Low	Low	Low
	Stratified zonal	Moderate	Moderate	Moderate
	Stratified rigorous	High	High	Very high
Secondary plant and control	Constant set-point tracking	Low	Low	Low
	Steady-state model	Moderate	Moderate	Moderate
	Dynamic-state model	High	High	Very high
Distribution	Close-coupled	Low	Low	Low
	Steady-state model	Moderate/high	Moderate/high	High
	Dynamic-state model	High	High	Very high
Primary plant and control	Sum of loads	Low/moderate	Low	Low
	Steady-state model	Moderate/high	Moderate/high	Moderate
	Dynamic-state model	High	High	Very high

Current interest in the integration of processes, methods and practices lies in the concept of open data transfer standards for the interoperability of software products realised through the development of Industry Foundation Classes (IFCs – see Karola *et al.* 2002; Bazjanac *et al.* 2002). An example of an integrated system is shown in Fig. 7.5.

Though part of the reason for the continuing segregation of methods is, without doubt, due to the high computational cost of developing fully integrated tools, there is another reason; that of the expertise level of the user. It is unusual for example to find an expert user in CFD applications who can readily turn to the problem of simulating plant and controls. It must also be expected that the high-rigour applications listed in Table 7.1 will require a higher level of expertise to apply these methods than would be the case with low-rigour applications.

This modelling practice does not merely hinge on the availability of a given tool for the job, but concomitantly, on the availability of expertise. Ellis & Mathews (2002), in reviewing advances made in HVAC system design tools, argue that the potential of many existing tools remains largely untapped. Visier (2000) argues for the use of 'virtual

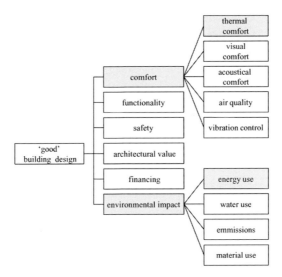

Fig. 7.3 Objective tree for building design processes. (Reprinted from *Building and Environment*, 37, de Wilde, P., Augenbroe, G., van der Voorden, M., Design analysis integration: Supporting the selection of energy saving building components, 807–16, © (2002), with permission from Elsevier.)

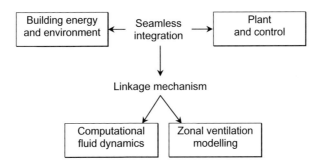

Fig. 7.4 Relationships between available program codes.

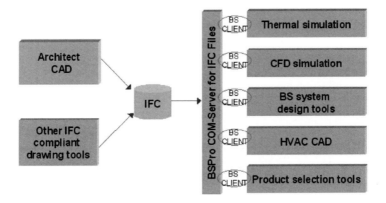

Fig. 7.5 Use of IFCs for interoperability. (Reprinted from *Energy and Buildings*, 34, Karola, A., Lahtela, H., Hänninen, R., Hitchcock, R., Chen, Q., Dajka, S. & Hagström, K., BSPro COM-Server – interoperability between software tools using industrial foundation classes, 901–07, © (2002), with permission from Elsevier.)

laboratories' for HVAC applications consisting of validated system models that can interact with real components for use in both design and testing of systems. The need for research into interfacing methods between virtual and real components is highlighted. Garde-Bentaleb *et al.* (2002) propose a novel method of expertise-integration among engineering and architectural specialisms through the use of expert rules (see Fig. 7.6). Augenbroe (2002) discusses the challenges presented to the building professions brought about by the maturation of building simulation tools, arguing for the need for better integration of simulation tools to bring about efficient integration of simulation expertise and the need for increased quality assurance. The issue of quality assurance and simulation tools is discussed below.

Integration of methods and tools – some specific examples

Approaches to integrating heating, lighting, ventilation and acoustic 'domains' are discussed by Citherlet *et al.* (2001) and Citherlet & Hand (2002). They favour the use of an integrated model approach (Fig. 7.7) over other methods such as model-sharing, model-exchange and model-coupling on the grounds of simplified data management and software maintenance and evolutionary development.

Clarke (2001) deals with the issue of domain integration – the integration of conduction, convection (including the conservation equations for momentum, etc), fabric moisture sorption and plant 'domains' within the ESP-r program code. Using a link to the Radiance program code, an illuminance mapping method is also included.

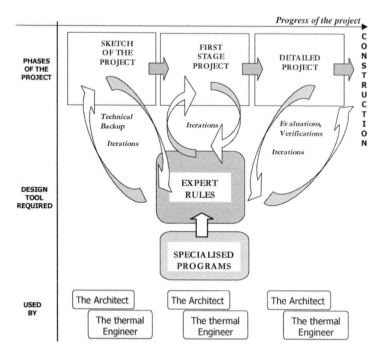

Fig. 7.6   Expert rules to support expertise-integration. (Reprinted from *Energy and Buildings*, 34, Garde-Bentaleb, F., Miranville, F., Boyer, H. & Depecker, P., Bringing scientific knowledge from research to the professional fields: The case of the thermal and airflow design of buildings in tropical climates, 511–21, © (2002), with permission from Elsevier.)

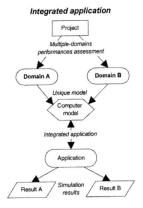

Integrated application

Fig. 7.7   Integrated model approach. (Reprinted from *Energy and Buildings*, 33, Citherlet, S., Clarke, J.A. & Hand, J., Integration in building physics simulation, 451–61, © (2001), with permission from Elsevier.)

Many other examples of methodological integration have resulted in stand-alone programs with library-based modularity for flexible model construction with varying levels of user expertise (Husaundee & Visier 1999; Gouda *et al*. 2002; Mathews *et al*. 1999; Strand *et al*. 2002; Crawley *et al*. 2001). As graphical modelling environments have improved, model construction has moved from menu- and spreadsheet-based input methods to information flow graphics (Fig.7.8a) and on to object-based component libraries that include drag-and-drop placement and connections of masked objects (Fig.7.8b).

Mathews *et al*. (1999) and Husaundee & Visier address the development of modular programs that include building envelope, plant and control specifically for application to control system modelling. Both developments make use of graphical environments for component selection and connection. Gouda *et al*. (2002) have also developed a similar tool in a third-party graphics environment specifically for controller design applications. Strand *et al*. (2002) and Crawley *et al*. (2001) report on the evolution of the programs BLAST and DOE-2, both of which contained inflexible generic plant and controls model options, into a new modular program EnergyPlus. The program has evolved as an integrated stand-alone modeller, as well as permitting links to third-part software (Fig.7.9).

Besides the facility to link the zonal ventilation program COMIS (Feustel 1999) and EnergyPlus (Crawley *et al*. 2001) – see Fig.7.9, Dorer & Weber 1999 describe the development of a linkage mechanism with the modular simulation program TRNSYS (undated) within a shared graphical user-interface by translating COMIS as a TRNSYS 'Type'.

For higher levels of energy and ventilation modelling sophistication, the coupling and integration of energy simulation and computational fluid dynamics (CFD) has received widespread attention (Bartak *et al*. 2002; Beausoleil-Morrison 2002; Zhai & Chen 2003; Negrão 1998; Zhai *et al*. 2002; Kim *et al*. 2001; Murakami *et al*. 2001).

Bartak *et al*. and Beausoleil-Morrison focus on integrative methods within the program ESP-r whilst Zhai *et al*. focus on linkage mechanisms between energy simulation programs. Though surface conditions and air diffusion are interrelated and it is therefore necessary to call the CFD calculations at each simulation time step, this is computationally very expensive. Figure 7.11 illustrates a bypass mechanism which therefore permits CFD calculations to be called at periods greater than the simulation time step – in effect a quasi-steady-state method as far as the CFD calculations are concerned (Zhai *et al*. 2002). Kim *et al*. (2001) and

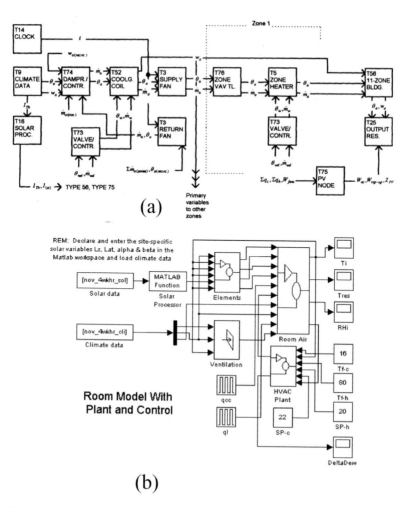

Fig. 7.8 Modular modelling examples. (a) Information flow using TRNSYS (Underwood 1996). (b) Masked components (Gouda *et al.* 2002).

Murakami *et al.* (2001) consider the coupling of a CFD model with surface radiation, HVAC control and comfort algorithms using a simpler energy model than is customary in energy simulation and resulting in, in effect, a modular integrated model (Fig. 7.12).

7.3 Validation and verification

Model *validation* is taken to mean the extent to which a model produces results that represent real world behaviour; whereas model *verification* describes the extent to which a model produces results that compare with some alternative method or standard. There is of course a fundamental difference; validation has a truth connotation whereas verification merely has a reference connotation. Put more simply by Blottner (see Rees *et al.* 2002c) verification can be thought of as 'solving the equations right' whereas validation as 'solving the right equations'.

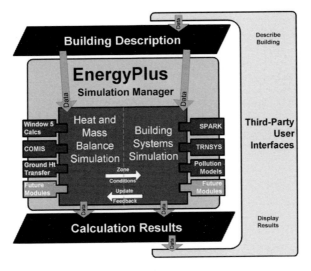

Fig. 7.9 EnergyPlus program with link mechanisms to other codes. (Reprinted from *Energy and Buildings*, 33, Crawley, D.B., Lawrie, L.K., Winklemann, F.C., Buhl, W.F., Huang, Y.J., Pedersen, C.C., Strand, R.K., Liesen, R.J., Fisher, D.E., Witte, M.J. & Glazer, J., Energy Plus: Creating a new-generation building energy simulation program, 319–31, © (2001), with permission from Elsevier.)

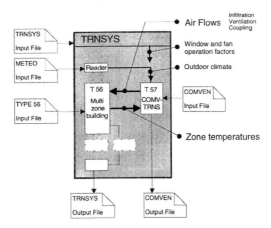

Fig. 7.10 Linkage of TRNSYS and COMIS. (Reprinted from *Energy and Buildings*, 30, Dorer, V. & Weber, A., Air, contaminant and heat transport models: Integration and application, 97–104, © (1999), with permission from Elsevier.)

Three general methods are now widely established for testing the performance of building energy calculation methods and programs using validation and verification approaches (Bowman & Lomas 1985; Bloomfield 1999; Neymark *et al.* 2002a,b) as summarised in Table 7.2.

Evidently, only the empirical method has the potential to offer a validation, the other two methods forming verification tests. Because each method has at least one serious weakness, the comprehensive testing of a calculation procedure, a piece of code or a program will tend to make use of a combination of methods rather than any single one.

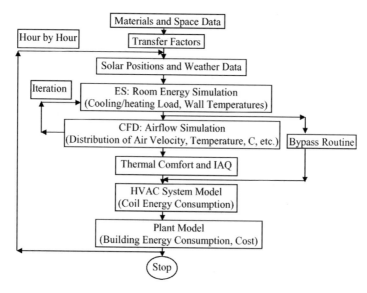

Fig. 7.11   Conditional coupling of energy simulation and CFD. (Reprinted from *Building and Environment*, 37, Zhai, Z., Chen, Q., Haves, P. & Klems, J.H., On approaches to couple energy simulation and computational fluid dynamics programs, 857–64, © (2002), with permission from Elsevier.)

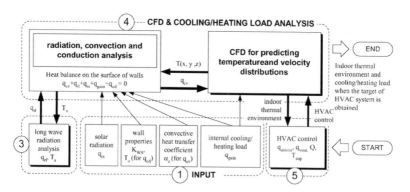

Fig. 7.12   CFD coupled with surface radiation, HVAC plant and control. (Reprinted from *Building and Environment*, 36, Murakami, S., Kato, S. & Kim, T., Indoor climate design based on CFD coupled simulation of convection, radiation, and HVAC control for attaining a given PMV value, 710–9, © (2001), with permission from Elsevier.)

Reviews on progress in the validation of computer programs for energy and environment in buildings can be found in Ahmad (1998) and Bloomfield (1999); and Chen & Srebric (2002) discuss issues for the verification, validation and standardisation specifically related to CFD. Amongst other things, these reviews give conclusions on the limited availability of robust and reliable validation test data sets and the need for such data to be documented and made available for re-use. As a consequence, much of the progress that has been made has been based on inter-program verification tests. The best of these have tended to use a small number of programs that have themselves been extensively verified and validated as the basis for the comparisons.

In many cases, those with the task of evaluating and testing programs for application to

Table 7.2 Procedures for model testing.

Method	Basis	Strengths	Weaknesses
Empirical *validation*	Results predicted by a modelling method are compared with measured experimental or field monitored data	A potential basis for 'truth' Flexible; wide range of problem types	Need to quantify all measurement uncertainties Expensive and time consuming
Analytical *verification*	Results predicted by a modelling method are compared with an exact known solution (usually mathematically derived)	Precise Easy to apply	Few instances where analytical solutions are derived Restricted to a sample of results; not comprehensive
Inter-model *comparison*	Results predicted by a modelling method are compared with an alternative modelling method	Useful for bug-fixing Any level of complexity (provided both models consistent) Comprehensive	Not a truth standard Model consistency 'calibration'

given problems will have a wide choice of available methodologies and software to choose from. Program testing therefore usually amounts to a two-stage process:

(1)   Selection of candidate programs for testing
(2)   Selection and application of an appropriate test method (Table 7.2)

Selection will usually amount to identifying candidate programs or methods that are fit in principle for the intended purpose. A generalisation of a procedure developed by Underwood *et al.* (1993) for the selection of programs to test their capabilities to model plant and controls is illustrated in Fig. 7.13.

- Stage 1: Identify the application area and set criteria against which programs are to be evaluated
- Stage 2: Submit each program to an outline check-list of features that address adminstrative and non-technical aspects of the evaluation criteria such as requirements for user training, cost, hardware needs, etc.
- Stage 3: Filter candidate programs to a short-list based on general technical features and existing use
- Stage 4: Submit the short-listed programs to a detailed checklist of features that address technical and modelling capabilities of each program
- Stage 5: Program testing

The filtering process will only be needed in situations where the number of initial candidate programs is large and includes a criterion that implies 'validation through use' by examining the size of any existing user community. Filters 1 and 2 are intended to test for any obvious technical deficiencies. At Stage 4 detailed characteristics of the short-listed programs are evaluated against a checklist of methodological capabilities, such as is given in Table 7.1,

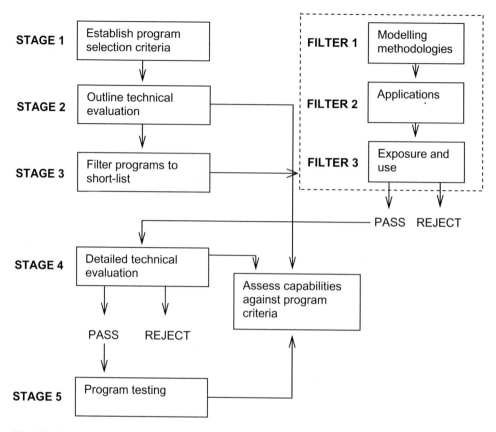

Fig. 7.13   Selection process for candidate programs to test.

drawn up against the application requirements in the program selection criteria. At Stage 5 each program is submitted to detailed testing (this stage will also need to include an assessment of the appropriate testing procedure – e.g. Table 7.2). Examples of progress with the application of the various tests are given in the following.

Analytical verification

Several applications of this test method to various aspects of building energy modelling have been reported (Stefanizzi *et al.* 1990; Bland 1992; Rees *et al.* 2002c; Neymark *et al.* 2002a,b; Ben-Nakhi & Aasem 2003).

Stefanizzi *et al.* and Bland carried out tests related to the specific modelling problems of conduction heat transfer and internal longwave radiation calculation respectively. 'Exact' solutions for the conduction case are based on an analytically solved differential equation for a homogenous slab subjected to a step change in one surface temperature with the other surface temperature held constant. For the longwave radiation case (Stefanizzi *et al.* 1990), analytically derived view factors for a variety of simple room geometries were used. Ben-Nakhi & Aasem report on an analytical test procedure for multi-layered conduction developed as a function that can be built-in to energy simulation programs.

Rees *et al*. (2002c) describe the development of a test suite for 16 analytical tests for cases involving surface convection, conduction, solar radiation, longwave radiation, infiltration rates and ground-coupled floors. Neymark *et al*. (2002a,b) report on the development and application of an analytical verification test to assess the ability of whole-building simulation programs to model unitary air conditioners. They have also developed a diagnostic procedure based on the International Energy Agency's building energy simulation test 'BESTEST' (Judkoff & Neymark 1999) inter-model comparison method.

Inter-program comparisons

Qualitative inter-methodological comparisons of the (predominantly North American) heat balance and radiant time-series cooling load calculation methods and the (predominantly UK) admittance method have been carried out by Rees *et al*. (2000a). In related work, quantitative comparisons based on peak load predictions of the heat balance method and admittance method were also carried out by Rees *et al*. (2000b). Other examples of comparison tests at specific levels of modelling capability or for specific methods are patchy. Davies *et al*. (2000) for example, deal with the specific case of earth-contact heat transfer by comparing two different model types; and Sowell & Haves (2001) compare the execution times for the research energy kernel 'SPARK' with the modular program HVACSIM+ (Clark 1985).

Significant work has been carried out on inter-program comparisons of whole building energy simulations mostly within the framework of BESTEST, including methods forming specific test cases in BESTEST (Judkoff & Neymark 1999; Hayez *et al*. 2001; Haddad & Beausoleil-Morrison 2001). Judkoff & Neymark propose a set of test cases ranging from simple tests through to realistic whole-building simulation test cases. The simple cases permit diagnostics of particular bugs, faulty algorithms, etc., whilst the more comprehensive tests

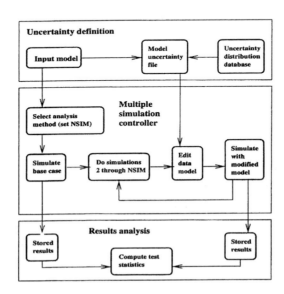

Fig. 7.14   Accommodating uncertainties in simulation modelling. (Reprinted from analysis *Energy and Buildings*, 33, MacDonald, I. & Strachan, P., Practical application of uncertainty, 219–27, © (2001), with permission from Elsevier.)

are designed for testing the capability of programs to model specific design analysis issues. Hayez *et al.* have developed HVAC-BESTEST for the specific purpose of testing the capabilities of programs to model unitary space cooling plant, whereas Haddad & Beausoleil-Morrison used the Home Energy Rating System 'HERS-BESTEST' for testing the performance of a program applied to domestic-scale applications.

Validation tests

The limited availability of robust and comprehensive test data has restricted the application of full validation tests to the results of energy simulation programs. Perhaps the most prominent of what has been done as far as whole-building energy simulation programs are concerned has taken place within the framework of various International Energy Agency annexes and tasks (e.g. Lomas *et al.* 1994).

Ventilation modelling programs have been dealt with separately since isothermal modelling capabilities can be relatively easily validated using meaurements with tracer gases (e.g. Upham *et al.* 2001). For a general review of the validation of multizone ventilation models, see Emmerich (2001).

A number of studies have focused on uncertainty in empirical validation tests both from the point of view of prediction uncertainties due to uncertain input data and from the point of view of measurement uncertainties in empirical validation. Aude *et al.* 2000, de Wit & Augenbroe 2002 and Macdonald & Strachan 2001 deal with energy simulation programs. Figure 7.14 illustrates the method used by Macdonald & Strachan to simulate with data containing uncertainties. Borchiellini & Fürbringer (1999) and Fürbringer & Roulet (1999) deal with uncertainties in multizone ventilation modelling programs.

References

Ahmad, Q.T. (1998) Validation of building thermal and energy models. *Building Services Engineering Research & Technology* 19(2) pp.61–66.

Aude, P., Tabary, L. & Depecker, P. (2000) Sensitivity analysis and validation of buildings' thermal models using adjoint-code method. *Energy and Buildings* 31 pp.267–83.

Augenbroe, G. (2002) Trends in building simulation. *Building and Environment* 37 pp.891–902.

Bartak, M., Beausoleil-Morrison, I., Clarke, J.A., *et al.* (2002) Integrating CFD and building simulation. *Building and Environment* 37 pp.865–71.

Bazjanac, V., Forester, J., Haves, P., Sucic, D. & Xu, P. (2002) HVAC component data modelling using Industry Foundation Classes. *Proceedings, System Simulation in Buildings – SSB '02, Liege.*

Beausoleil-Morrison, I. (2002) The adaptive conflation of computational fluid dynamics with whole-building thermal simulation. *Energy and Buildings* 34 pp.857–71.

Ben-Nakhi, A.E. & Aasem, E.O. (2003) Development and integration of a user friendly validation module within whole building simulation. *Energy Conversion Management* 44 pp.53–64.

Bland, B.H. (1992) Conduction in dynamic thermal models: Analytical tests for validation. *Building Services Engineering Research & Technology* 13(4) pp.197–208.

Bloomfield, D.P. (1999) An overview of validation methods for energy and environmental software. *ASHRAE Transactions* 105(2) pp.685–93.

Borchiellini, R. & Fürbringer, J-M. (1999) An evaluation exercise of a multizone air flow model. *Energy and Buildings* 30 pp.35–51.

Bowman, N.T. & Lomas, K.J. (1985) Empirical validation of dynamic thermal computer models of buildings. *Building Services Engineering Research & Technology* 6(4) pp.153–62.

Chen, Q. & Srebric, J. (2002) A procedure for verification, validation and reporting of indoor environment CFD analysis. *International Journal of Heating, Ventilating, Air-conditioning and Refrigerating Research* 8(2) pp.201–16.

Citherlet, S., Clarke, J.A. & Hand, J. (2001) Integration in building physics simulation. *Energy and Buildings* 33 pp.451–61.

Citherlet, S. & Hand, J. (2002) Assessing energy, lighting, room acoustics, occupant comfort and environmental impacts performance of building with a single simulation program. *Building and Environment* 37 pp.845–56.

Clark, D.R. (1985) *HVACSIM+ Building Systems and Equipment Simulation Program Reference Manual – NBSIR 84–2996*. National Institute of Standards and Technology, Gaithersburg, MD.

Clarke, J.A. (2001) Domain integration in building simulation. *Energy and Buildings* 33 pp.303–8.

Crawley, D.B., Lawrie, L.K., Winklemann, F.C., *et al.* (2001) EnergyPlus: Creating a new-generation building energy simulation program. *Energy and Buildings* 33 pp.319–31.

Davies, M., Zoras, S. & Adjali, M.H. (2000) Development and inter-model comparative testing of a novel earth-contact heat-transfer method. *Building Services Engineering Research & Technology* 21(1) pp.19–26

Dorer, V. & Weber, A. (1999) Air, contaminant and heat transport models: Integration and application. *Energy and Buildings* 30 pp.97–104.

Ellis, M.W. & Mathews, E.H. (2002) Needs and trends in building and HVAC system design tools. *Building and Environment* 37 pp.461–70.

Emmerich, S.J. (2001) Validation of multizone IAQ modelling of residential-scale buildings: A review. *ASHRAE Transactions* 107(2) pp.619–28.

Feustel, H.E. (1999) COMIS – an international multizone air-flow and contaminant transport model. *Energy and Buildings* 30 pp.3–18.

Fürbringer, J.-M. & Roulet, C.-A. (1999) Confidence of simulation results: Put a sensitivity analysis module in your model – the IEA-ECBS Annex 23 experience of model evaluation. *Energy and Buildings* 30 pp.61–71.

Garde-Bentaleb, F., Miranville, F., Boyer, H. & Depecker, P. (2002) Bringing scientific knowledge from research to the professional fields: The case of the thermal and airflow design of buildings in tropical climates. *Energy and Buildings* 34 pp.511–21.

Gouda, M.M., Underwood, C.P. & Danaher, S. (2002) Modelling the robustness properties of HVAC plant under feedback control. *Proceedings Sixth International SSB Conference, Liege.*

Haddad, K.H. & Beausoleil-Morrison, I. (2001) Results of HERS-BESTEST on an energy simulation computer program. *ASHRAE Transactions* 107(2) pp.713–21.

Hayez, S., Dalibart, C., Guyon, G. & Feburie, J. (2001) HVAC BESTEST: Clim2000 and CA-SIS results. *Proceedings, Seventh International IBPSA Conference, Rio* 2 pp.1127–33.

Husaundee, A.M.I. & Visier, J-C. (1999) The building HVAC system in control engineering – a modelling approach in a widespread graphical environment. *ASHRAE Transactions* 105(1) pp.319–29.

Judkoff, R. & Neymark, J.S. (1999) Adaptation of the BESTEST intermodel comparison method for proposed ASHRAE Standard 140P: Method of test for building energy simulation programs. *ASHRAE Transactions* 105(2) pp.721–36.

Karola, A., Lahtela, H., Hänninen, R., Hitchcock, R., Chen Q., Dajka, Q. & Hagström, K. (2002) BSPro COM-Server – interoperability between software tools using industrial foundation classes. *Energy and Buildings* 34 pp.901–07.

Kim, T., Kato, S. & Murakami, S. (2001) Indoor cooling/heating analysis based on coupled simulation of convection, radiation and HVAC control. *Building and Environment* 36 pp.901–8.

Lomas, K.J., Eppel, H., Martin, C. & Bloomfield, D. (1994) Empirical validation of thermal building simulation programs using test room data. *Final Report for IEA Annex 21/Task 12*. DeMontfort University, Leicester.

MacDonald, I. & Strachan, P. (2001) Practical application of uncertainty analysis. *Energy and Buildings* 33 pp.219–27.

Mathews, E.H., van Heerden, E. & Arndt, D.C. (1999) A tool for integrated HVAC, building, energy and control analysis part 1: Overview of QUICKcontrol. *Building and Environment* 34 pp.429–49.

Murakami, S., Kato, S. & Kim, T. (2001) Indoor climate design based on CFD coupled simulation of convection, radiation, and HVAC control for attaining a given PMV value. *Building and Environment* 36 pp.710–9.

Negrão, C.O.R. (1998) Integration of computational fluid dynamics with building thermal and mass flow simulation. *Energy and Buildings* 27 pp.155–65.

Neymark, J,, Judkoff, R., Knabe, G. & Le, H-T. (2002a) An analytical verification procedure for testing the ability of whole-building energy simulation programs to model space conditioning equipment at typical design conditions. *ASHRAE Transactions* 108(2) pp.1144–54.

Neymark, J., Judkoff, R., Knabe, G., Le, H.-T., Dürig, M., Glass, A. & Zweifel, G. (2002b) Applying the building energy simulation test (BESTEST) diagnostic method to verification of space conditioning equipment models used in whole-building energy simulation programs. *Energy and Buildings* 34 pp.917–31.

Rees, S.J., Spitler, J.D., Davies, M.G. & Haves, P. (2000a) Qualitative comparison of North American and UK cooling load calculation methods. *International Journal of Heating, Ventilating, Air-conditioning and Refrigerating Research* 6(1) 75–99.

Rees, S.J., Spitler, J.D., Holmes, M.J. & Haves, P. (2000b) Comparison of peak load predictions and treatment of solar gains in the admittance and heat balance load calculation procedures. *Building Services Engineering Research & Technology* 21(2) pp.125–38.

Rees, S.J., Xiao, D. & Spitler, J.D. (2002c) An analytical verification test suite for building fabric models in whole building energy simulation programs. *ASHRAE Transactions* 108(1) pp.30–42.

Sowell, E.F. & Haves, P. (2001) Efficient solution strategies for building energy system simulation. *Energy and Buildings* 33 pp.309–17.

Stefanizzi, P., Wilson, A. & Pinney, A. (1990) Internal long-wave radiation exchange in buildings: Comparison of calculation methods: II Testing of algorithms. *Building Services Engineering Research & Technology* 11(3) pp.87–96.

Strand, R.K., Fisher, D.E., Liesen, R.J. & Pedersen, C.O. (2002) Modular HVAC simulation and the future integration of alternative cooling systems in a new building energy simulation program. *ASHRAE Transactions* 108(2) pp.1107–17.

TRNSYS (undated) A Transient Simulation Program. University of Wisconsin Solar Energy Laboratory, Madison, WI.

Underwood, C.P. (1996) Photovoltaic-clad buildings – integration with variable volume air conditioning. *Proceedings, CIBSE/ASHRAE Conference, Harrogate* 1 pp.376–86.

Underwood, C.P., Potter, S.E., Wiltshire, T.J., Wright, A.J., Bloomfield, D. & Pinney, A.A. (1993) The capabilities of computer programs to predict plant and control system/operation in buildings. *Proceedings, CIBSE National Conference, Manchester* pp.389–98.

Upham, R.D., Yuill, G.K. & Bahnfleth, W.P. (2001) A validation study of multizone airflow and contaminant migration simulation programs as applied to tall buildings. *ASHRAE Transactions* 107(2) pp.629–44.

Visier, J-C. (2000) Virtual laboratories used for climatic engineering. *International Journal of Heating, Ventilating, Air Conditioning and Refrigerating Research* 6(4) pp.285–7.

de Wilde, P., Augenbroe, G., van der Voorden, M. (2002) Design analysis integration: Supporting the selection of energy saving building components. *Building and Environment* 37 pp.807–16.

de Wit, S. & Augenbroe, G. (2002) Analysis of uncertainty in building design evaluations and its implications. *Energy and Buildings* 34 pp.951–8.

Zhai, Z. & Chen, Q. (2003) Solution characters of iterative coupling between energy simulation and CFD programs. *Energy and Buildings* 35 pp.493–505.

Zhai, Z., Chen, Q., Haves, P. & Klems, J.H. (2002) On approaches to couple energy simulation and computational fluid dynamics programs. *Building and Environment* 37 pp.857–64.

Appendix A

Derivation of Ramp Function Responses from Transfer Functions

The transfer functions $B(s)$ and $D(s)$ are:

$$B(s) = \frac{\sinh\left(\sqrt{s/\alpha} \cdot L\right)}{k\sqrt{s/\alpha}}$$

(A.1)

$$D(s) = \cosh\left(\sqrt{s/\alpha} \cdot L\right)$$

(A.2)

Roots of B(s)

To have $B(s) = 0$ for $s \neq 0$ requires:

$$\sinh\left(\sqrt{s/\alpha} \cdot L\right) = 0$$

Since $\sinh(iz) = i \sin(z)$, where $i = \sqrt{-1}$, let:

$$z = \frac{1}{i}\sqrt{s/\alpha} \cdot L$$

Hence, $\sin h\left(\sqrt{s/\alpha} \cdot L\right) = 0$ requires:

$$\frac{1}{i}\sqrt{s/\alpha} \cdot L = m\pi \text{ for } m = 1, 2, \dots$$

Therefore, the roots for $B(s)$, β_m, such that $B(s = \beta_m) = 0$ are:

$$\beta_m = -\frac{m^2\pi^2\alpha}{L^2} \text{ for } m = 1, 2, 3, \dots$$

(A.3)

Response to unit ramp input

From Equations A.1 and A.2

$$B(0) = \lim_{s \to 0} \frac{\sinh\left(\sqrt{s/\alpha} \cdot L\right)}{k\sqrt{s/\alpha}} = \lim_{s \to 0} \frac{\dfrac{L}{2\sqrt{s\alpha}} \cosh\left(\sqrt{s/\alpha} \cdot L\right)}{\dfrac{k}{2\sqrt{s\alpha}}} = \frac{L}{k} \tag{A.4}$$

$$D(0) = 1 \tag{A.5}$$

Therefore:

$$\frac{1}{B(0)} = \frac{D(0)}{B(0)} = \frac{k}{L} \tag{A.6}$$

Consider:

$$\frac{d}{ds}\left(\frac{1}{B(s)}\right) = -\frac{1}{B(s)^2} \frac{dB(s)}{ds} \tag{A.7}$$

$$\frac{d}{ds}\left(\frac{D(s)}{B(s)}\right) = \frac{1}{B(s)} \frac{dD(s)}{ds} - D(s)\frac{1}{B(s)^2}\frac{dB(s)}{ds} \tag{A.8}$$

From Equation A.2:

$$\frac{dD(s)}{ds} = \frac{L}{2\sqrt{s\alpha}}\sinh\left(\sqrt{s/\alpha} \cdot L\right)$$

$$\frac{dD(s)}{ds}\bigg|_{s=0} = \lim_{s \to 0} \frac{L\sinh\left(\sqrt{s/\alpha} \cdot L\right)}{2\sqrt{s\alpha}} = \lim_{s \to 0} \frac{\dfrac{L^2}{2\sqrt{s\alpha}}\cosh\left(\sqrt{s/\alpha} \cdot L\right)}{\sqrt{\alpha/s}} \tag{A.9}$$

Hence:

$$\frac{dD(s)}{ds}\bigg|_{s=0} = \lim_{s \to 0} \frac{L^2}{2\alpha}\cosh\left(\sqrt{s/\alpha} \cdot L\right) = \frac{L^2}{2\alpha} \tag{A.10}$$

It follows from the result in Equation A.6 and the above that:

$$\frac{1}{B(s)}\frac{dD(s)}{ds}\bigg|_{s=0} = \frac{kL}{2\alpha} \tag{A.11}$$

Differentiating Equation A.1 yields:

$$\frac{dB(s)}{ds} = \frac{1}{k\sqrt{s/\alpha}}\frac{L}{2\sqrt{s\alpha}}\cosh\left(\sqrt{s/\alpha}\cdot L\right) - \frac{1}{k^2(s/\alpha)}\frac{k}{2\sqrt{s\alpha}}\sinh\left(\sqrt{s/\alpha}\cdot L\right)$$

$$= \frac{L}{2ks}\cosh\left(\sqrt{s/\alpha}\cdot L\right) - \frac{1}{2ks\sqrt{s/\alpha}}\sinh\left(\sqrt{s/\alpha}\cdot L\right)$$

$$= \frac{L\sqrt{s/\alpha}\cosh\left(\sqrt{s/\alpha}\cdot L\right) - \sinh\left(\sqrt{s/\alpha}\cdot L\right)}{2ks\sqrt{s/\alpha}} \qquad\qquad (A.12)$$

Hence:

$$\left.\frac{dB(s)}{ds}\right|_{s=0} = \lim_{s\to 0}\frac{L\sqrt{s/\alpha}\cosh\left(\sqrt{s/\alpha}\cdot L\right) - \sinh\left(\sqrt{s/\alpha}\cdot L\right)}{2ks\sqrt{s/\alpha}}$$

$$= \lim_{s\to 0}\frac{\sqrt{s/\alpha}\dfrac{L^2}{2\sqrt{s\alpha}}\sinh\left(\sqrt{s/\alpha}\cdot L\right) + \dfrac{L}{2\sqrt{s\alpha}}\cosh\left(\sqrt{s/\alpha}\cdot L\right) - \dfrac{L}{2\sqrt{s\alpha}}\cosh\left(\sqrt{s/\alpha}\cdot L\right)}{\dfrac{6}{2}k\sqrt{s/\alpha}}$$

$$= \lim_{s\to 0}\frac{L^2}{6k}\frac{\sinh\left(\sqrt{s/\alpha}\cdot L\right)}{\sqrt{s\alpha}}$$

$$= \lim_{s\to 0}\frac{L^2}{6k}\frac{\dfrac{L}{2\sqrt{s\alpha}}\cosh\left(\sqrt{s/\alpha}\cdot L\right)}{\dfrac{1}{2}\sqrt{\dfrac{\alpha}{s}}}$$

Hence:

$$\left.\frac{dB(s)}{ds}\right|_{s=0} = \frac{L^3}{6k\alpha} \qquad\qquad (A.13)$$

From the above results:

$$\left.\frac{d}{ds}\left(\frac{1}{B(s)}\right)\right|_{s=0} = -\frac{k^2}{L^2}\frac{L^3}{6k\alpha} = -\frac{kL}{6\alpha} \qquad\qquad (A.14)$$

Also:

$$\left.\frac{d}{ds}\left(\frac{D(s)}{B(s)}\right)\right|_{s=0} = \frac{k}{L}\frac{L^2}{2\alpha} - \frac{k^2}{L^2}\frac{L^3}{6k\alpha} = \frac{1}{3}\frac{kL}{\alpha} \qquad\qquad (A.15)$$

Applying the above results to Equation 1.103:

$$\xi_{0,0}(t) = L^{-1}\left\{\frac{D(s)}{s^2 B(s)}\right\}$$

$$= \frac{D(0)}{B(0)}t + \left[\frac{d}{ds}\left(\frac{D(s)}{B(s)}\right)\right]_{s=0} + \sum_{m=1}^{\infty} \frac{D(\beta_m)}{\beta_m^2 (dB(s)/ds)_{s=\beta_m}}\exp(\beta_m t)$$

$$= \frac{k}{L}t + \frac{kL}{3\alpha} + \sum_{m=1}^{\infty} \frac{2k\sqrt{\beta_m/\alpha}\cosh\left(\sqrt{\beta_m/\alpha}\cdot L\right)}{\beta_m\left(L\sqrt{\beta_m/\alpha}\cosh\left(\sqrt{\beta_m/\alpha}\cdot L\right) - \sinh\left(\sqrt{\beta_m/\alpha}\cdot L\right)\right)}\exp(\beta_m t) \quad \text{(A.16)}$$

Note that:

$$\beta_m = -\frac{m^2\pi^2\alpha}{L^2} \text{ for } m = 1, 2, 3, \ldots$$

$$\sqrt{\frac{\beta_m}{\alpha}} = \sqrt{-\frac{m^2\pi^2}{L^2}} = \frac{m\pi}{L}i$$

Therefore, the following term in Equation A.16 can be evaluated:

$$\frac{2k\sqrt{\beta_m/\alpha}\cosh\left(\sqrt{\beta_m/\alpha}\cdot L\right)}{\beta_m\left(L\sqrt{\beta_m/\alpha}\cosh\left(\sqrt{\beta_m/\alpha}\cdot L\right) - \sinh\left(\sqrt{\beta_m/\alpha}\cdot L\right)\right)}$$

$$= -\frac{2k\dfrac{m\pi}{L}i\cosh(m\pi i)}{\dfrac{m^2\pi^2\alpha}{L^2}(m\pi i\cosh(m\pi i) - \sinh(m\pi i))}$$

$$= -\frac{kL}{\alpha}\frac{2}{m^2\pi^2}\left(\frac{m\pi i\cosh(m\pi i)}{m\pi i\cosh(m\pi i) - \sinh(m\pi i)}\right) \quad \text{(A.17)}$$

Note that:

$$\cos(x) = \cosh(ix)$$
$$\sin(x) = \frac{1}{i}\sinh(ix)$$
$$\sin(m\pi) = 0$$
$$\cos(m\pi) = (-1)^m$$

The right hand side of Equation A.17, therefore, becomes:

$$= -\frac{kL}{\alpha}\frac{2}{m^2\pi^2}\left(\frac{m\pi i\cos(m\pi)}{m\pi i\cos(m\pi) - i\sin(m\pi)}\right)$$

$$= -\frac{kL}{\alpha}\frac{2}{m^2\pi^2}$$

Hence, Equation 1.103 can be simplified to:

$$\xi_{0,0} = \frac{k}{L}t + \frac{kL}{3\alpha} - \frac{kL}{\alpha}\frac{2}{\pi^2}\sum_{m=1}^{\infty}\frac{1}{m^2}\exp\left(-\frac{m^2\pi^2\alpha}{L^2}t\right) \tag{A.18}$$

Similarly, Equation 1.104 can be simplified as follows:

$$\xi_{1,0}(t) = L^{-1}\left\{\frac{1}{s^2 B(s)}\right\}$$

$$= \frac{1}{B(0)}t + \left[\frac{d}{ds}\left(\frac{1}{B(s)}\right)\right]_{s=0} + \sum_{m=1}^{\infty}\frac{1}{\beta_m^2 (dB(s)/ds)_{s=\beta_m}}\exp(\beta_m t)$$

$$= \frac{k}{L}t - \frac{kL}{6\alpha} + \sum_{m=1}^{\infty}\frac{2k\sqrt{\beta_m/\alpha}}{\beta_m\left(L\sqrt{\beta_m/\alpha}\cosh\left(\sqrt{\beta_m/\alpha}\cdot L\right) - \sinh\left(\sqrt{\beta_m/\alpha}\cdot L\right)\right)}\exp(\beta_m t) \tag{A.19}$$

Hence:

$$\xi_{1,0} = \frac{k}{L}t - \frac{kL}{6\alpha} - \frac{kL}{\alpha}\frac{2}{\pi^2}\sum_{m=1}^{\infty}\frac{(-1)^m}{m^2}\exp\left(-\frac{m^2\pi^2\alpha}{L^2}t\right) \tag{A.20}$$

Appendix B

Derivation of Response Factors for Two-Layered Walls

The equations that govern the heat transfer at the surfaces of each layer are:

$$\begin{Bmatrix} q_1(n) \\ q_2(n) \end{Bmatrix} = \sum_{j=0}^{n} \begin{bmatrix} X_A(j) & -Y_A(j) \\ Y_A(j) & -Z_A(j) \end{bmatrix} \begin{Bmatrix} T_1(n-j) \\ T_2(n-j) \end{Bmatrix} \tag{B.1}$$

$$\begin{Bmatrix} q_2(n) \\ q_3(n) \end{Bmatrix} = \sum_{j=0}^{n} \begin{bmatrix} X_B(j) & -Y_B(j) \\ Y_B(j) & -Z_B(j) \end{bmatrix} \begin{Bmatrix} T_2(n-j) \\ T_3(n-j) \end{Bmatrix} \tag{B.2}$$

Equating the heat transfer rate across plane 2 from the above equations:

$$\sum_{j=0}^{\infty} Y_A(j)T_1(n-j) - Z_A T_2(n-j) = \sum_{j=0}^{n} X_B(j)T_2(n-j) - Y_B T_3(n-j) \tag{B.3}$$

From Equation B.3:

$$\sum_{j=0}^{\infty} (X_B(j) + Z_A(j))T_2(n-j) = \sum_{j=0}^{n} Y_A(j)T_1(n-j) + Y_B(j)T_3(n-j)$$

Let:

$$S(j) = X_B(j) + Z_A(j) \tag{B.4}$$

$$\sum_{j=0}^{\infty} S(j)T_2(n-j) = \sum_{j=0}^{n} Y_A(j)T_1(n-j) + Y_B(j)T_3(n-j) \tag{B.5}$$

Solving for $T_2(n)$ from Equation B.5:

$$T_2(n) = \frac{1}{S(0)} \left\{ \sum_{j=0}^{n} Y_A(j)T_1(n-j) + Y_B(j)T_3(n-j) - \sum_{j=1}^{n} S(j)T_2(n-j) \right\} \tag{B.6}$$

Note that instead of up to time step n, Equation B.5 that applies up to time step $n-k$ will be:

$$\sum_{j=0}^{n-k} S(j)T_2(n-k-j) = \sum_{j=0}^{n-k} Y_A(j)T_1(n-k-j) + Y_B(j)T_3(n-k-j)$$

The above equation can be re-written as:

$$\sum_{j=k}^{n} S(j-k)T_2(n-j) = \sum_{j=k}^{n} Y_A(j-k)T_1(n-j) + Y_B(j-k)T_3(n-j) \tag{B.7}$$

Therefore, $T_2(n-k)$ can be expressed as:

$$T_2(n-k) = \frac{1}{S(0)}\left\{ \sum_{j=k}^{n} Y_A(j-k)T_1(n-j) + Y_B(j-k)T_3(n-j) - \sum_{j=k+1}^{n} S(j-k)T_2(n-j) \right\} \tag{B.8}$$

From Equations B.1 and B.2:

$$\begin{aligned}
q_1(n) &= X_A(0)T_1(n) + X_A(1)T_1(n-1) + \ldots + X_A(k)T_1(n-k) + \ldots + X_A(n)T_1(0) \\
&\quad - Y_A(0)T_2(n) - Y_A(1)T_2(n-1) - \ldots - Y_A(k)T_2(n-k) - \ldots - Y_A(n)T_2(0)
\end{aligned} \tag{B.9}$$

$$\begin{aligned}
q_3(n) &= Y_B(0)T_2(n) + Y_B(1)T_2(n-1) + \ldots + Y_B(k)T_2(n-k) + \ldots + Y_B(n)T_2(0) \\
&\quad - Z_B(0)T_3(n) - Z_B(1)T_3(n-1) - \ldots - Z_B(k)T_3(n-k) - \ldots - Z_B(n)T_3(0)
\end{aligned} \tag{B.10}$$

$T_2(n-k)$ for $k = 0 \ldots n$, as shown in Equation B.8, can now be substituted into Equations B.9 and B.10 to yield expressions for $q_1(n)$ and $q_3(n)$ in terms of $T_1(n \ldots 0)$ and $T_3(n \ldots 0)$, as shown below:

$$\begin{Bmatrix} q_1(n) \\ q_3(n) \end{Bmatrix} = \sum_{j=0}^{n} \begin{bmatrix} X(j) & -Y(j) \\ Y(j) & -Z(j) \end{bmatrix} \begin{bmatrix} T_1(n-j) \\ T_3(n-j) \end{bmatrix} \tag{B.11}$$

where $X(j)$, $Y(j)$ and $Z(j)$ denote the response factors for the two-layered wall.

To ease collection of coefficients for $T_1(n \ldots 0)$ and $T_3(n \ldots 0)$ after the substitution, Equation B.8 is first expanded as follows:

$$\begin{aligned}
T_2(n-k) &= (1/S(0))\{ Y_A(0)T_1(n-k) + Y_A(1)T_1(n-k-1) + \ldots + Y_A(n-k)T_1(0) \\
&\quad + Y_B(0)T_3(n-k) + Y_B(1)T_3(n-k-1) + \ldots + Y_B(n-k)T_3(0) \\
&\quad - S(1)T_2(n-k-1) - S(2)T_2(n-k-2) - \ldots - S(n-k)T_2(0)\}
\end{aligned} \tag{B.12}$$

The $T_2(n-k-1)$ term can also be expanded as follows:

$$
\begin{aligned}
T_2(n-k-1) = (1/S(0))\{ &Y_A(0)T_1(n-k-1) + Y_A(1)T_1(n-k-2) + \ldots + Y_A(n-k-1)T_1(0) \\
+ &Y_B(0)T_3(n-k-1) + Y_B(1)T_3(n-k-2) + \ldots + Y_B(n-k-1)T_3(0) \\
- &S(1)T_2(n-k-2) - S(2)T_2(n-k-3) - \ldots - S(n-k-1)T_2(0)\} \quad \text{(B.13)}
\end{aligned}
$$

Equations B.12 and B.13 show that the coefficients of $T_1(n-k-m)$ and $T_3(n-k-m)$ in Equation B.12, for $m = 0 \ldots n-k$, include not only $(1/S(0))Y_A(m)$ and $(1/S(0))Y_B(m)$, but expansion of the $S(i)T_2(n-k-i)$ $(i=1 \ldots m)$ terms in the equation will generate further coefficients for the same temperature terms. Each expansion of the $S(i)T_2(n-k-i)$ term will yield $(S(i)/S(0))Y_A(m-i)$ and $(S(i)/S(0))Y_B(m-i)$, which are backward shifted by i time steps compared to $Y_A(m)$ and $Y_B(m)$. Furthermore, expansion of each $T_2(n-k-i)$ generates a further list of T_2 terms which should also be expanded until all coefficients for the T_1 and T_3 terms of the time step under concern have been found.

In the light of the above, the combined response factors for the two-layered wall can be determined from the response factors of individual layers as follows:

For $j = 0$:

$$
X(0) = X_A(0) - Y_A(0)\frac{Y_A(0)}{S(0)} = X_A(0) - \frac{Y_A(0)^2}{X_B(0) - Z_A(0)} \quad \text{(B.14a)}
$$

$$
Y(0) = Y_A(0)\frac{Y_B(0)}{S(0)} = \frac{Y_A(0)Y_B(0)}{X_B(0) - Z_A(0)} \quad \text{(B.14b)}
$$

$$
Z(0) = Z_B(0) - Y_B(0)\frac{Y_B(0)}{S(0)} = Z_B(0) - \frac{Y_A(0)^2}{X_B(0) - Z_A(0)} \quad \text{(B.14c)}
$$

For $j = 1$:

$$
X(1) = X_A(1) - Y_A(1)\frac{Y_A(0)}{S(0)} - \left(\frac{Y_A(0)}{S(0)} \left(Y_A(1) - \frac{S(1)}{S(0)}Y_A(0) \right) \right) \quad \text{(B.15a)}
$$

$$
Y(1) = Y_A(1)\frac{Y_B(0)}{S(0)} - \left(\frac{Y_A(0)}{S(0)} \left(Y_B(1) - \frac{S(1)}{S(0)}Y_B(0) \right) \right) \quad \text{(B.15b)}
$$

$$
Z(1) = Z_B(1) - Y_B(1)\frac{Y_B(0)}{S(0)} - \left(\frac{Y_B(0)}{S(0)} \left(Y_B(1) - \frac{S(1)}{S(0)}Y_B(0) \right) \right) \quad \text{(B.15c)}
$$

In general:

$$X(j) = X_A(j) - Y_A(j)\frac{Y_A(0)}{S(0)} - \left(\frac{Y_A(j-1)}{S(0)}\left(Y_A(1) - \frac{S(1)}{S(0)}Y_A(0)\right)\right)$$
$$- \left(\frac{Y_A(j-2)}{S(0)}\left(Y_A(2) - \frac{S(1)}{S(0)}\left(Y_A(1) - \frac{S(1)}{S(0)}Y_A(0)\right) - \frac{S(2)}{S(0)}Y_A(0)\right)\right)$$
$$- \left(\frac{Y_A(j-3)}{S(0)}\left(\begin{matrix} Y_A(3) - \frac{S(1)}{S(0)}\left(Y_A(2) - \frac{S(1)}{S(0)}\left(Y_A(1) - \frac{S(1)}{S(0)}Y_A(0)\right) - \frac{S(2)}{S(0)}Y_A(0)\right) \\ -\left(\frac{S(2)}{S(0)}\left(Y_A(1) - \frac{S(1)}{S(0)}Y_A(0)\right)\right) - \frac{S(3)}{S(0)}Y_A(0) \end{matrix}\right)\right)$$
$$- \dots \tag{B.16a}$$

$$Y(j) = Y_A(j)\frac{Y_B(0)}{S(0)} - \left(\frac{Y_A(j-1)}{S(0)}\left(Y_B(1) - \frac{S(1)}{S(0)}Y_B(0)\right)\right)$$
$$- \left(\frac{Y_A(j-2)}{S(0)}\left(Y_B(2) - \frac{S(1)}{S(0)}\left(Y_B(1) - \frac{S(1)}{S(0)}Y_B(0)\right) - \frac{S(2)}{S(0)}Y_B(0)\right)\right)$$
$$- \left(\frac{Y_A(j-3)}{S(0)}\left(\begin{matrix} Y_B(3) - \frac{S(1)}{S(0)}\left(Y_B(2) - \frac{S(1)}{S(0)}\left(Y_B(1) - \frac{S(1)}{S(0)}Y_B(0)\right) - \frac{S(2)}{S(0)}Y_B(0)\right) \\ -\left(\frac{S(2)}{S(0)}\left(Y_B(1) - \frac{S(1)}{S(0)}Y_B(0)\right)\right) - \frac{S(3)}{S(0)}Y_B(0) \end{matrix}\right)\right)$$
$$- \dots \tag{B.16b}$$

$$Z(j) = Z_B(j) - Y_B(j)\frac{Y_B(0)}{S(0)} - \left(\frac{Y_B(j-1)}{S(0)}\left(Y_B(1) - \frac{S(1)}{S(0)}Y_B(0)\right)\right)$$
$$- \left(\frac{Y_B(j-2)}{S(0)}\left(Y_B(2) - \frac{S(1)}{S(0)}\left(Y_B(1) - \frac{S(1)}{S(0)}Y_B(0)\right) - \frac{S(2)}{S(0)}Y_B(0)\right)\right)$$
$$- \left(\frac{Y_B(j-3)}{S(0)}\left(\begin{matrix} Y_B(3) - \frac{S(1)}{S(0)}\left(Y_B(2) - \frac{S(1)}{S(0)}\left(Y_B(1) - \frac{S(1)}{S(0)}Y_B(0)\right) - \frac{S(2)}{S(0)}Y_B(0)\right) \\ -\left(\frac{S(2)}{S(0)}\left(Y_B(1) - \frac{S(1)}{S(0)}Y_B(0)\right)\right) - \frac{S(3)}{S(0)}Y_B(0) \end{matrix}\right)\right)$$
$$- \dots \tag{B.16c}$$

Index